A Level
Salters Advanced
Chemistry
for OCR

B

David Goodfellow • Mark Gale

OXFORD
UNIVERSITY PRESS

Great Clarendon Street, Oxford, OX2 6DP, United Kingdom

Oxford University Press is a department of the University of Oxford. It furthers the University's objective of excellence in research, scholarship, and education by publishing worldwide. Oxford is a registered trade mark of Oxford University Press in the UK and in certain other countries

British Library Cataloguing in Publication Data
Data available

978-0-19-833292-3

10 9 8 7 6 5 4 3 2 1

Paper used in the production of this book is a natural, recyclable product made from wood grown in sustainable forests. The manufacturing process conforms to the environmental regulations of the country of origin.

Printed in Great Britain by Bell and Bain Ltd., Glasgow

Acknowledgements

Cover: Image Source / Alamy

Artwork by Q2A Media

Although we have made every effort to trace and contact all copyright holders before publication this has not been possible in all cases. If notified, the publisher will rectify any errors or omissions at the earliest opportunity.

AS/A Level course structure

This book has been written to support students studying for OCR A Level Chemistry B (Salters). The content is arranged by chemical idea. The content covered is shown in the contents list, which also shows you the page numbers for the main topics within each chapter.

AS exam

A level exam

Year 1 content

1 Elements of life
2 Developing fuels
3 Elements from the sea
4 The ozone story
5 What's in a medicine

Year 2 content

6 The chemical industry
7 Polymers and life
8 Oceans
9 Developing metals
10 Colour by design

A Level exams will cover content from Year 1 and Year 2 and will be at a higher demand. You will also carry out practical activities throughout your course.

Contents by chemical idea

| How to use this book | v |

1 Measuring amounts of substance 2
1.1 Amount of substance (EL 6) 2
1.2 Balanced equations (EL 6) 5
1.3 Using equations to work out reacting masses (EL 9) 7
1.4 Concentrations of solutions (EL 9) 8
1.5 Calculations involving gases (DF 8, OZ 1) 10
Experimental techniques 12
Practice questions 15

2 Atomic structure 17
2.1 A simple model of the atom (EL 1) 17
2.2 Nuclear fusion (EL 1) 18
2.3 Shells, sub-shells, and orbitals (EL 3) 19
Practice questions 21

3 Bonding, shapes, and sizes 23
3.1 Chemical bonding (EL 5) 23
3.2 Shapes of molecules (EL 5) 25
Practice questions 27

4 Energy changes and chemical reactions 28
4.1 Energy out, energy in (DF 1) 28
4.2 Enthalpy cycles (DF 2) 31
4.3 Bond enthalpies (DF 4) 33
4.4 Ions in solution (O1) 34
4.5 Enthalpy and entropy (O5) 38
Experimental techniques 41
Practice questions 42

5 Structure and properties 46
5.1 Ionic substances in solution (EL 7) 46
5.2 Bonding, structure, and properties (EL 7) 48
5.3 Bonds between molecules: temporary and permanent dipoles (OZ 6) 50
5.4 Bonds between molecules: hydrogen bonding (OZ 7) 52
5.5 Protein structure (PL 5) 53
5.6 Molecular recognition (PL 7) 55
5.7 Bonding dyes to fibres (CD 6) 57
Experimental techniques 59
Practice questions 60

6 Radiation and matter 63
6.1 Light and electrons (EL 2, OZ 2) 63
6.2 What happens when radiation interacts with matter? (OZ 2) 65
6.3 Radiation and radicals (OZ 3) 66
6.4 Infrared spectroscopy (WM 3) 69
6.5 Mass spectrometry (EL 1, WM 4) 70
6.6 Further mass spectrometry (PL9) 72
6.7 Nuclear magnetic resonance (NMR) spectroscopy (PL 9) 74
6.8 Using combined spectroscopic techniques to deduce the structure of organic molecules (PL9) 78
6.9 Coloured organic molecules (CD 1) 80
Practice questions 82

7 Equilibrium in chemistry 85
7.1 Chemical equilibrium (ES 4) 85
7.2 Equilibrium constant K_c (ES 4) 86
7.3 Equilibrium constant and concentrations (ES 4) 87
7.4 Le Châtelier's principle (ES 7) 88
7.5 Equilibrium constant K_c, temperature, and pressure (CI 2) 89

7.6 Solubility equilibria (O4) 91
7.7 Gas–liquid chromatography (CD 8) 92
Experimental techniques 94
Practice questions 95

8 Acids and bases 97
8.1 Acids, bases, alkalis, and neutralisation (EL 9) 97
8.2 Strong and weak acids and pH (O2) 99
8.3 Buffer solutions (O3) 102
Experimental techniques 104
Practice questions 105

9 Redox 107
9.1 Oxidation and reduction (ES 2) 107
9.2 Electrolysis as redox (ES 3) 110
9.3 Redox reactions, cells, and electrode potentials (DM 4) 111
9.4 Rusting and its prevention (DM 5) 115
Experimental techniques 117
Practice questions 119

10 Rates of reactions 121
10.1 Factors affecting reaction rates (OZ 4) 121
10.2 The effect of temperature on rate (OZ 4) 122
10.3 Catalysts (DF 5) 123
10.4 How do catalysts work? (OZ 5) 124
10.5 Rates of reactions (CI 3) 125
10.6 Rate equation of a reaction (CI 4) 126
10.7 Finding the order of reaction with experiments (CI 5) 128
10.8 Enzymes (PL 6) 131
Experimental techniques 133
Practice questions 135

11 The periodic table 137
11.1 Periodicity (EL 4) 137
11.2 The s-block: Groups 1 and 2 (EL 8) 139
11.3 The p-block: Group 7 (EL 8, ES 6) 141
11.4 The d-block (DM1, DM3, DM6) 144
11.5 Nitrogen chemistry (CI1) 147
Experimental techniques 149
Practice questions 150

12 Organic chemistry: frameworks 152
12.1 Alkanes (DF 3, DF 9) 152
12.2 Alkenes (DF 6, DF 7) 154
12.3 Structural isomerism and E/Z isomerism (DF 9) 157
12.4 Arenes (CD 2, CD3) 158
12.5 Reactions of arenes (CD 4) 161
12.6 Azo dyes (CD 5) 164
Practice questions 166

13 Organic chemistry: modifiers 168
13.1 Naming organic compounds: Alkanes, haloalkanes, and alcohols (DF 9) 168
13.2 Haloalkanes (OZ 6, OZ 8) 169
13.3 Alcohols (WM 1) 171
13.4 Carboxylic acids and phenols (WM 2) 173
13.5 Carboxylic acids (PL 1) 175
13.6 Amines (PL 2) 176
13.7 Hydrolysis of amides and esters (PL 3) 178
13.8 Amino acids, peptides, and proteins (PL 4) 179
13.9 Oils and fats (CD 7) 182

13.10	Aldehydes and ketones (CD 9)	183
	Experimental techniques	185
	Practice questions	188

14 Chemistry in industry — 191

14.1	Risks and benefits of chlorine (ES 5)	191
14.2	Atom economy (ES 6)	193
14.3	Operation of a chemical manufacturing process (CI 6)	194
	Practice questions	197

15 Human impacts — 197

15.1	Atmospheric pollutants (DF 10)	199
15.2	Alternatives to fossil fuels (DF 11)	200
15.3	Radiation in, radiation out (O2)	201
	Practice questions	204

16 Molecules in living systems — 206

| 16.1 | DNA and RNA (PL 8) | 206 |
| | Practice questions | 210 |

17 Organic synthesis — 212

17.1	Functional group reactions (CD 10)	212
17.2	Classification of organic reactions (CD 11)	214
	Practice questions	216

Answers to summary questions	**217**
Answers to practice questions	**230**
Synoptic questions	**237**
Answers to synoptic questions	**241**
For reference	**244**
Periodic table	**245**

How to use this book

This book contains many different features. Each feature is designed to support and develop the skills you will need for your examinations, as well as foster and stimulate your interest in chemistry.

This book is structured by chemical idea.

 Worked example

Step-by-step worked solutions.

Common misconception

Common student misunderstandings clarified.

 Go further

Familiar concepts in an unfamiliar context.

Maths skill

A focus on maths skills.

Model answers

Sample answers to exam-style questions.

Summary Questions

1   These are short questions at the end of each topic.

2   They test your understanding of the topic and allow you to apply the knowledge and skills you have acquired.

3   The questions are ramped in order of difficulty. Lower-demand questions have a paler background, with the higher-demand questions having a darker background. Try to attempt every question you can, to help you achieve your best in the exams.

Specification references

→   At the beginning of each topic there is a specification reference to allow you to monitor your progress.

Key term

Pulls out key terms for quick reference.

Revision tips

Revision tips contain prompts to help you with your understanding and revision.

Synoptic link

These highlight the key areas where topics relate to each other. As you go through your course, knowing how to link different areas of chemistry together becomes increasingly important. Many exam questions, particularly at A Level, will require you to bring together your knowledge from different areas.

1 What mass of sodium hydroxide, NaOH is needed to form 200 cm^3 of a 0.25 mol dm^{-3} solution?

A 2000 g

B 2 g

C 32 g

D 10 g

(1 mark)

2 Choose the phrase that correctly completes the following definition:

The relative isotopic mass is the mass of one atom of an isotope compared to....

A The mass of a carbon-12 atom.

B 12 g of carbon-12 atoms.

C one-twelfth of 12 g of carbon-12 atoms.

D one-twelfth of the mass of a carbon-12 atom.

(1 mark)

3 1.40 g of ethene, C_2H_4, reacts with excess steam, H_2O, to form ethanol, C_2H_5OH. The equation for the reaction is:

$C_2H_4 + H_2O \rightarrow C_2H_5OH$

1.06 g of ethanol are formed in the reaction. The % yield of this process is given by the equation:

A $\dfrac{1.06}{1.40} \times 100$

B $\dfrac{1.06}{\frac{(1.40 \times 28)}{46}} \times 100$

C $\dfrac{1.06}{\frac{(1.40 \times 46)}{28}} \times 100$

D $\dfrac{1.40}{\frac{(1.06 \times 28)}{46}} \times 100$

(1 mark)

4 A compound that contains C, H, and O atoms only has the following % composition by mass: C: 40% O: 53%.

The relative molecular mass of the compound was found to be 60.

The molecular formula of the compound is:

A CH_2O

B CO

C $C_2H_4O_2$

D C_2O_2

(1 mark)

5 Barium chloride solution was added to a solution of sodium sulfate. A white precipitate of barium sulfate was formed. The following are possible equations to describe what has happened. Which of these are correct?

   1 $BaCl(aq) + NaSO_4(aq) \rightarrow NaCl(aq) + BaSO_4(s)$

   2 $Na^+(aq) + Cl^-(aq) \rightarrow NaCl(s)$

   3 $Ba^{2+}(aq) + SO_4^{2-}(aq) \rightarrow BaSO_4(s)$

Practice questions including questions that cover practical and maths skills.

Experimental techniques

Specification reference: EL (b) (ii), EL (c) (ii), DF (a)

Weighing a solid

Masses of solids are normally measured on an electronic balance masses to two or three decimal places.

Weighing out a known mass of solid

1 Place a weighing bottle or similar container on the balance.

2 Add the required mass of solid.

3 Reweigh the weighing bottle + solid.

4 Transfer the solid to the reaction vessel or volumetric flask.

5 Reweigh the weighing bottle without the solid.

Heating to constant mass

In some experiments you need to measure the mass of a solid rem after thermal decomposition. Thermal decomposition of hydrated carried out in a crucible supported on a pipe-clay triangle.

Weighing to constant mass ensures that decomposition is complete:

1 Weigh an empty crucible.

2 Add the hydrated salt and weigh the crucible + hydrated salt.

3 Heat the crucible strongly for several minutes.

4 Allow the crucible to cool and weigh the crucible and

Dedicated experiential techniques pages, detailing all of the practical skills you need to know for your examinations.

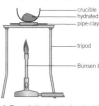

▲ Figure 1 Heating a hydrated salt in

crucible
hydrated
pipe-clay

tripod

Bunsen b

1.1 Amount of substance

Isotopes

Isotopes are atoms of the same element. They have the same atomic number (number of protons), but they have different mass numbers (different numbers of neutrons). Most elements exist naturally as a mixture of isotopes.

For example, there are two common isotopes of chlorine:

- $^{35}_{17}Cl$ has 17 protons and 18 neutrons.
- $^{37}_{17}Cl$ has 17 protons and 20 neutrons.

Relative atomic mass (A_r)

The **relative atomic mass** (A_r) is the average of the relative isotopic masses of the naturally occurring isotopes of an element, taking into account their abundances.

So the A_r value of an element tells you the average mass of an atom of that element compared to one-twelfth of an atom of ^{12}C. A_r values have no units.

> ### 🖩 Worked example: Calculating a relative atomic mass
>
> Silicon, Si, exists as three naturally occurring isotopes. The % abundances of the three isotopes are:
>
> $^{28}_{14}Si$: 92.2%, $^{29}_{14}Si$: 4.7%, $^{30}_{14}Si$: 3.1%. Calculate the relative atomic mass of silicon to 2 decimal places.
>
> **Step 1:** Multiply the % abundance by the relative mass of each isotope.
>
> **Step 2:** Divide by 100.
>
> **Step 3:** Round the answer to 2 decimal places.
>
> $$\frac{(92.2 \times 28)+(4.7 \times 29)+(3.1 \times 30)}{100} = 28.109 = 28.11 \text{ to 2 d.p.}$$

Relative formula mass (M_r)

The relative formula mass (M_r) is the sum of the relative atomic masses (A_r) for each atom in the formula. For a simple molecular compound, it is sometimes called the relative molecular mass. M_r values have no units.

Avogadro constant and the mole

The mole is the unit of **amount of substance**:

- To obtain 1 mole of a substance, weigh out the A_r or M_r exactly in grams.
- The Avogadro constant, $6.02 \times 10^{23} \text{mol}^{-1}$ (given the symbol N_A), is the number of particles in 1 mole of a substance.

Calculations involving amount of substance

Calculating amount of a substance (number of moles)

To work out the amount in moles of a substance, use:

Amount in moles $= \dfrac{\text{mass in grams}}{A_r}$ for an atomic element

amount in moles $= \dfrac{\text{mass in grams}}{M_r}$ for a molecule or compound.

Empirical and molecular formulae

Empirical formula

To find the empirical formula of a compound, you simply need to calculate the number of moles of each element in the compound.

 Worked example: Calculating empirical formula

A sample of aluminium oxide contains 0.310 g of aluminium and 0.276 g of oxygen. Calculate the empirical formula.

	Al ($A_r = 27.0$)	O ($A_r = 16.0$)
mass / g	0.310	0.276
moles / mol	$\frac{0.310}{27.0} = 0.011\,48$	$\frac{0.276}{16.0} = 0.017\,25$
ratio (divide each number by the smallest of the two numbers)	$\frac{0.01148}{0.01148} = 1.00$	$\frac{0.01725}{0.01148} = 1.50$ (to 3 s.f.)

The ratio of Al to O is 1:1.5. This is not a whole number ratio; the simplest whole number ratio would be 2:3.

The formula is Al_2O_3.

If the masses are given as percentages (e.g. 52.9% Al and 47.1% O), then simply treat these numbers as masses (52.9 g and 47.1 g) to start the calculation.

Molecular formula

The molecular formula may be the same as the empirical formula, or it may be a whole number multiple of the empirical formula.

You can use the M_r value of the molecule to find the molecular formula.

 Worked example: Molecular formula

The empirical formula of a hydrocarbon molecule is CH_2. The M_r value is found to be 84.

Step 1: $M_r(CH_2) = 14$.

Step 2: Number of empirical formula units needed to give an M_r of 84 $= \frac{84}{14} = 6$.

Step 3: $CH_2 \times 6 = C_6H_{12}$.

The molecular formula is C_6H_{12}.

Water of crystallisation

Some ionic salts are hydrated. They contain water of crystallisation, which can be driven off when the salt is heated.

Experimental data can be used to find the formula of the ionic salt.

 Worked example: Finding the formula of a hydrated salt

Hydrated cobalt(II) chloride has the formula $CoCl_2 \bullet \times H_2O$.

1.173 g of hydrated cobalt chloride is heated to drive off the water of crystallisation. The mass remaining is 0.641 g.

Step 1: Calculate the mass of water driven off: 1.173 − 0.641 = 0.532 g.

Step 2: Calculate the number of moles of water ($M_r = 18$): $\frac{0.532}{18} = 0.029\,56$.

Step 3: Calculate the number of moles of anhydrous salt ($M_r = 129.9$): $\frac{0.641}{129.9} = 0.004\,93$.

Step 4: Calculate the ratio $\frac{\text{moles water}}{\text{moles anhydrous salt}}$: $\frac{0.2956}{0.00493} = 5.996$.

The formula of the hydrated salt = $CoCl_2 \bullet 6H_2O$.

Synoptic link

Relative atomic masses are calculated from the abundances of the different isotopes of an element as measured by a mass spectrometer. The use of a mass spectrometer is described in Topic 6.5, Mass spectrometry.

Revision tip

You should use the A_r values (to 1 d.p.) from the periodic table to calculate the M_r value of a compound.

Key terms

Avogadro constant: The number of particles in 1 mole of a substance. Unit: mol^{-1}.

Mole: An amount of substance containing the same number of particle as there are in 12 g of ^{12}C. Unit: mol.

Revision tip

If an element exists as molecules, e.g. oxygen (O_2), look carefully at the question to see whether you are being asked to calculate the number of moles of atoms or molecules. The relative masses of these particles are different!

Key term

Empirical formula: The simplest whole number ratio of atoms of each element in a compound.

Revision tip

If the ratio from a calculation is not a whole number (e.g. 1:1.67), then do not be tempted to round up to make the whole number ratio 1:2. Find a pair of whole numbers that match this ratio (in this case 3:5).

Key term

Molecular formula: the actual number of atoms of each element in a compound molecule.

Synoptic link

Details of the experimental method used to find the formula of hydrated salts can be found in Experimental Techniques.

Percentage composition

You can work out the percentage by mass of an element in a compound using the formula of the compound.

 Worked example: Working out the percentage by mass of an element

Calculate the percentage by mass of nitrogen in ammonium sulfate, $(NH_4)_2SO_4$.

Step 1: $M_r (NH_4)_2SO_4 = 2 \times (14.0 + 1.0 + 1.0 + 1.0 + 1.0) + 32.1 + (4 \times 16.0) = 132.1$.

Step 2: Mass of nitrogen in 1 mole of $(NH_4)_2SO_4 = 14.0 + 14.0 = 28.0$.

$$\% \text{ by mass} = \frac{\text{mass of element in 1 mole}}{M_r \text{ of compound}} \times 100 = \frac{28.0}{132.1} \times 100 = 21.2\%$$

Summary questions

1   What is the M_r value of:
    a   aluminium sulfate, $Al_2(SO_4)_3$ (*1 mark*)
    b   naphthalene $C_{10}H_8$?   (*1 mark*)
2   Calculate the % by mass of C in benzene, C_6H_6.   (*1 mark*)

3   Relative isotopic masses have whole number values when measured to 2 decimal places, whereas relative atomic masses are often not whole numbers. Explain why.   (*2 marks*)
4   The % by mass of C in CO_2 is 27%. However, 33% of the atoms in CO_2 are C. Explain the difference in these numbers.   (*2 marks*)

5   Bromine and boron both have two naturally occurring isotopes:
    a   The % abundance of the two isotopes of bromine are: ^{79}Br: 50.52%, ^{81}Br: 49.48%. Calculate the relative atomic mass of bromine to 2 d.p.   (*1 mark*)
    b   The two isotopes of boron are ^{10}B and ^{11}B. The relative atomic mass of boron is 10.8 to 1.d.p. Calculate the % abundance of each of the two B isotopes.   (*1 mark*)
6   A compound containing C, H, and O only has a composition of 73.47% C and 10.20% H. The M_r is found to be 196. What is the molecular formula of the compound?   (*2 marks*)

1.2 Balanced equations
Specification reference: EL(d)

Writing formulae

You can use the charges on ions to write the formulae of ionic substances.

 Worked example: Writing formula of ionic compounds

Write the formula of calcium chloride.

Step 1: Charges on ions: Ca^{2+}, Cl^-.

Step 2: To balance the charges, 2 Cl^- ions are needed for 1 Ca^{2+}.

Step 3: The formula is $CaCl_2$.

 Worked example: Writing the formula of more complex ionic compounds

Write the formula of aluminium sulfate:

Step 1: Charges on ions: Al^{3+}, SO_4^{2-}.

Step 2: To balance the charges, 1½ SO_4^{2-} are needed for 1 Al^{3+}.

Step 3: The simplest whole number ratio is $2Al^{3+} : 3SO_4^{2-}$.

Step 4: The formula is written as $Al_2(SO_4)_3$.

Full chemical equations

Once the correct formulae have been written for the substances in a chemical equation, you can add balancing numbers to produce a full balanced equation.

 Worked example: Writing a balanced equation

A piece of calcium is added to water. The calcium reacts to form a solution of calcium hydroxide and hydrogen is given off.

Step 1: Write the word equation: calcium + water $\rightarrow$ calcium hydroxide + hydrogen.

Step 2: Put in the formulae: $Ca + H_2O \rightarrow Ca(OH)_2 + H_2$.

Step 3: Balance the equation using large numbers before the formulae:

$$Ca + 2H_2O \rightarrow Ca(OH)_2 + H_2.$$

Check:

	Left	Right
Ca	1	1
O	2	2
H	4	4

Step 4: Add the state symbols if the question asks for them:

$$Ca(s) + 2H_2O\ (l) \rightarrow Ca(OH)_2\ (aq) + H_2\ (g).$$

Revision tip
You will need to learn some formulae of some common substances.

Water	H_2O
Carbon dioxide	CO_2
Hydrogen	H_2
Hydrochloric acid	HCl
Sulfuric acid	H_2SO_4

Revision tip
There are some formulae of common ions that you should also learn:

Zinc	Zn^{2+}		
		Hydroxide	OH^-
		Carbonate	CO_3^{2-}
		Sulfate	SO_4^{2-}
		Nitrate(V)	NO_3^-
		Ammonium	NH_4^+
Silver	Ag^+	Hydrogen carbonate	HCO_3^-

Revision tip
The charges of simple ions of elements in the s- and p-block of the periodic table follow a regular pattern. You can use this pattern to check the charges on simple ions such as Ba^{2+} and Cl^-.

Revision tip
You should use these state symbols: (s) = solid, (l) = liquid, (aq) = aqueous solution, (g) = gas.

Ionic equations

The term *ionic equation* is sometimes used to describe reactions involving ionic compounds in solution. Any ions that do not change during the reaction are left out, making it easier to see what has happened.

 Worked example: Writing an ionic equation for a precipitation reaction

Silver nitrate solution is added to a solution containing chloride ions. A precipitate of silver chloride is formed. Write an ionic equation for the reaction, including state symbols.

Step 1: Write the formula of the precipitate: $AgCl$.

Step 2: Show the ions that combine to form this precipitate:

$$Ag^+ + Cl^- \longrightarrow AgCl.$$

Step 3: Add the state symbols:

$$Ag^+ (aq) + Cl^- (aq) \longrightarrow AgCl(s).$$

Summary questions

1   What is the usual charge on an ion formed by an atom of these elements?
    a   sulfur (Group 6)
    b   rubidium (Group 1).   (*2 marks*)

2   Write the formula of these ionic compounds:
    a   barium hydroxide
    b   aluminium oxide
    c   sodium sulfate.   (*3 marks*)

3   a   A piece of calcium oxide reacts with hydrochloric acid to form water and a solution of calcium chloride. Write a full balanced equation for this process.   (*2 marks*)
    b   Barium chloride solution is added to a solution containing sulfate ions. A white precipitate of barium sulfate is formed. Write an ionic equation, including state symbols.   (*2 marks*)

1.3 Using equations to work out reacting masses

Specification reference: EL (b) (i)

Calculating reacting masses

You can convert between masses and moles using these relationships:

$$\text{Moles} = \frac{\text{mass}}{A_r \text{ or } M_r} \qquad \text{Mass} = \text{moles} \times M_r \text{ (or } A_r\text{)}$$

 Worked example: Calculating reacting masses

What mass of aluminium is needed to completely react with 7.98 g of iron(III) oxide, Fe_2O_3?

Formulae and balancing numbers taken from the equation	Fe_2O_3	2Al
M_r or A_r value	159.6	27.0
Mass	7.98	$0.100 \times 27 = 2.70$ g
Amount in moles	$\frac{7.98}{159.6} = 0.0500$	0.100

Step 1: Convert the mass of Fe_2O_3 into moles using:

$$\text{moles} = \frac{\text{mass}}{A_r \text{ or } M_r}.$$

Step 2: Use the ratio of the balancing numbers from the equation to work out the moles of Al: $2 \times 0.0500 = 0.100$ mol.

Step 3: Convert the moles of Al into a mass using: mass = moles $\times M_r$ (or A_r).

Percentage yield

Reacting mass calculations allow you to calculate the expected yield of a reaction.

 Worked example: % yield

7.98 g of iron(III) oxide, Fe_2O_3, are reacted with excess aluminium. 4.62 g of iron are formed. Calculate the % yield (to 3 s.f.)

Step 1: Calculate the expected yield:

Formulae and balancing numbers taken from the equation	Fe_2O_3	2Fe
M_r or A_r value	159.6	55.8
Mass	7.98	$0.100 \times 55.8 = 5.58$ g
Amount in moles	$\frac{7.98}{159.6} = 0.0500$	0.100

Step 2: Compare the actual yield to the expected calculated yield:

$$\frac{4.62}{5.58} \times 100 = 82.645\ldots$$

Step 3: Give the answer to 3 significant figures: 82.6%.

Key term

Expected yield: The mass of a product that can be formed from a given mass of reactant.

The percentage yield of a reaction = $\frac{\text{actual yield}}{\text{expected yield}} \times 100$.

Common misconception: Converting moles

Be careful when converting moles of product to mass of product (Step 3): do not try and use the A_r value of 2Al (54.0). The balancing number (2) has already been taken account in the second step.

Summary questions

1 When carbon is burnt in oxygen, it forms carbon dioxide $(C + O_2 \rightarrow CO_2)$. Calculate the mass of carbon dioxide formed from 10 g of carbon.

   (1 mark)

2 Uranium hexafluoride (UF_6) is formed in a reaction between uranium and chlorine trifluoride (ClF_3): $U + 2ClF_3 \rightarrow UF_6 + Cl_2$. What mass of ClF_3 is needed to form 100 g of UF_6? *(3 marks)*

3 In a reaction to form UF_6, 1 kg of uranium is reacted with 1 kg of ClF_3. 1.36 kg of UF_6 are formed. Calculate the % yield of UF_6 in this reaction. *(4 marks)*

Calculating concentrations

Concentrations can be measured in grams per cubic decimetre ($g\,dm^{-3}$).

Example

A solution containing 40 g of sodium hydroxide in $2\,dm^3$ of solution has a concentration of $\frac{40}{2} = 20g\,dm^{-3}$

A more common unit for concentration is *moles per cubic decimetre* ($mol\,dm^{-3}$).

1 mole of a substance, dissolved and then made up to a volume of $1\,dm^3$ of solution, has a concentration of $1\,mol\,dm^{-3}$.

> **Worked example: Converting from $g\,dm^{-3}$ to $mol\,dm^{-3}$**
>
> A solution of sodium hydroxide, NaOH, has a concentration of $36\,g\,dm^{-3}$. What is the concentration in $mol\,dm^{-3}$?
>
> **Step 1:** Find the M_r value for NaOH: $M_r = 40$.
>
> **Step 2:** Convert 36 g into moles: $\frac{36}{40} = 0.90\,mol$.
>
> The concentration is $0.90\,mol\,dm^{-3}$.

Moles = volume (in dm^3) × concentration (in $mol\,dm^{-3}$)

OR, if the volume is given in cm^3:

$$\text{Moles} = \frac{\text{volume (in } cm^3)}{1000} \times \text{concentration (in } mol\,dm^{-3})$$

> **Worked example: Calculating concentrations from masses and volumes**
>
> 5.85 g of NaCl ($M_r = 58.5$) is dissolved and made up to $250\,cm^3$. What is the concentration in $mol\,dm^{-3}$?
>
> **Step 1:** Calculate the number of moles of NaCl: $\frac{5.85}{58.8} = 0.100\,mol$.
>
> **Step 2:** Calculate the volume in dm^3: $\frac{250}{1000} = 0.250\,dm^3$.
>
> **Step 3:** Rearrange the expression linking moles, volume and concentration:
>
> $$n = c \times v, \text{ so } c = \frac{n}{v}.$$
>
> **Step 4:** Substitute the numbers into this expression:
>
> $$c = \frac{0.100}{0.250} = 0.400\,mol\,dm^{-3}$$

Diluting solutions

Solutions are diluted by taking a small volume of the original solution and diluting it to a greater volume using distilled water.

> **Worked example**
>
> $10.0\,cm^3$ of a $1.00\,mol\,dm^{-3}$ solution of HCl was transferred to a volumetric flask and made up to $250\,cm^3$ with water. What is the concentration of the diluted solution?
>
> **Step 1:** Calculate the dilution factor $= \frac{10.0}{250}$
>
> **Step 2:** Use this to calculate the new concentration of the diluted solution $= \frac{10.0}{250} \times 1.00$
>
> $$= 0.0400\,mol\,dm^{-3}$$

Revision tip

Volumes can be measured in dm^3 or cm^3. If volumes are given in cm^3, you will need to convert it to dm^3.

$1\,dm^3 = 1000\,cm^3$ or $1\,cm^3 = \frac{1}{1000}\,dm^3$

So to convert cm^3 to dm^3, you need to divide by 1000.

Revision tip

You need to be able to rearrange the expressions linking moles, volume, and concentration. Practise doing this frequently!

Synoptic link

Details of the experimental method used to make up and dilute solutions can be found in Experimental Techniques.

Revision tip

You can calculate the new concentration of the diluted solution by using the dilution factor:

Dilution factor = $\frac{\text{volume of original solution}}{\text{new volume of diluted solution}}$

So the new concentration of the diluted solution = old concentration × dilution factor.

Titrations

A titration is a method for finding the concentration of a solution by reacting a known volume with another solution of known concentration. Both the reacting volumes are measured as precisely as possible.

Using the balanced equation for the reaction, the unknown concentration of the solution can be found.

Synoptic link

You will need to be able to write equations for the reactions of common acids and alkalis. Look at Topic 1.2, Balanced equations and Topic 8.1, Acids, bases, alkalis, and neutralisation, to help you.

 Worked example: Finding the concentration of an alkali

In a titration, 25.0 cm³ of potassium hydroxide solution was pipetted into a conical flask. A 0.020 mol dm⁻³ solution of sulfuric acid was added from a burette. An indicator in the solution changed colour when 27.9 cm³ of sulfuric acid had been added. What is the concentration of the potassium hydroxide?

Formulae and balancing ratio	H_2SO_4	2KOH
volume / cm³	27.9 (0.0279 dm³)	25.0 (0.0250 dm³)
concentration / mol dm⁻³	0.020	$\frac{1.116 \times 10^{-3}}{0.0250} = 0.045$
moles	5.58×10^{-4}	1.116×10^{-3}

Step 1: Write out the balanced equation for the reaction, in order to find the ratio of moles:

$$H_2SO_4 + 2KOH \rightarrow K_2SO_4 + 2H_2O.$$

Step 2: Calculate the number of moles of H_2SO_4:

$$0.020 \times 0.0279 = 5.58 \times 10^{-4} \text{ mol}$$

Step 3: Use the balancing ratio to find the number of moles of KOH:

$$2 \times 5.58 \times 10^{-4} = 1.116 \times 10^{-3} \text{ mol}$$

Step 4: Convert this number to a concentration using the volume of KOH:

$$c = \frac{n}{v} = \frac{1.116 \times 10^{-3}}{0.0250} = 0.045 \text{ mol dm}^{-3}$$

Maths skill: Avoiding rounding errors

In multi-step calculations, like these titration calculations, it is important to avoid rounding your answers at an early stage. Keep the number in your calculator as you go through each stage of the calculation, and only round down to the appropriate number of significant figures at the end of the calculation.

Summary questions

1   What is the concentration of these solutions of NaCl:
   a   0.200 mol in 250 cm³
   b   2.0 × 10⁻³ mol in 50 cm³
   c   2.5 mol in 20 dm³.   *(3 marks)*

2   28.6 g of HCl was dissolved in water and made up to 200 cm³. 10 cm³ of this solution was then removed and diluted to 250 cm³ by adding distilled water:
   a   What is the concentration in mol dm⁻³ of the original solution?
   b   What is the concentration of the diluted solution?   *(3 marks)*

3   20.0 cm³ of a solution of KOH was analysed in a titration against 0.0125 mol dm⁻³ sulfuric acid (H_2SO_4). An average titre of 16.73 cm³ of sulfuric acid was recorded. Calculate the concentration of the KOH in g dm⁻³.   *(4 marks)*

 Go further

In some questions there may be extra steps in your calculation after you find the concentration of the reacting solution. For example, the solution used in the titration may have been diluted and you may need to use the dilution factor to find the original concentration of the undiluted solution.

The potassium hydroxide in the worked example was taken from an alkaline battery. 3.57 cm³ of the battery fluid was made up to 250 cm³ with distilled water for use in the titration. Calculate the concentration of potassium hydroxide in the original battery fluid.

1.5 Calculations involving gases

Specification reference: DF (a) OZ (i)

Molar volumes of gases

One mole of any gas occupies the same volume as any other gas, as long as the conditions are the same.

This volume is called the molar volume, and for room temperature and pressure (around 20 °C and 1 atmosphere pressure, or 10 100 Pa), this is 24.0 dm³.

Using molar volumes of gases

At RTP (room temperature and pressure), the amount in moles of a gas is given by:

$$\text{Moles} = \frac{\text{volume in dm}^3}{24.0}; \text{ so volume in dm}^3 = \text{moles} \times 24.0.$$

Calculating reacting masses and volumes

 Worked example: Calculating the volume of gas produced in a reaction

Sodium hydrogencarbonate, $NaHCO_3$, decomposes to produce carbon dioxide gas when it is heated:

$$2 NaHCO_3 \rightarrow Na_2CO_3 + H_2O + CO_2$$

Calculate the volume of CO_2 (measured at RTP) that would be produced from 2.69 g of sodium hydrogen carbonate.

Formulae and balancing ratio	$2NaHCO_3$	CO_2
M_r	84.0	
Mass in g	2.69 g	
Volume of gas in dm³		$0.0160 \times 24.0 = 0.384$ dm³
moles	0.0320	0.0160

Step 1: Write out the formulae and balancing ratio of the substances.

Step 2: Calculate the number of moles of $NaHCO_3 = \frac{\text{mass}}{M_r} = \frac{2.69}{84} = 0.0320$ mol.

Step 3: Use the balancing ratio to find the moles of $CO_2 = \frac{0.0320}{2} = 0.0160$ mol.

Step 4: Convert the moles of CO_2 into a volume of gas using the volume of gas = moles × molar volume.

Step 5: molar volume = $0.0160 \times 24.0 = 0.384$ dm³.

You may be asked to convert this into a volume in cm³:

0.384 dm³ $= 0.384 \times 1000 = 384$ cm³.

If the question involves only gases, you can use a simpler method. There is no need to convert the volumes into moles, because equal volumes of gases contain the same number of moles.

The ratio of the reacting volumes is the same as the ratio of moles.

> **Revision tip**
>
> The molar gas volume at RTP is provided on the data sheet.

> **Revision tip**
>
> If you are given a volume of a gas in cm³, you will need to convert it to dm³ to use this expression:
>
> 1000 cm³ = 1 dm³.

> **Revision tip**
>
> To convert between m³ and dm³ (or cm³), you need to know that 1 m³ = 1000 dm³.
>
> So 1 m³ = 1 000 000 cm³.

> **Revision tip**
>
> The value of R, the gas constant, is given in your data sheet.

Worked example: Calculating reacting volumes of gases

When propane is burnt in a good supply of air, it undergoes complete combustion:

$$C_3H_8(g) + 5O_2(g) \rightarrow 3CO_2(g) + 4H_2O(l)$$

4.5 dm³ of propane is burnt in air. Calculate the volume of oxygen needed.

Step 1: Find the ratio of moles oxygen: moles propane (5:1).

Step 2: Multiply the volume of propane by this ratio: 4.5 × 5 = 22.5 dm³.

The ideal gas equation

The connection between volume, pressure and temperature for a gas is given by the ideal gas equation:

$$pV = nRT$$

p = pressure in Pa V = volume in m³ n = number of moles

T = temperature in K R = the gas constant: 8.314 J mol⁻¹ K⁻¹.

By rearranging the ideal gas equation, you can predict the pressure or volume of a gas, given all the other conditions.

 Worked example: Calculating the volume of a gas at non-standard conditions

34.0 g of ammonia are heated to 373 K at a pressure of 150 000 Pa. Calculate the volume, in dm³, of this mass of ammonia.

Step 1: Rearrange the ideal gas equation to make V the subject:

$$V = \frac{nRT}{p}$$

Step 2: Calculate n, the number of moles of ammonia:

$$M_r(NH_3) = 17. \text{ So } n = \frac{34}{17} = 2.$$

Step 3: Substitute the values of n, R, T, and p into the equation:

$$V = \frac{2 \times 8.314 \times 373}{150\,000} = 0.0413 \text{ m}^3.$$

Step 4: Convert m³ to dm³:

$$1 \text{ m}^3 = 1000 \text{ dm}^3, \text{ so } V = 0.0413 \times 1000 = 41.3 \text{ dm}^3$$

Concentrations of gases

Percentage by volume

In mixtures of gases, such as the air in the atmosphere, the concentration of each gas is often given as the percentage by volume.

Parts per million

If a gas is present in a low concentration – less than 1% by volume – the concentration might be given in parts per million (ppm).

Maths skill: Rearranging equations

The ideal gas equation contains 5 different terms. You may be asked to rearrange it to make p, V, n, or T the subject of the equation. Practise rearranging the equation so that you are confident in rearranging in all these different ways.

Revision tip

To convert between % and ppm, remember that 100% = 1 000 000 ppm. So 1% = 10 000 ppm. To convert from % to ppm, multiply by 10 000. To convert from ppm to %, divide by 10 000.

Summary questions

1   a   The concentration by volume of neon in the atmosphere is 18.2 ppm. Express this as a %.

   b   The carbon dioxide concentration is now greater than 0.04%. Convert this concentration into ppm. *(2 marks)*

2   Hydrogen peroxide (H_2O_2) decomposes slowly to form oxygen gas:
$$2H_2O_2(aq) \rightarrow 2H_2O(l) + O_2(g)$$
   What volume of oxygen gas is formed if 2.40 g of hydrogen peroxide decomposes at RTP? *(3 marks)*

3   A sample of oxygen has a volume of 360 dm³ at a temperature of 400 K and 200 000 Pa. How many moles of oxygen molecules are present? *(3 marks)*

Experimental techniques

Weighing a solid

Masses of solids are normally measured on an electronic balance that records masses to two or three decimal places.

Weighing out a known mass of solid

1   Place a weighing bottle or similar container on the balance.

2   Add the required mass of solid.

3   Reweigh the weighing bottle + solid.

4   Transfer the solid to the reaction vessel or volumetric flask.

5   Reweigh the weighing bottle without the solid.

Heating to constant mass

In some experiments you need to measure the mass of a solid remaining after thermal decomposition. Thermal decomposition of hydrated salts can be carried out in a crucible supported on a pipe-clay triangle.

Weighing to constant mass ensures that decomposition is complete:

1   Weigh an empty crucible.

2   Add the hydrated salt and weigh the crucible + hydrated salt.

3   Heat the crucible strongly for several minutes.

4   Allow the crucible to cool and weigh the crucible and contents.

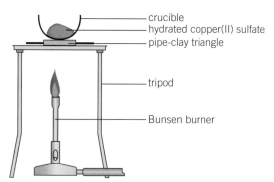

▲ **Figure 1** *Heating a hydrated salt in a crucible.*

5   Heat strongly for a further minute and reweigh the crucible and contents again.

6   If the mass has changed, then repeat this last step until two successive weighings are identical.

Volumetric equipment

Pipettes and burettes are described as volumetric equipment, because they can be used to measure volumes of liquids accurately.

Pipettes

Pipettes are used to precisely measure out fixed volumes of liquids (e.g. $10\,cm^3$ or $25\,cm^3$).

Liquid is sucked up into the pipette until it reaches the mark on the neck of the pipette.

The level of the liquid is viewed at eye level.

Burettes

Burettes are used to measure the volume of solution used to reach the end point of a reaction in a titration, or to measure out the volumes of solutions needed to produce a soluble salt.

> ### Key term
>
> **Thermal decomposition:** A reaction in which heat causes a compound to break down into simpler substances.

> ### Synoptic link
>
> You can see how to perform calculations using the data from thermal decomposition of hydrated salts in Topic 1.3, Using equations to work out reacting masses.

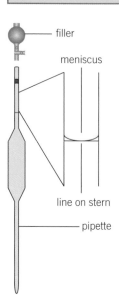

▲ **Figure 2** *Measuring the volume of liquid in a pipette*

Readings on burettes should be recorded to the nearest 0.05 cm³.

Titrations

This technique is used to determine the concentration of a solution by reacting it with another solution of known concentration.

Carrying out an acid/alkali titration

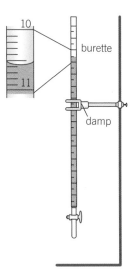

> ### Go further
>
> Other types of reactions can be used as the basis of a titration. Sodium thiosulfate can be added to a solution of iodine using a starch indicator, or potassium manganate(VII) solution can be added to solutions of reducing agents, such as Fe²⁺ ions.
>
> The end point of a thiosulfate / iodine titration occurs when a blue-black colour is seen; the end point of a manganate titration occurs when a pink colour is seen. What substances are responsible for these colours?

When the reaction is complete, it is signalled by the colour change of an indicator. The main stages are:

1 Pipette a known volume of the solution of unknown concentration into a clean conical flask.

2 Place the flask on a white tile – this improves the visibility of any colour changes.

3 Add a few drops of a suitable indicator – do not add too much, because the indicator reacts with acids and alkalis and so will interfere with your results.

4 Rinse a clean burette with the solution of known concentration.

5 Fill the burette with this solution.

6 Take an initial reading of the volume from the burette, using the lowest point of the meniscus of the liquid.

7 Swirl the conical flask and add the liquid from the burette 1 cm³ at a time until the expected colour change occurs. Keep swirling to mix the contents of the flask throughout the process.

8 Rinse the conical flask with distilled water and repeat steps 1, 2, 3, 5, and 6. This time, run in the solution from the burette up to 1 cm³ before the colour change obtained in step 7. Then, swirling thoroughly, add more solution dropwise until the desired colour change is observed. This gives an accurate result.

9 Repeat step 8 until you get three titre values within 0.1 cm³ of each other. These are called concordant results. The average of these titres is used in the calculations.

Preparing solutions

A standard solution can be made up either from a solid or from accurate dilution of another standard solution.

Making a standard solution from a solid

1 Weigh out the mass of solute required (see Weighing a solid, above).

2 Transfer the solid into a beaker and dissolve by stirring it with some distilled water.

3 Transfer this solution to a volumetric flask of a suitable volume.

4 Rinse the beaker and stirring rod with more distilled water and add these washings to the flask.

Synoptic link

Calculations using titration data are described in Topic 1.4, Concentrations of solutions.

Revision tip

To describe the colour change at the end point of a titration, you will need to know the colours of some common indicators. Phenolphthalein is magenta in alkali and colourless in acid. Methyl orange is red in acid and yellow in alkali.

Key terms

Titre: The volume of liquid delivered from a burette when the end-point of a reaction is reached.

Standard solution: A solution whose concentration is accurately known.

Synoptic link

Calculating the concentration of a diluted solution is described in Topic 1.4, Concentrations of solutions.

5 Add distilled water to the flask to make it up to the mark. Use a dropping pipette for the last cm³.

6 Invert the flask several times to ensure complete mixing.

Making a standard solution by dilution

A standard solution is diluted down by taking a small volume of the original solution and diluting it to a greater volume using distilled water.

1 A volumetric pipette is used to measure out the required volume of the original solution.

2 This is transferred to a volumetric flask of a suitable volume.

3 The flask is made up to the mark as described above.

Measuring volumes of gases

The volume of a gas produced in a chemical reaction can be measured using either a gas syringe or an inverted burette.

Gas syringe

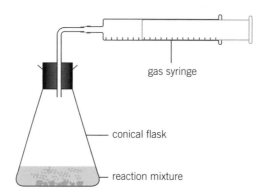

▲ **Figure 4** *Collecting a gas using a gas syringe*

● Up to 100 cm³ of gas can be collected.

● The volume can be measured to the nearest 1 cm³.

Using an inverted burette or measuring cylinder

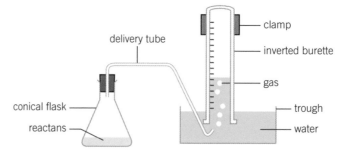

▲ **Figure 5** *Collecting a gas using a measuring cylinder*

● A burette can collect up to 50 cm³ of gas; measuring cylinders can collect much greater volumes.

● This method is not suitable if the gas is soluble in water.

Chapter 1 Practice questions

1 What mass of sodium hydroxide, NaOH is needed to form 200 cm^3 of a 0.25 mol dm^{-3} solution?

A 2000 g

B 2 g

C 32 g

D 10 g

(1 mark)

2 Choose the phrase that correctly completes the following definition:

The relative isotopic mass is the mass of one atom of an isotope compared to....

A The mass of a carbon-12 atom.

B 12 g of carbon-12 atoms.

C one-twelfth of 12 g of carbon-12 atoms.

D one-twelfth of the mass of a carbon-12 atom.

(1 mark)

3 1.40 g of ethene, C_2H_4, reacts with excess steam, H_2O, to form ethanol, C_2H_5OH. The equation for the reaction is:

$C_2H_4 + H_2O \rightarrow C_2H_5OH$

1.06 g of ethanol are formed in the reaction. The % yield of this process is given by the equation:

A $\frac{1.06}{1.40} \times 100$

B $\frac{1.06}{\frac{(1.40 \times 28)}{46}} \times 100$

C $\frac{1.06}{\frac{(1.40 \times 46)}{28}} \times 100$

D $\frac{1.40}{\frac{(1.06 \times 28)}{46}} \times 100$

(1 mark)

4 A compound that contains C, H, and O atoms only has the following % composition by mass: C: 40% O: 53%.

The relative molecular mass of the compound was found to be 60.

The molecular formula of the compound is:

A CH_2O

B CO

C $C_2H_4O_2$

D C_2O_2

(1 mark)

5 Barium chloride solution was added to a solution of sodium sulfate. A white precipitate of barium sulfate was formed. The following are possible equations to describe what has happened. Which of these are correct?

1 $BaCl(aq) + NaSO_4(aq) \rightarrow NaCl(aq) + BaSO_4(s)$

2 $Na^+(aq) + Cl^-(aq) \rightarrow NaCl(s)$

3 $Ba^{2+}(aq) + SO_4^{2-}(aq) \rightarrow BaSO_4(s)$

A 1, 2, and 3

B Only 1 and 2

C Only 2 and 3

D Only 3

(1 mark)

6 Hydrochloric acid in a sample of toilet cleaner was analysed.
25 cm³ of the toilet cleaner was dissolved in water to make a solution of 250 cm³.
10.0 cm³ samples from this **diluted solution** were extracted using a volumetric pipette and titrated against 0.500 mol dm⁻³ sodium hydroxide. The average titre of sodium hydroxide was 15.75 cm³.
The equation for the reaction between sodium hydroxide and hydrochloric acid is:

$NaOH (aq) + HCl(aq) \rightarrow NaCl(aq) + H_2O(l)$.

a Calculate the average number of moles of sodium hydroxide used in the titration. *(1 mark)*

b Write down the number of moles of hydrochloric acid in a 10.0 cm³ sample. *(1 mark)*

c Calculate the concentration of hydrochloric acid, in mol dm⁻³, in the diluted solution to 3 s.d. *(2 marks)*

d What was the concentration, in g dm⁻³, of hydrochloric acid in the original, undiluted, solution of toilet cleaner? *(2 marks)*

e A technician made up 200 cm³ of the standard 0.500 mol dm⁻³ sodium hydroxide solution from a stock solution of 2 mol dm⁻³ sodium hydroxide. Describe how the technician could carry out this procedure, giving details of the equipment that would be used to ensure that the solution was made up to a high degree of accuracy. *(4 marks)*

7 Magnesium carbonate is a common mineral found on Earth.
When it is heated it forms carbon dioxide gas:

$$MgCO_3(s) \rightarrow MgO(s) + CO_2(s)$$

2.00 g of magnesium carbonate was heated. The carbon dioxide gas was collected and its volume was measured.

a Draw a labelled diagram showing how this experiment could be carried out. *(3 marks)*

b Calculate the volume of gas that is formed in this experiment, measured at room temperature and pressure. *(2 marks)*

c Magnesium carbonate has also been detected on the surface of Mars. Scientists expect that any magnesium carbonate present on the surface of Mars will be decomposing slowly to form carbon dioxide, which will be released into the Martian atmosphere.

Calculate the volume, in dm³, that would be occupied by 1 mole of carbon dioxide in the Martian atmosphere. Assume that the pressure in the Martian atmosphere is 600 Pa and that the temperature is 210 K. *(3 marks)*

2.1 A simple model of the atom

Specification reference: EL(g), EL(a)

Inside the atom

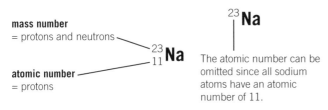

mass number
= protons and neutrons

atomic number
= protons

$^{23}_{11}$Na

^{23}Na

The atomic number can be omitted since all sodium atoms have an atomic number of 11.

▲ **Figure 1** *A nuclear symbol*

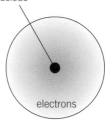

protons and neutrons in nucleus

electrons

▲ **Figure 2** *The nuclear model of the atom*

The nuclear symbol tells you how many protons, electrons, and neutrons there are in an atom of an element:

- number of protons = atomic number (bottom number)
- number of electrons in a neutral atom = atomic number (bottom number)
- number of neutrons = mass number – atomic number (top number – bottom number).

So a sodium atom has 11 protons, 11 electrons, and 12 neutrons.

Atomic models

This model of an atom, known as the nuclear model, has developed over time. Models are tested using experimental investigations and are revised when observations are made that are not predicted by the model.

The succession of models

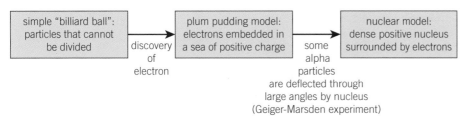

| simple "billiard ball": particles that cannot be divided | | plum pudding model: electrons embedded in a sea of positive charge | | nuclear model: dense positive nucleus surrounded by electrons |

discovery of electron

some alpha particles are deflected through large angles by nucleus (Geiger-Marsden experiment)

▲ **Figure 3** *Timeline for development of the atomic model*

Evidence from atomic spectra and the patterns of ionisation enthalpy led to a more sophisticated model of the atom known as the Bohr model.

The electrons are arranged in shells (also called energy levels).

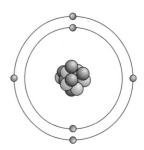

▲ **Figure 4** *The Bohr (electron shell) model of the atom*

Summary questions

1 The electron was discovered in 1897. Suggest one way in which this discovery caused the accepted model of the atom to change. (*1 mark*)

2 How many protons, neutrons and electrons are there in:
   a $^{3}_{1}$H   **b** $^{47}_{20}$Ca   **c** $^{23}_{11}$Na   (*3 marks*)

3 In the Geiger-Marsden experiment, alpha particles (helium nuclei) were fired at a thin sheet of gold atoms. Most of the alpha particles passed through the gold sheet but some of the alpha particles bounced back. Describe the features of the atomic model that were suggested by this observation. In your answer you should make it clear how the experimental evidence links with the model. (*2 marks*)

2.2 Nuclear fusion

Nuclear fusion

Under certain conditions lighter nuclei can join together to form a single heavier nucleus.

High temperatures and/or pressures are required to provide the energy needed to overcome the repulsion between two positive nuclei.

The formation of new elements

Nuclear fusion is common in the centre of stars, where the necessary conditions of high temperature and pressure exist. In most stars, helium is being formed from hydrogen by nuclear fusion.

Heavier elements are formed in some particularly hot or heavy stars.

Nuclear equations

You need to be able to write nuclear equations for fusion reactions:

- In a nuclear equation, the total mass numbers and atomic numbers of the products and reactants must balance.
- Some fusion reactions produce one or more neutrons. The nuclear symbol for a neutron is ^1_0n; this should be included in the equation as a product.

 Worked example: Writing nuclear equations

Write a nuclear equation for the fusion of a $^1_1$H nucleus with a $^2_1$H nucleus to form a single heavier nucleus.

Step 1: Write out the equation, leaving out the details of the product nucleus:

$$^1_1H + {}^2_1H \longrightarrow {}^?_??$$

Step 2: Add up the total mass number and atomic number on the left-hand side of the equation. These will be the mass number and atomic number of the product nucleus:

$$^1_1H + {}^2_1H \longrightarrow {}^3_2?$$

Step 3: Look up the symbol of the element with atomic number = 2 (He)

$$^1_1H + {}^2_1H \longrightarrow {}^3_2He$$

Summary questions

1. Nuclear fusion reactions require high temperatures and pressure. Why are these conditions necessary? *(2 marks)*

2. Complete these nuclear equations for fusion reactions by identifying particle X:

   a. $^2_1H + {}^3_1H \longrightarrow {}^4_2He + X$

   b. $^{13}_6C + {}^4_2He \longrightarrow X + {}^1_0n$ *(2 marks)*

3. A fusion reaction occurs between two ^3_2He to produce a different helium nucleus and two identical hydrogen nuclei. Write the nuclear equation for this reaction. *(1 mark)*

2.3 Shells, sub-shells, and orbitals

Specification reference: EL (e), EL (f)

- Electrons exist in shells and these are designated as $n = 1$, $n = 2$, $n = 3$, etc. The further away a shell is from the nucleus, the larger its n number.
- These shells are sub-divided into **sub-shells**, labelled *s*, *p*, *d*, and *f*.
- Notice that the 4*s* sub-shell is at a lower energy than the 3*d*. This affects the order in which electrons fill up the sub-shells.

Orbitals

- Each sub-shell is further divided into atomic orbitals – each atomic orbital can hold a maximum of two electrons. These two electrons must have opposite (or paired) spins.
- *s*-orbitals and *p*-orbitals have different shapes.
- Orbitals are often represented by boxes. Drawing arrows in these boxes is a good way of representing the filling up of orbitals by electrons.

▲ **Figure 3** *A pair of electrons in an orbital. Electrons have a property known as 'spin' and the direction of the arrows show that the electrons have opposite spins*

Distribution of electrons in atomic orbitals

Shell	Description	Sub-shells
First	$n = 1$ has only one sub-shell	*s*
Second	$n = 2$ has two sub-shells	*s* and *p*
Third	$n = 3$ has three sub-shells	*s*, *p*, and *d*
Fourth	$n = 4$ has four sub-shells	*s*, *p*, *d*, and *f*

- An *s* sub-shell has 1 orbital holding a maximum of 2 electrons.
- A *p* sub-shell has 3 orbitals holding a maximum of 6 electrons.
- A *d* sub-shell has 5 orbitals holding a maximum of 10 electrons.

1. The orbitals are filled in order of increasing energy.
2. Where there is more than one orbital at the same energy, the orbitals are first occupied singly by electrons. When each orbital is singly occupied, then electrons pair up in orbitals.
3. Electrons in singly occupied orbitals have parallel spins.
4. Electrons in doubly occupied orbitals have opposite spins.

Representing electron distribution

The way that electrons are distributed between sub-shells is called the electron configuration of an atom or ion. You could be asked to write out an electronic configuration or to draw out the arrangement of electrons in boxes.

▲ **Figure 4** *The electronic configuration of N and Na atoms, showing electrons in boxes*

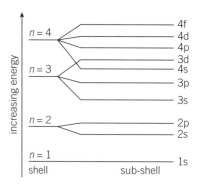

▲ **Figure 1** *Shells $n = 1$ to $n = 4$ and sub-shells*

s-orbital

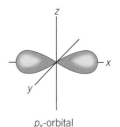

p_x-orbital

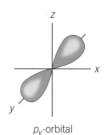

p_y-orbital

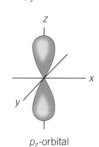

p_z-orbital

▲ **Figure 2** *The shapes of s- and p-orbitals*

🖩 Worked example: Electron configuration of a magnesium atom

Step 1: Sodium has an atomic number of 11, so there are 11 protons and 11 electrons in an Na atom.

Step 2: The first 2 electrons go into the lowest available sub-shell: $1s^2$.

Step 3: The next 2 electrons go into the next lowest sub-shell: $2s^2$.

Step 4: The next 6 electrons go into the next lowest sub-shell: $2p^6$.

Step 5: The total number of electrons added so far is 10; a Na atom has 11 electrons, so just one more needs to go into the next lowest sub-shell: $3s^1$.

Step 6: The electron configuration is $1s^2 2s^2 2p^6 3s^1$.

s, *p*, and *d* blocks

The periodic table is divided into blocks. The name of the block tells you the type of sub-shell occupied by the outermost (highest energy) electrons.

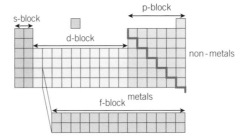

▲ **Figure 5** *The periodic table, showing the s, p, d, and f blocks*

🖩 Worked example: The outer sub-shell structure of arsenic (As)

Step 1: Arsenic is in the *p*-block. The outer sub-shell will be a *p* sub-shell.

Step 2: Arsenic is in the 3rd column of the *p*-block. The *p* sub-shell contains 3 electrons.

Step 3: Arsenic is in the 4th period. The outer sub-shell configuration is $4p^3$.

Summary questions

1 The orbitals in a 3*p* sub-shell can be represented as boxes, as shown in the following diagram. Copy and complete the diagram to show the arrangement of the 3*p* electrons in a sulfur atom. *(2 marks)*

3*p*

2 Write out the full electronic configuration of:
   a an Al atom   b an O^{2-} ion   c a Ca^{2+} ion.   *(3 marks)*

3 a Complete the electron configuration of a V atom: [Ar] ___ ___.
   b Write out the electron configuration of the outer electron sub-shell of a Polonium (Po) atom.   *(2 marks)*

Chapter 2 Practice questions

1 The symbol for an ion of lithium is $_3^7\text{Li}^+$.

Which of the rows in the table correctly describes the numbers of the different sub-atomic particles in this ion? *(1 mark)*

	Protons	Neutrons	Electrons
A	3	4	2
B	3	4	3
C	4	3	3
D	4	3	2

2 Ernest Rutherford proposed a new model of the atom in 1911. In this model, a small dense nucleus is surrounded by a cloud of negative electrons. The evidence that best supports this model is:

A The jumps observed in the pattern of successive ionisation enthalpies of atoms

B The discovery of the electron

C The deflection of alpha particles through large angles by gold nuclei

D The lines seen in atomic emission spectra. *(1 mark)*

3 Which of the following equations represents a fusion reaction?

1 $2\,_6^{12}\text{C}\,_0^1\text{n} \rightarrow\ _{12}^{23}\text{Mg}$ 2 $_3^7\text{Li}\,_1^1\text{H} \rightarrow 2\,_2^4\text{He}$, 3 $_{15}^{32}\text{P} \rightarrow\ _{-1}^{0}\text{e} +\ _{16}^{32}\text{S}$

A 1, 2, and 3

B Only 1 and 2

C Only 2 and 3

D Only 3 *(1 mark)*

4 Which of the following statements about *p*-orbitals is true?

1 Each *p*-orbital can hold 6 electrons.

2 *p*-orbitals have a spherical shape.

3 Ground state electrons in helium atoms do not occupy *p*-orbitals.

A 1, 2, and 3

B Only 1 and 2

C Only 2 and 3

D Only 3 *(1 mark)*

5 Fusion reactions occur in centre of the Sun and other stars. High temperatures and pressures are necessary for fusion reactions.

a i State what is meant by the term fusion reaction. *(2 marks)*

ii Explain why high temperatures and pressures are necessary for fusion reactions to occur. *(2 marks)*

b In the Sun, the most common fusion process involves isotopes of hydrogen. In other stars, helium nuclei are also involved in fusion processes.

i Complete this equation for the fusion process that occurs in the Sun: _____ + _____ $\rightarrow\ _2^3\text{He}$. *(2 marks)*

ii Identify the nucleus X formed as a result of the fusion of helium nuclei: $3\,_2^4\text{He} \rightarrow \text{X}$. *(2 marks)*

iii Suggest why this process occurs in some stars, but is not the main fusion process in the Sun. *(1 mark)*

6 This question is about the electron configurations of some atoms and ions.

a The diagram below shows an incomplete diagram to show the arrangement of electrons in an atom of silicon (Si).

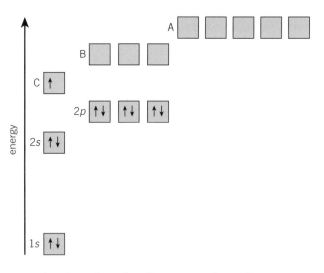

i Complete the diagram to show the arrangement of electrons in an atom of silicon. *(3 marks)*

ii How many electrons can occupy the sub-shell labelled C on the diagram? *(1 mark)*

iii Identify the sub-shells labelled A and B in this diagram.
A: _____ B: _____. *(2 marks)*

iv State one difference between the shape of an *s*-orbital and a *p*-orbital. *(1 mark)*

b Complete the table to show the electron configurations of some atoms and ions. *(4 marks)*

Species	Electron configuration
Na⁺	....
....	$1s^22s^22p^5$
Cr	[Ar] ....

c Antimony (Sb) is in the *p*-block of the periodic table.

i What does this information tell you about the arrangement of electrons in an antimony atom? *(1 mark)*

ii Write down the configuration of the outer sub-shell in an antimony atom. *(1 mark)*

3.1 Chemical bonding

Specification reference: EL (i)

The types of bonding

The type of bond depends on the two atoms involved in the bond.

▼ **Table 1**

	Metal	Non-metal
Metal	metallic bonding	ionic bonding
Non-metal	ionic bonding	covalent bonding

Ionic bonding

The metal atom transfers electron(s) to the non-metal atom. This results in the formation of charged ions. The ions that are formed often have full outer shells, which makes them particularly stable.

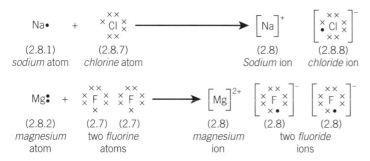

▲ **Figure 1** *Dot-and-cross diagrams for the formation of ionic compounds*

Ionic lattices and electrostatic attraction

- The cations (positive ions) and anions (negative ions) produced are held together in a giant ionic lattice.
- There is an electrostatic attraction between the cations and anions.

Covalent bonding

The two non-metal atoms involved in a covalent bond *share* one or more pairs of electrons. If two pairs of electrons are shared, then a double bond is formed.

Dative covalent bonds

Some molecules contain **dative covalent bonds**.

▲ **Figure 2** *Dative covalent bonds in molecules and ions*

Covalent bonds and electrostatic attraction

- There is an electrostatic attraction between the positive nuclei of the two atoms in the bond and the shared pair of negative electrons.
- This is greater than the repulsion between the two nuclei.

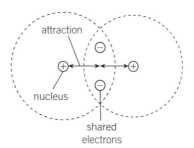

▲ **Figure 3** *The electrostatic attractions in a covalent bond*

Revision tip

Note that you only need to draw the outermost electrons.

Revision tip

It is important to make it clear that the ions are completely separate from each other, with no shared electrons. Drawing a bracket round the ion helps to clarify this. Do not forget the ionic charge!

Synoptic link

The structure of a giant ionic lattice is described in more detail in Topic 5.2, Bonding, structure, and properties.

▼ **Table 2** *Dot-and-cross structures of covalent molecules*

Molecular formula	Dot-and-cross diagram	Structural formula
H_2	H×H	H—H
NH_3	H×N with H above and below	H—N with H above and below
H_2O	H×O with H and dots	H—O with H
O_2	O×O	O=O
N_2	N×N	N≡N
CO_2	O×O×O	O=C=O

Revision tip

To draw a *dot-and-cross* diagram for a covalent molecule you need to first make sure that you include the correct number of electrons from each atom.

Key term

Dative covalent bond: A type of covalent bond in which both electrons come from the same atom.

Summary questions

1 What type of bonding is most likely between:
   a lithium and bromine, **b** phosphorus and oxygen? *(2 marks)*

2 Draw a *dot-and-cross* diagram to show the arrangement of outer shell electrons in the ionic compound magnesium oxide, MgO. *(2 marks)*

3 In N_2O there is a central N atom bonded to a second N atom and an oxygen atom. Draw a *dot-and-cross* diagram to suggest a possible arrangement of electrons in this molecule. *(2 marks)*

3.2 Shapes of molecules

Specification reference: EL (k)

Electron pair repulsion

- The shape of a molecule depends on the number of **groups** of electrons in the outer shell of the central atom.
- A group of electrons could be a bonding pair (a single bond), two bonding pairs (a double bond), three bonding pairs (a triple bond), or a lone pair.
- Groups of electrons repel each other.
- Groups of electrons will arrange themselves so as to be as far apart in space as possible, to minimise this repulsion.

The arrangement of electron pairs in molecules

You need to learn the 3-dimensional arrangements taken up by electron pairs in atoms with 2, 3, 4, 5, and 6 electron pairs in the outer shell. You will then be able to use the terms in this table to describe the shape of the molecule. You also need to know the bond angles that result from these arrangements.

Revision tip

Be careful how you describe electrons; you can describe the repulsion of electron pairs if there are no double or triple bonds in a molecule. If double or triple bonds are present, you should discuss the repulsion of electron groups.

Revision tip

The most common arrangement for electron pairs is tetrahedral, because many atoms in molecules have 8 electrons in their outer shell. You should learn any common exceptions to this, such as BF_3 and SF_6.

Number of pairs of electrons	2	3	4	5	6
Arrangement of electrons	Linear	Triangular planar	Tetrahedral	Trigonal bipyramid	Octahedral
Example	Beryllium chloride, $BeCl_2$	Boron trifluoride, BF_3	Methane, CH_4	Phosphorus pentachloride, PCl_5	Sulfur hexafluoride, SF_6
Diagram	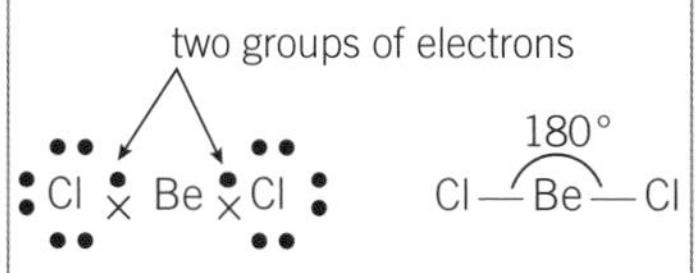 two groups of electrons $Cl \times Be \times Cl$ $180°$ $Cl—Be—Cl$	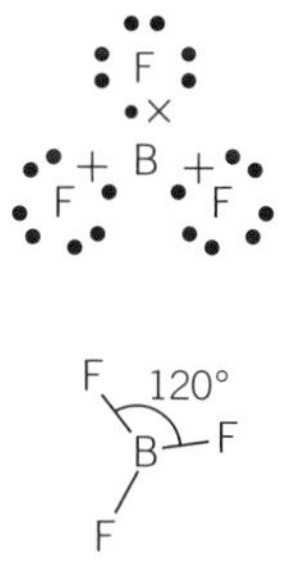 F, B, F, F $120°$ B—F	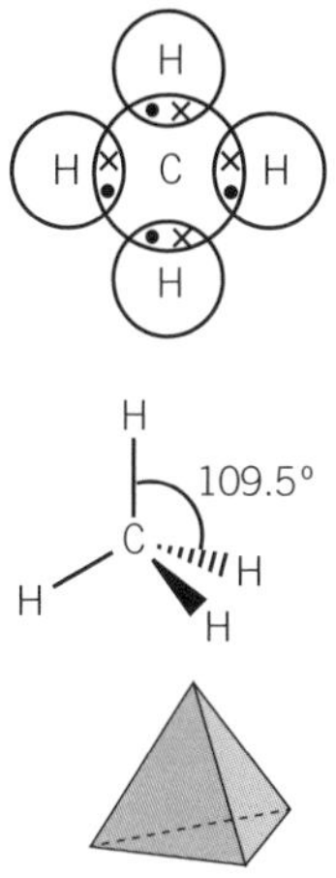 H, C, H, H $109.5°$	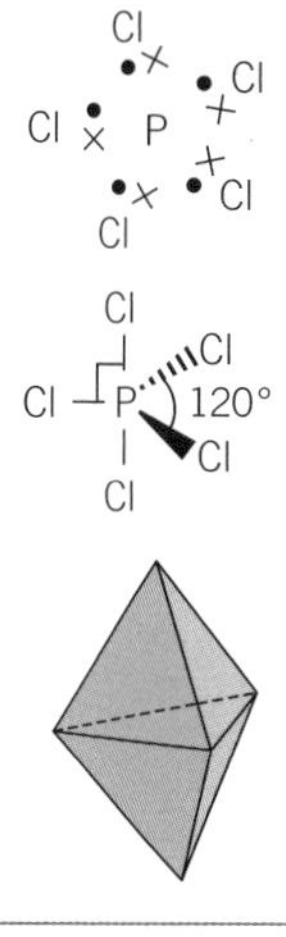 Cl, P, Cl, Cl $120°$	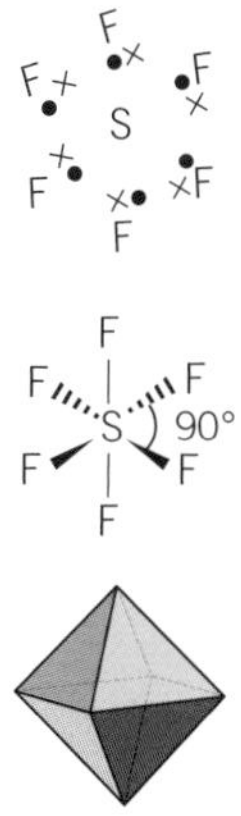 F, S, F $90°$
Bond angle	180°	120°	109.5	90 and 120°	90°

The effect of lone pairs

The shape of the molecule

If the molecule consists of a mixture of lone pairs and bonding pairs, you will need to use some different words to describe the shape of the molecule:

Key term

Lone pair: A pair of electrons in the outer shell of an atom that are not part of a covalent bond.

Pairs of electrons in outer shell	3 bonding pairs 1 lone pair	2 bonding pairs 2 lone pairs
Shape	Pyramidal	Bent (or V-shaped)
Example	Ammonia, NH_3	Water, H_2O
Diagram	ammonia	water

The bond angles

Lone pairs repel more strongly than bonding pairs of electrons. The effect of this is to force the bonding pairs slightly closer together, which makes the bond angle slightly smaller. On average, each lone pair present decreases the bond angle by about 2.5°.

Model answer: Use the idea of electron pair repulsion to describe the shape of a hydrogen sulfide molecule, H_2S. In your answer, you should predict and explain a value for the bond angle.

Electron pairs repel each other and get as far away as possible to minimise repulsion. The central S atom in the H_2S molecule has 4 electron pairs in its outer shell. These electron pairs take up a tetrahedral arrangement, but two of the electron pairs are lone pairs, so the shape of the molecule is V-shaped.

The angle between electron pairs in a tetrahedral arrangement is 109.5°, but the presence of two lone pairs decreases the angle between the bonding pairs by 5° (2 × 2.5°). So the bond angle is likely to be 104.5°.

Molecules with double and triple bonds

You can think of a double or triple bond as a single group of electrons. This often means that there are only 2 or 3 groups of electrons around a central atom.

Groups of electrons	2 double bonds	1 triple bond, 1 bonding pair	1 double bond, 2 bonding pairs
Shape	Linear	Linear	Triangular planar
Example	Carbon dioxide, CO_2	Ethyne, C_2H_2	Methanal, CH_2O
Diagram	$O=C=O$ 180°	$H-C\equiv C-H$ 180°	each 120°

Summary questions

1. Name the 3-dimensional arrangement taken up by (a) 6 pairs of electrons (b) 3 pairs of electrons. Assuming that all the pairs are bonding pairs, suggest the bond angle in each case. *(4 marks)*

2. Draw a *dot-and-cross* diagram for the molecule phosphorus trichloride, PCl_3. Use this diagram and the principle of electron pair repulsion to describe the shape of a PCl_3 molecule. In your answer you should predict and explain the bond angle in the molecule. *(7 marks)*

3. a Suggest a *dot-and-cross* diagram for a sulfate ion, SO_4^{2-}, which contains an S atom surrounded by 4 O atoms, two of which have negative charges.

   b Draw a 3-dimensional diagram of the shape of a sulfate ion. In your answer you should predict and explain the bond angle in the molecule. *(5 marks)*

1 In which of these compounds is the bonding most likely to be ionic?

   A NO_2

   B MgH_2

   C HCl

   D BF_3 *(1 mark)*

2 Covalent bonds cause a force of attraction between atoms. The best explanation of this is:

   A The atoms now have a full outer shell and so are more stable.

   B There is electrostatic attraction between nuclei and the electron pair.

   C The two electrons in the bond attract one another.

   D There is electrostatic attraction between positive and negatively charged atoms. *(1 mark)*

3 Which of these molecules is most likely to contain a dative covalent bond, in order to achieve noble gas structures for all the atoms in the molecule?

   A CO_2

   B CO

   C N_2

   D H_2CO *(1 mark)*

4 The *dot-and-cross* diagram of the molecule BF_3 is shown.

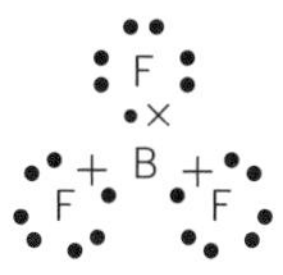

   Which of these statements about the molecule are true?

   1 It has a pyramidal structure.

   2 The bond angle will be 120°.

   3 The molecule will be planar. *(1 mark)*

   A 1,2, and 3

   B Only 1 and 2

   C Only 2 and 3

   D Only 3 *(1 mark)*

5 What name is given to the shape of a molecule of sulfur hexafluoride, with 6 electron pairs in the outer shell of the sulfur atom?

   A Octahedral   **B** Tetrahedral

   C Hexahedral   **D** Hexagonal *(1 mark)*

6 Ammonium chloride, NH_4Cl is an ionic compound used as a fertiliser. The two ions present in the compound are the ammonium ion, NH_4^+, and the chloride ion, Cl^-. At room temperature, they form a 3-dimensional ionic lattice.

   a i Draw out a diagram to show the arrangement of these ions in a layer of the ionic lattice. *(2 marks)*

   ii Describe the forces that hold this lattice together at room temperature. *(2 marks)*

   b The ammonium ion contains covalent bonds. The *dot-and-cross* diagram of the ammonium ion is shown.

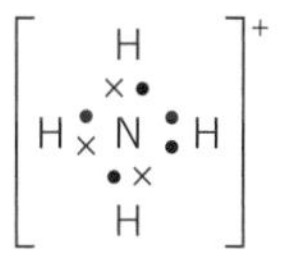

   i Explain how this *dot-and-cross* diagram shows the presence of a dative covalent bond in the ammonium ion. *(2 marks)*

   ii Use the *dot-and-cross* diagram to predict the shape and bond angle of the ammonium ion. *(5 marks)*

   c In the soil, ammonium ions can be broken down into nitrogen gas, N_2, by certain types of bacteria.

   Draw a *dot-and-cross* diagram to show the arrangement of electrons in a molecule of nitrogen gas. *(2 marks)*

4.1 Energy out, energy in

Specification reference: DF (e), DF (d), DF(f), DF(a)

Enthalpy changes

The enthalpy change, $\Delta_r H$, for a reaction is the quantity of energy transferred to or from the surroundings when the reaction is carried out in an open container.

Exothermic and endothermic reactions

- An exothermic reaction gives out energy from the system to the surroundings. The temperature of the surroundings increases, $\Delta_r H$ is negative.
- An endothermic reaction takes energy into the system from the surroundings. The temperature of the surroundings decreases, $\Delta_r H$ is positive.

Bond breaking and bond forming

- Breaking chemical bonds requires energy and so is an endothermic process.
- Forming chemical bonds releases energy and so is an exothermic process.

The overall effect of bond breaking and forming

Many reactions involve both bond breaking:

- If the energy released by forming bonds is greater than the energy required to break bonds, then the reaction will be **exothermic.**
- If the energy released by forming bonds is less than the energy required to break bonds, then the reaction will be **endothermic.**

Standard enthalpy changes

Several types of enthalpy change are particularly important. These are given specific names and symbols. If an enthalpy change is measured under standard conditions, it is known as a standard enthalpy change:

- (standard) enthalpy change of reaction ($\Delta_r H$), $2NaOH(aq) + H_2SO_4(aq) \rightarrow Na_2SO_4(aq) + 2H_2O(l)$; $\Delta_r H^\ominus = -115\,kJ\,mol^{-1}$
- $\Delta_c H$ (standard) enthalpy change of combustion ($\Delta_c H$), $CH_4(g) + 2O_2(g) \rightarrow CO_2(g) + 2H_2O(l)$; $\Delta_c H = -890\,kJ\,mol^{-1}$
- (standard) enthalpy change of formation ($\Delta_f H$), $H_2(g) + \frac{1}{2}O_2(g) \rightarrow H_2O(l)$; $\Delta_f H = -286\,kJ\,mol^{-1}$
- (standard) enthalpy change of neutralisation ($\Delta_{neut} H$), $H^+(aq) + OH^-(aq) \rightarrow H_2O(l)$; $\Delta_{neut} H = -58\,kJ\,mol^{-1}$

Enthalpy changes and moles

You can calculate the amount of energy released or taken in during a reaction by using values for enthalpy changes of reaction, $\Delta_r H$, and data about the number of moles of the reactants.

🖩 Worked example: Enthalpy change for an acid-base reaction

Sodium hydroxide solution reacts with sulfuric acid in an exothermic reaction:

$2NaOH(aq) + H_2SO_4(aq) \rightarrow Na_2SO_4(aq) + 2H_2O(l)$ $\Delta_r H^\ominus = -115\,kJ\,mol^{-1}$

Calculate the energy released when $100\,cm^3$ of $2.00\,mol\,dm^{-3}$ NaOH reacts with excess H_2SO_4.

Step 1: Calculate the number of moles of NaOH $= \dfrac{100}{1000} \times 2.00 = 0.200\,mol$.

Step 2: The $\Delta_r H^\ominus$ value gives the enthalpy change for the number of moles of the equation: number of moles of NaOH in the equation $= 2$.

Step 3: Find the ratio between the number of moles that actually react and number of moles in the equation: $\frac{0.2}{2} = 0.1$.

Step 4: Multiply the $\Delta_r H$ by this ratio: $-115 \times 0.1 = -11.5\,\text{kJ}$.
11.5 kJ of energy will be released in this reaction (released because the sign is negative).

Calculating enthalpy changes from experimental results

Experiments to measure enthalpy changes usually involve transferring energy to or from water.

The amount of energy transferred to or from a mass of water is given by the equation:

E, energy transferred = m (mass of water) × c (specific heat capacity of water) × ΔT (temperature change of the water).

 Worked example: Calculating enthalpy changes of combustion

0.880 g of heptane (C_7H_{16}) undergo complete combustion and the energy released is used to heat 250 cm^3 of water. The temperature of the water rises by 19.0 °C. Calculate the $\Delta_c H$ of heptane in kJ mol^{-1}. Give your answer to 3 significant figures.

Step 1: Use E = m c ΔT to calculate the energy transferred to the water:
E = 250 × 4.18 × 19 = 19 855 J.

(250 cm^3 water is assumed to have a mass of 250 g, because the density of water is 1.00 g cm^{-3}.)

Step 2: Convert the energy released into kJ = $\frac{19855}{1000}$ = 19.855 kJ.

Step 3: Calculate the number of moles of heptane burnt: moles = $\frac{\text{mass}}{M_r} = \frac{0.880}{100}$
= 0.0880 mol.

Step 4: Scale the energy released to find the energy that would be released if 1 mole was burnt: = $\frac{\text{energy released in experiment}}{\text{number of moles burnt}} = \frac{19.855}{0.0880}$ = 2 256.25 kJ mol^{-1}.

Step 5: Write down the $\Delta_c H$ value with the correct sign, units and significant figures:
$\Delta_c H = -2\,260$ kJ mol^{-1}.

 Worked example: Calculating enthalpy changes in solution

An excess of zinc is added to 100 cm^3 of 1.00 mol dm^{-3} copper(II) sulfate. The maximum temperature rise measured was 36.0 °C.

Calculate $\Delta_r H$ for this reaction in kJ per mole of copper sulfate. Give your answer to 3 s.f.

Step 1: Use E = m c ΔT to calculate the energy transferred to the water:
E = 100 × 4.18 × 36.0 = 15 048 J
(100 cm^3 solution is assumed to contain 100 g water).

Step 2: Convert the energy released into kJ: = $\frac{15048}{1000}$ = 15.048 kJ.

Step 3: Calculate the number of moles of copper sulfate reacted:
moles = $\frac{100}{1000} \times 1$ = 0.1 mol.

Step 4: Scale the energy released to find the energy that would be released if one mole was burnt: = $\frac{\text{energy released in experiment}}{\text{number of moles reacted}} = \frac{15.048}{0.1}$ = 150.48 kJ mol^{-1}.

Step 5: Write down the $\Delta_r H$ value with the correct sign, units and significant figures:
$\Delta_r H = -150$ kJ mol^{-1}.

Revision tip

Make sure that you are careful about the wording when you are comparing bond breaking and bond forming. One process requires energy and the other releases energy. So answers such as 'bond breaking requires more energy than bond forming' are not correct.

Key terms

Standard conditions: 1 atmosphere pressure (101 kPa), 298 K (25°C) and a concentration of 1 mol dm^{-3} for any solutions.

Standard enthalpy change of reaction ($\Delta_r H$): The enthalpy change for a reaction (described by an equation) that occurs between the number of moles of the reactants specified in the equation.

Standard enthalpy change of combustion $\Delta_c H$: The enthalpy change when 1 mole of a substance burns completely in oxygen under standard conditions.

Standard enthalpy change of formation, $\Delta_f H$: The enthalpy change when 1 mole of a substance is formed from its constituent elements in their standard states under standard conditions.

(Standard) enthalpy change of neutralisation ($\Delta_{neut} H$): The enthalpy change when 1 mole of hydrogen ions react with one mole of hydroxide ions to form 1 mole of water under standard conditions and in solutions with a concentration of 1 mol dm^{-3}.

Synoptic link

You need to be able to describe how to obtain results to find energy transferred in chemical reactions. You can find this information in Experimental techniques.

Revision tip

The value of c, the specific heat capacity of water, is $4.18\,\mathrm{J\,g^{-1}\,K^{-1}}$.

 Go further

In some experiments, energy is transferred to a solution of an ionic compound, such as copper sulfate solution. The volume of the solution is measured. What assumptions are necessary to complete the calculation for the amount of energy transferred? Do you think these assumptions are justified?

Summary questions

1. All combustion reactions are exothermic.
    a State what you can conclude about:
       i the sign of $\Delta_r H$ for a combustion reaction
       ii the temperature change that will occur in the surroundings when combustion occurs. *(2 marks)*
    b Explain why combustion reactions are exothermic, using ideas about bonding in your answer. *(2 marks)*

2. $1.60\,\mathrm{g}$ of methanol (CH_3OH) is burnt in a spirit burner and used to heat $150\,\mathrm{cm^3}$ of water. The maximum temperature rise recorded was $41.0\,^\circ C$. Calculate a value for $\Delta_c H$ of methanol, in $\mathrm{kJ\,mol^{-1}}$. Give your answer to 3 s.f. *(4 marks)*

3. Nitrogen triiodide, NI_3, decomposes in an exothermic reaction:
   $2NI_3(s) \rightarrow N_2(g) + 3I_2(g)\ \Delta_r H^\ominus = -290\,\mathrm{kJ\,mol^{-1}}$.
   Calculate the energy released by this reaction if 1 kg of NI_3 decomposes. *(3 marks)*

4.2 Enthalpy cycles

Specification reference: DF (g)

Hess' Law

Hess' Law states that as long as the starting and finishing points are the same, the enthalpy change for a chemical reaction will always be the same, no matter how you go from start to finish. It is useful for calculating unknown enthalpy changes from ones for which data is available.

Hess' Law calculations from $\Delta_f H$ data

The key to Hess' Law calculations is to construct a triangular enthalpy cycle, to show the relationship between the various enthalpy changes. If $\Delta_f H$ data is given, then the substances at the bottom of the triangle will be elements in their standard states.

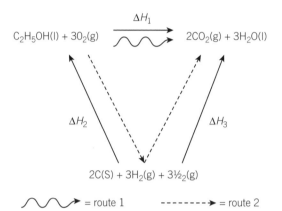

▲ **Figure 1** *Enthalpy cycle for Hess' Law calculation using $\Delta_f H$ data*

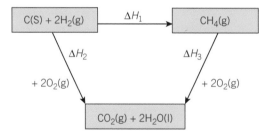

▲ **Figure 2** *Enthalpy cycle for Hess' Law calculation using $\Delta_c H$ data*

> **Revision tip**
> Remember that if an equation is reversed, the sign for ΔH is also reversed.

> **Revision tip**
> Remember to multiply $\Delta_f H$ by the number of moles in the equation.

> **Revision tip**
> You can calculate an enthalpy change for a reaction without drawing an enthalpy cycle by remembering that $\Delta_r H = \Delta_f H(\text{products}) - \Delta_f H (\text{reactants})$.

> 🖩 **Worked example: Hess' Law calculation using $\Delta_f H$ data**
>
> Use the data provided to calculate a value for the enthalpy change represented by the equation:
>
> $$C_2H_5OH(l) + 3O_2(g) \rightarrow 2CO_2(g) + 3H_2O(l)$$
>
> | Compound | $\Delta_f H(\text{kJ mol}^{-1})$ |
> |---|---|
> | $C_2H_5OH(l)$ | −277 |
> | $CO_2(g)$ | −394 |
> | $H_2O(l)$ | −286 |
>
> **Step 1:** Construct an enthalpy cycle showing the reactants and products of the reaction as well as elements in their standard states. Make sure that the arrows point in the correct directions to represent the enthalpy changes.
>
> **Step 2:** Choose suitable labels to represent each arrow $(\Delta H_1, \Delta H_2, \text{etc.})$.
>
> **Step 3:** Calculate ΔH_2: $\Delta H_2 = \Delta_f H (C_2H_5OH(l)) = -277 \text{ kJ mol}^{-1}$.

> **Common misconception: Missing data for $\Delta_f H$?**
>
> Students often worry that they cannot find data to complete Hess' Law calculations. But, of course, the enthalpy change of formation of an element in its standard state must be zero, by definition, because forming an element from the same element involves no chemical change. So $\Delta_f H (O_2)$ can be left out of the calculation.

Revision tip

Make sure that the arrows in your enthalpy cycle point in the correct direction. If $\Delta_f H$ data is given, elements will appear at the bottom of the cycle and the arrows will point up; if $\Delta_c H$ data is given, combustion products will appear at the bottom of the cycle and arrows will point down.

Maths skill: Checking the sign

Think carefully about your answer, to check whether it is mathematically sensible. If the reaction you are dealing with is a combustion reaction, this will be exothermic. Your answer should have a negative sign. If the sign does not seem sensible, then check your working, particularly where you are handling numbers with negative signs. It is easy to accidently leave some of these signs out of your calculation.

Revision tip

Always include a sign in any value of ΔH that you give in an answer even if the value is positive.

Step 4: Calculate ΔH_3: $\Delta H_3 = 2 \times \Delta_f H\ (CO_2(g)) + 3 \times \Delta_f H\ (H_2O(l)) = (2 \times -394) + (3 \times -286) = -1646\ \text{k mol}^{-1}$.

Step 5: Use Hess' Law to write down a relationship between $(\Delta H_1, \Delta H_2,$ and $\Delta H_3)$:
$\Delta H_1 = \Delta H_3 - \Delta H_2$.

Step 6: Plug the calculated values for ΔH_2 and ΔH_3 into the relationship to find ΔH_1:
$\Delta H_1 = -1646 - (-277) = -1369\ \text{kJ mol}^{-1}$.

 Worked example: Hess' Law calculation using $\Delta_c H$ data

Calculate the enthalpy change for the reaction $C(s) + 2H_2(g) \rightarrow CH_4(g)$.

Substance	$\Delta_c H$ / kJ mol^{-1}
$C(s)$	−394
$H_2(g)$	−286
$CH_4(g)$	−890

Step 1: Construct an enthalpy cycle showing the reactants and products of the reaction as well as combustion products (CO_2 and H_2O). Make sure that the arrows point in the correct directions to represent the enthalpy changes.

Step 2: Add in the correct number of oxygen molecules to make the combustion equations balance, as shown in the cycle above.

Step 3: Choose suitable labels to represent each arrow $(\Delta H_1, \Delta H_2,$ etc.).

Step 4: Calculate ΔH_2: $\Delta H_2 = \Delta_c H(C(s)) + 2 \times \Delta_c H(H_2) = -394 + (2 \times -286)$
$= -966\ \text{kJ mol}^{-1}$.

Step 5: Calculate ΔH_3: $\Delta H_3 = \Delta_c H\ (CH_4(g)) = -890\ \text{kJ mol}^{-1}$.

Step 6: Use Hess' Law to write down a relationship between $(\Delta H_1, \Delta H_2,$ and $\Delta H_3)$:
$\Delta H_1 = \Delta H_2 - \Delta H_3$.

Step 7: Plug the calculated values for ΔH_2 and ΔH_3 into the relationship to find ΔH_1:
$\Delta H_1 = -966 - (-890) = -76\ \text{kJ mol}^{-1}$.

Summary questions

1 What $\Delta_f H$ data would you need in order to calculate the enthalpy change for the reaction: $C_2H_4(g) + H_2(g) \rightarrow C_2H_6(g)$? *(2 marks)*

2 Hydrazine, N_2H_4 reacts with oxygen to form N_2 and H_2O:
$N_2H_4(l) + O_2(g) \rightarrow N_2(g) + 2H_2O(l)$.
Use the following $\Delta_f H$ values to calculate a value for the $\Delta_r H$ of the reaction of hydrazine with oxygen:
$\Delta_f H$ values / kJ mol^{-1}: $N_2H_4(l) + 50.6$, $H_2O(l) -285.8$. *(4 marks)*

3 The enthalpy change of formation of propane, $C_3H_8(g)$ can be found using $\Delta_c H$ data for propane and the elements carbon and hydrogen.
a Draw out a fully labelled Hess's Law cycle that would enable you to calculate the $\Delta_f H$ of propane by this method.
b Use the following data to calculate a value for the $\Delta_f H$ of propane:
$\Delta_c H\ [C(s)] = -394\ \text{kJ mol}^{-1}$, $\Delta_c H[H_2(g)] = -286\ \text{kJ mol}^{-1}$
$\Delta_c H\ [C3H_8(g)] = -2\ 219\ \text{kJ mol}^{-1}$. *(6 marks)*

4.3 Bond enthalpies

Specification reference: DF (e), DF (g)

Bond enthalpies

Bond enthalpy is a measure of the strength of a covalent bond; the greater the bond enthalpy, the stronger the bond. Short bonds are stronger than long bonds, so they have a greater bond enthalpy.

Bond enthalpies and enthalpy changes of reaction

You can use bond enthalpy data to calculate a value for the enthalpy change of a reaction involving covalent molecules.

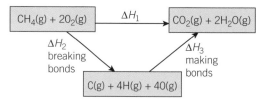

▲ **Figure 1** *Using Hess' Law to find the enthalpy change of a reaction:* $\Delta H_1 = \Delta H_2 + \Delta H_3$

🖩 Worked example: Enthalpy change from bond enthalpy data

Calculate the enthalpy change of combustion for ethane (C_2H_6) using the bond enthalpy data from the table in Table 1:

$$C_2H_6 + 3\tfrac{1}{2}O_2 \rightarrow 2CO_2 + 3H_2O$$

Step 1: Draw out diagrams to show the structures of the reactants and products.

Step 2: List the bonds broken and formed:

Bonds broken	Bonds formed
$1 \times C-C$ (347) = +347	
$6 \times C-H$ (413) = +2478	$4 \times C=O$ (805) = 3220
$3.5 \times O=O$ (498) = +1743	$6 \times O-H$ (464) = 2784
Total = **+4568**	Total = **−6004**
(+ sign because bond breaking)	(− sign because bond making)

Step 3: Multiply the number of bonds broken and formed by the bond enthalpies to calculate the total enthalpy change for bond breaking and bond forming.

Step 4: Add together the enthalpy changes for bond breaking and bond forming:

$$\Delta H = +4\,568 - 6\,004 = -1\,436\ \text{kJ mol}^{-1}.$$

Problems with this calculation

The value obtained by this calculation is not the same as the data book value for $\Delta_r H$. There are two main reasons for this:

1. Average bond enthalpies are used in the calculation; the actual bond enthalpies in particular molecules will be slightly different.

2. Bond enthalpy data are for gaseous molecules; some molecules may be in a liquid state at 298 K.

Common misconception: Adding up the bonds

There are a number of common errors to beware of:

- Miscounting the number of C–C bonds in a molecule: C_2H_6 has 1 C–C bond, not 2.

- In a combustion reaction, forgetting to include the bonds in the oxygen molecules.

- Treating the bonds in CO_2 as C–O, not C=O.

Key term

Bond enthalpy: The energy required to break 1 mole of a particular bond, averaged over a range of different gaseous compounds.

▼ **Table 1**

Bond	Bond enthalpy / kJ mol⁻¹
C–C	+347
C–H	+413
O=O	+498
O–H	+464
C=O (in CO₂)	+805
C–O	+358
H–H	+436
C=C	+612

Summary questions

1 Describe how C–C and C=C bonds differ in:
   a bond strength,   b bond length.   (*2 marks*)

2 a Calculate a value for the $\Delta_c H$ of methane, CH₄ using bond enthalpy data.
   b Suggest why this value differs from the data book value for $\Delta_c H$ of methane.   (*4 marks*)

3 Cyclopropane, C₃H₆, has a structure in which the three carbon atoms are arranged in the form of a ring. The $\Delta_c H^\ominus$ of cyclopropane is −2 091 kJ mol⁻¹. Use this value and other bond enthalpy data from this page to deduce a value for the bond enthalpy of C–C in cyclopropane. Comment on your answer.   (*7 marks*)

4.4 Ions in solution

Revision tip

Notice that although you can imagine the dissolving process involves the breaking up of an ionic lattice – an endothermic process – the lattice enthalpy refers to the formation of an ionic lattice and is therefore **exothermic**.

Key terms

Enthalpy change of solution, $\Delta_{sol}H$: The enthalpy change when 1 mole of solute dissolves to form a solution, e.g.

$CaCl_2(s) + aq \rightarrow Ca^{2+}(aq) + 2Cl^-(aq)$

Lattice enthalpy, $\Delta_{LE}H$: The enthalpy change when 1 mole of an ionic solid is formed from gaseous ions, e.g.

$Ca^{2+}(g) + 2Cl^-(g) \rightarrow CaCl_2(s)$

Enthalpy change of hydration, $\Delta_{hyd}H$: The enthalpy change when 1 mole of a gaseous ion is hydrated by forming bonds to water molecules, e.g.

$Ca^{2+}(g) + aq \rightarrow Ca^{2+}(aq)$

Hydrated ion: An ion bonded to water molecules, e.g.

$Ca^{2+}(aq)$

▼ **Table 1** Enthalpy of hydration values for some cations

Ion	$\Delta_{hyd}H/kJ\,mol^{-1}$
Li^+	−520
Na^+	−406
K^+	−320
Rb^+	−296
Mg^{2+}	−1926
Ca^{2+}	−1579
Sr^{2+}	−1446
Al^{3+}	−4680

Dissolving ionic compounds

Many (but not all) ionic compounds are soluble in water. The ionic lattice breaks up and new bonds are formed between water molecules and the separate ions.

Enthalpy level diagrams are drawn to illustrate the energy changes involved in dissolving processes.

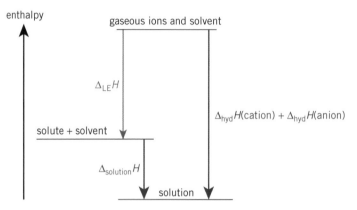

▲ **Figure 1** An enthalpy level diagram for the dissolving of an ionic compound

Enthalpy changes involved in dissolving

Figure 1 shows the relationship between the lattice enthalpy, the enthalpy change of solution, and the enthalpy changes of hydration of the positive and negative ions.

Hydrated ions

Hydrated ions are surrounded by water molecules that bond to them by ion–dipole bonds.

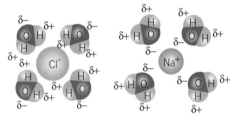

▲ **Figure 2** Hydrated chloride and sodium ions

Factors affecting the enthalpy of hydration of ions

Enthalpy of hydration will be more negative if

- the ionic charge is greater
- the ionic radius is smaller.

Alternatively, you can say that the enthalpy of hydration is more negative if the **charge density** of an ion is greater. Increased charge density means that an ion attracts more water molecules, and forms stronger ion–dipole bonds. So the energy released by bond forming is greater.

Factors affecting the lattice enthalpy

Lattice enthalpy is also affected by ionic charge and size of ions, following a similar pattern to enthalpy of hydration.

Lattice enthalpy becomes more negative if:

- the ions in the lattice are more highly charged
- the ions in the lattice are smaller.

So lattice energy is more negative if the ions in an ionic lattice have a greater charge density.

Synoptic link

Charge density was introduced to explain the patterns in the thermal decomposition of carbonates in Topic 11.2, The s-block: Groups 1 and 2.

Synoptic link

Calculations involving these enthalpy changes may make use of Hess' Law. This was introduced in Topic 4.2, Enthalpy cycles.

▼ **Table 2** *Enthalpy of lattice enthalpy values for some ionic compounds*

Compound	$\Delta_{LE}H/\text{kJ mol}^{-1}$	Compound	$\Delta_{LE}H/\text{kJ mol}^{-1}$
Li_2O	−2806	MgO	−3800
Na_2O	−2488	CaO	−3419
K_2O	−2245	SrO	−3222
LiF	−1047	MgF_2	−2961
NaF	−928	CaF_2	−2634
KF	−826	Al_2O_3	−15 916

Model answer: Calculating enthalpy changes involved in the dissolving process

An enthalpy level diagram for the dissolving of calcium hydroxide is shown in Figure 3.

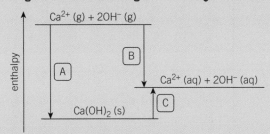

▲ **Figure 3** *Enthalpy cycle for the dissolving of calcium hydroxide*

Use the data below to calculate the enthalpy change of solution of calcium hydroxide, $\Delta_{sol}H$ [Ca(OH)$_2$]

$\Delta_{LE}H[Ca(OH)_2] = -2506\,\text{kJ mol}^{-1}$

$\Delta_{hyd}H[Ca^{2+}(aq)] = -1579\,\text{kJ mol}^{-1}$

$\Delta_{hyd}H[OH^-(aq)] = -460\,\text{kJ mol}^{-1}$

1 Identify the data that tells you the value of enthalpy change A. This is the lattice enthalpy, so A = −2506 kJ mol⁻¹.

2 Identify the data that tells you the value of enthalpy change B. This is the sum of the enthalpy changes of hydration of 1Ca²⁺ + 2OH⁻, so B = −1579 + 2 × −460 = −2499 kJ mol⁻¹

3 Use Hess's Law to write down the relationship between A, B, and C. A = B − C, so C (the enthalpy of solution) = B − A

4 Calculate C: C = −2499 − (−2506) = +7 kJ mol⁻¹

> Make sure that you give the value a sign, and that the sign is consistent with the diagram, which shows an endothermic reaction.

Measuring enthalpy changes of solution experimentally

Lattice enthalpies and enthalpy changes of hydration cannot be measured directly in experiments. However enthalpy changes of solution can be measured using the method described in Chapter 4 Experimental techniques.

A known mass of solid is dissolved in a known mass of solvent and the temperature change is measured.

The energy exchanged with the water is calculated using the equation $E = mc\Delta T$.

This number is then scaled up to find the energy exchanged when 1 mole of solute dissolves.

The relationship between enthalpy change of solution and solubility

If the energy released by forming bonds in the hydration of ions can compensate for the energy required to break up the lattice, then an ionic solute will be soluble.

So enthalpy changes of solution are either exothermic or very slightly endothermic.

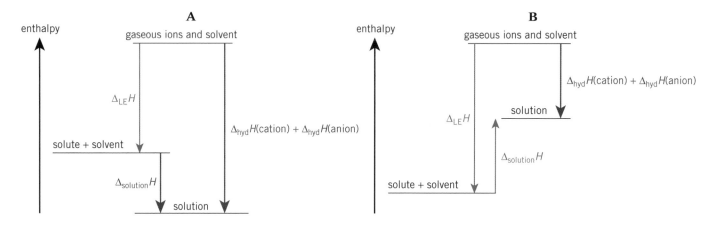

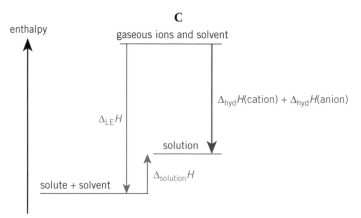

▲ **Figure 4** *Enthalpy cycles for exothermic and endothermic dissolving processes (soluble and insoluble). Cycle A (exothermic) is likely to be soluble, Cycle B (very endothermic) will be insoluble and Cycle C (slightly endothermic) may be soluble*

Model answer: Explaining the solubility of ionic substances

Many ionic substances are soluble in water. Explain why.

- Ionic bonds are broken when an ionic substance dissolves. Some hydrogen bonds also break between water molecules.

- Ion–dipole bonds form between water molecules and the free ions.

- The strength of the bonds formed is similar to the strength of the bonds broken.

- So the energy released by bond formation is sufficient to compensate for the energy required to the break the bond between ions.

> Describe the bonds broken during dissolving.

> Describe the bonds that can form when the substance dissolves.

> Compare the strength of the bonds broken and formed. Be careful to use precise language – common errors are to be imprecise and to say things like 'more energy is needed to make bonds than to break them' **OR** 'more bonds are broken than made'.

> Relate strength of bonds to energy changes.

You may also be asked to discuss why ionic substances do not dissolve in non-polar solvents such as cyclohexane. The difference here is that the ion–dipole bonds formed between the solvent molecules and the ions are much weaker than the bonds that are broken.

Model answer: Explaining why ionic compounds are insoluble in non-polar solvents

Sodium chloride does not dissolve in cyclohexane. Explain why.

- In order to dissolve, ionic bonds would need to break between the ions in the sodium chloride lattice. Instantaneous dipole–induced dipole bonds would also need to break between the cyclohexane molecules.

- Because cyclohexane molecules do not have a permanent dipole, only weak ion–dipole bonds could form between cyclohexane molecules and the free ions.

- The strength of the bonds that could form is much weaker than the bond that would need to break.

- So the energy released by bond formation is not sufficient to compensate for the energy required to the break the bond between ions.

> Imagine that the compound does dissolve and describe the bonds broken during dissolving.

> Describe the bonds that could form if the substance dissolves.

> Compare the strength of the bonds broken and formed.

> Relate strength of bonds to energy changes.

Summary questions

1 Magnesium chloride, $MgCl_2$, dissolves in water in an exothermic reaction. Draw out a diagram to show the energy changes involved in the dissolving process. Label the substances involved and the enthalpy changes. *(7 marks)*

2 Calcium chloride is soluble in water but does not dissolve in the organic solvent cyclohexane. Explain why. *(4 marks)*

3 a Use the diagram you drew for question 1, and the data below, to calculate a value for the enthalpy of hydration per mole of chloride ions:

$\Delta_{LE}H[MgCl_2]^- = -2526\,kJ\,mol^{-1}$

$\Delta_{hyd}H[Mg^{2+}(aq)] = -1926\,kJ\,mol^{-1}$

$\Delta_{sol}H[MgCl_2(s)] = -155\,kJ\,mol^{-1}$ *(4 marks)*

b The enthalpy of hydration of the Ca^{2+} ion is $-1579\,kJ\,mol^{-1}$. Discuss why this value is less negative than that for the Mg^{2+} ion. *(2 marks)*

Revision tip

You can also use the same ideas to explain why two covalent liquids will mix. Simply list the type of intermolecular bonds broken and formed and compare the strength of these bonds.

Synoptic link

A complete explanation for solubility of solutes requires discussion of entropy changes. Entropy is discussed in Topic 4.5, Enthalpy and entropy.

4.5 Enthalpy and entropy

Specification reference: O (d), (e), (f), (g)

Key term

Entropy: A measure of the number of ways in which molecules and energy quanta can be arranged, shown by the symbol: S.

Entropy
Entropy of particles and feasible processes

A feasible process is one that will occur without any energy input – although it may occur very slowly.

Many feasible processes involve particles becoming more disordered – liquids or gases tend to mix and solids tend to dissolve.

Entropy is a measure of the disorder of these particles or of energy quanta. It is calculated by working out the number of ways of arranging the particles or energy quanta.

▼ **Table 1** *When water changes state, the entropy of the particles increases*

State		Entropy S/$J\,K^{-1}\,mol^{-1}$
increasing disorder	$H_2O(s)$	+41
	$H_2O(l)$	+70
	$H_2O(g)$	+189

Differences in the entropy of solids, liquids, and gases

Particles in a solid are rigidly fixed in place, whereas particles in a gas are free to move around and take up many different positions. This means that there are more ways of arranging the particles.

In general gases have a much higher molar entropy than liquids, which in turn have a higher molar entropy than solids.

Synoptic link

The terms 'system' and 'surroundings' were used when discussing enthalpy changes in Topic 4.1, Energy out, energy in.

Entropy changes

The total change in entropy associated with a process is given the symbol $\Delta_{tot}S$.

This is the sum of two entropy changes:

- The entropy change in the system, $\Delta_{sys}\,S$, which is due to the change in the number of ways of arranging particles.
- The entropy change in the surroundings , $\Delta_{surr}\,S$, which is due to the change in the number of ways of arranging energy quanta that are exchanged with the surroundings.

$$\text{So } \Delta_{tot}S = \Delta_{sys}\,S + \Delta_{surr}\,S$$

Revision tip

The entropy change in the system is sometimes just referred to as ΔS, the entropy change of the reaction.

Calculating entropy
Calculating $\Delta_{sys}S$ and $\Delta_{surr}S$

$\Delta_{sys}S + \Delta_{surr}S$ can be calculated given appropriate data.

- $\Delta_{sys}S$ is calculated using values of molar entropy, S:

$$\Delta_{sys}S = \Sigma S \text{ (products)} - \Sigma S \text{ (reactants)}$$

Revision tip

The symbol Σ means 'sum of'. Remember to take into account the number of moles of each substance when adding up the entropy values.

- $\Delta_{surr}S$ varies with temperature, but can be calculated from the temperature in K and the enthalpy change for the reaction in $J\,mol^{-1}$:

$$\Delta_{surr}S = -\frac{\Delta H}{T}$$

Total entropy change and feasible reactions

If $\Delta_{tot}S$ is positive, at a certain temperature, then you can deduce that the reaction is feasible at that temperature.

Notice that if a feasible reaction is reversible, then the backwards reaction will have a negative $\Delta_{tot}S$ and will therefore not be feasible at this temperature.

If $\Delta_{tot}S = 0$ at a certain temperature then you can conclude that a reversible process will reach equilibrium where neither the forward or backward reaction is favoured.

Predicting the effect of changing temperature

Because changing temperature affects the magnitude of $\Delta_{tot}S$, it can also affect whether or not a reaction is feasible.

 Worked example: Calculating a value for $\Delta_{tot}S$

Calculate $\Delta_{tot}S$ at 333 K for the reaction:

$$2NH_3(g) + 2O_2(g) \rightarrow N_2O(g) + 3H_2O(g) \qquad \Delta H = -552\,kJ\,mol^{-1}$$

$H_2O(g)$	$S^{\ominus} = +189\,J\,K^{-1}\,mol^{-1}$
$N_2O(g)$	$S^{\ominus} = +220\,J\,K^{-1}\,mol^{-1}$
$NH_3(g)$	$S^{\ominus} = +192\,J\,K^{-1}\,mol^{-1}$
$O_2(g)$	$S^{\ominus} = +205\,J\,K^{-1}\,mol^{-1}$

Step 1: Calculate $\Delta_{sys}S$ by adding up the molar entropies of reactants and products, remembering to multiply by the number of moles present in the equation:

$\Delta_{sys}S = [220 + (3 \times 189)] - [(2 \times 192) + (2 \times 205)] = 787 - 794$
$= -7\,J\,K^{-1}\,mol^{-1}$

Step 2: Calculate $\Delta_{surr}S$ using the equation $\Delta_{surr}S = \dfrac{-\Delta H}{T}$

$\Delta H = -552\,000\,J$ and $T = 333\,K$, so $\Delta_{surr}S = \dfrac{-(-552\,000)}{333} = \dfrac{552\,000}{333}$
$= +1658\,J\,K^{-1}\,mol^{-1}$

3 Add together the values of $\Delta_{sys}S$ and $\Delta_{surr}S$.

$\Delta_{tot}S = (-7) + (1658) = +1651\,J\,K^{-1}\,mol^{-1}$

 Worked example: Predicting the sign of $\Delta_{sys}S$

Predict and explain the sign of $\Delta_{sys}S$ for this reaction:

$$Ca(s) + 2H_2O(l) \rightarrow Ca(OH)_2(s) + H_2(g)$$

Step 1: Compare the number of moles of gas on the two sides of the equation. In this equation there are 1 mole of gas on the RHS of the equation and none on the LHS.

Step 2: Decide whether entropy increases or decreases. There are more ways of arranging the particles in a gas than a solid, so the entropy increases.

Step 3: Relate this to the sign of $\Delta_{sys}S$: $\Delta_{sys}S = S\,(products) - S\,(reactants)$, so $\Delta_{tot}S$ is positive.

Revision tip

You will probably be given a value for ΔH in $kJ\,mol^{-1}$. Be sure to convert this into $J\,mol^{-1}$.

Revision tip

If you know that a reaction is feasible, you can work backwards to deduce that $\Delta_{tot}S$ is positive. This may in turn help you to deduce something about $\Delta_{sys}S$ or $\Delta_{surr}S$; for example, if you know that $\Delta_{surr}S$ is negative then $\Delta_{sys}S$ must be positive or the process could never be feasible.

Revision tip

You may be asked to describe what effect changing temperature has on the feasibility of a reaction, or to calculate the temperature at which a reaction becomes feasible.

Maths skill: Rearranging equations

To find T at which $\Delta_{tot}S$ is zero, you will need to be confident in rearranging equations.

Calculate the temperature in kelvin at which $\Delta_{tot}S$ is zero for the reaction:

$$N_2(g) + 3H_2(g) \rightarrow 2NH_3(g) \text{ where } \Delta H = -92.4 \text{ kJ mol}^{-1} \text{ and } \Delta_{sys}S = -198.3 \text{ J K}^{-1} \text{ mol}^{-1}$$

$$0 = \Delta_{sys}S + \Delta_{surr}S$$

$$0 = -198.3 + -\frac{(-92\,400)}{T} = -198.3 + \frac{92\,400}{T}$$

1 Simplify the equation by moving $\Delta_{sys}S$ onto the LHS of the equation (add 198.3 to both sides):

$$198.3 = \frac{92\,400}{T}$$

2 Rearrange the equation so that T is no longer at the bottom of an expression (multiply both sides by T):

$$198.3T = 92\,400$$

3 Rearrange the equation to find T:

$$T = \frac{924\,00}{198.3}$$

$$T = 466.0 \text{ K}$$

Summary questions

1 Predict and explain the sign of $\Delta_{sys}S$ for this reaction:
   $$N_2(g) + 3H_2(g) \rightarrow 2NH_3(g)$$
   (4 marks)

2 Calculate the value of $\Delta_{tot}S$ at 298 K for the decomposition of magnesium nitrate and comment on the significance of this value:
   $$2Mg(NO_3)_2(s) \rightarrow 2MgO(s) + 4NO_2(g) + O_2(g) \quad \Delta H = +510.8 \text{ kJ mol}^{-1}$$
   Values of S in J K^{-1} mol^{-1}: $Mg(NO_3)_2(s) = +164.0$, $MgO(s) = +26.9$, $NO_2(g) = +240.0$, $O_2(g) = +205.0$
   (6 marks)

3 Magnesium carbonate is stable at room temperature, but decomposes when heated. Deduce the signs of $\Delta_{sys}S$, $\Delta_{surr}S$, and $\Delta_{tot}S$ at 298 K.
   (4 marks)

Experimental techniques

Specification reference: DF (f)

Measuring energy transferred in a reaction

You can carry out experiments in which energy is transferred to or from a known mass of water.

Determining enthalpy changes of combustion of flammable liquids

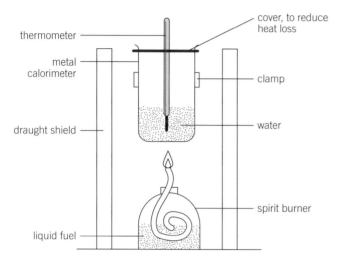

▲ **Figure 1** *Simple apparatus for measuring* $\Delta_c H$

Record the initial temperature of the water and the maximum temperature reached when a known volume of water is heated by the complete combustion of a measured mass of fuel. The energy transferred = m × c × ΔT. You can now calculate the enthalpy change for the combustion of 1 mole of the fuel used.

Determining enthalpy changes for reactions in solution

This method can be used for reactions between two solutions or between a solid and a solution. The energy is transferred to or from the water in the solution. The reaction is carried out in an insulated container, such as a polystyrene cup, fitted with a lid.

Record the volumes and initial temperature of the solution(s) and the maximum or minimum temperature reached during the reaction. The energy transferred = m × c × ΔT. You can now calculate the enthalpy change for the reaction of the number of moles of reactants specified in the equation.

Using cooling curves for reactions involving solutions

You can adapt the experiment to correct for heat loss by taking measurements of the temperature over several minutes and plotting a cooling curve. This produces a more accurate value for ΔT.

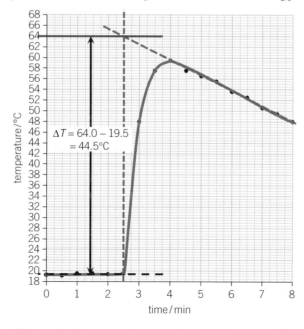

▲ **Figure 3** *Using a cooling curve to calculate a corrected value for ΔT (reaction started after 2½ minutes)*

Revision tip

You need to be aware that this method often produces values for $\Delta_c H$ that are less negative than the databook value.

You should be able to explain this by thinking about heat losses to the surroundings and incomplete combustion of the fuel.

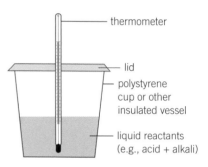

▲ **Figure 2** *Apparatus to determine ΔrH for a reaction in solution*

Synoptic link

The method for calculating enthalpy changes using data from experiments is described in Topic 4.1, Energy out, energy in.

Revision tip

You should be aware of the assumptions that are made in carrying out calculations using this method:

1 The density of the solution is the same as water.

2 The specific heat capacity of the solution is the same as water.

Common misconception: Heat loss and heat gain

Heat loss to the surroundings is a common source of error in experiments to determine enthalpy changes. However, if a reaction is endothermic, the source of error is actually heat gained from the surroundings.

41

1 For which of these processes is the enthalpy change certain to be endothermic?

A $Cl_2(g) \rightarrow 2Cl(g)$

B $Cl_2(g) + CH_4(g) \rightarrow CH_3Cl(g) + HCl(g)$

C $NH_3(g) + BF_3(g) \rightarrow NH_3BF_3(g)$

D $H_2(g) + Cl_2(g) \rightarrow 2HCl(g)$ *(1 mark)*

2 The combustion of methane is exothermic. The best explanation of this is:

A The reaction involves only bond formation.

B The energy released by bond forming is greater than the energy released by bond breaking.

C The energy released by bond forming is greater than the energy required for bond breaking.

D The energy required to form bonds is less than the energy released by breaking bonds. *(1 mark)*

3 Which of these processes represents the reaction that occurs during the measurement of the enthalpy change of formation of AgCl(s)?

A $Ag^+ (aq) + Cl^-(aq) \rightarrow AgCl(s)$

B $Ag(s) + \frac{1}{2}Cl_2(g) \rightarrow AgCl(s)$

C $2Ag(s) + Cl_2(g) \rightarrow 2AgCl(s)$

D $Ag(g) + Cl(g) \rightarrow AgCl(s)$ *(1 mark)*

4 A known mass of methanol is burned in a plentiful supply of oxygen. The energy released is transferred to water and the results are used to calculate the $\Delta_c H$ of methanol. The value obtained is much less negative than the databook value for $\Delta_c H^\ominus$. The most likely explanation of this is:

A Heat transferred from the surroundings to the water.

B Incomplete combustion of the fuel.

C Heat transferred from the water to the surroundings.

D Carrying out the reaction under non-standard conditions. *(1 mark)*

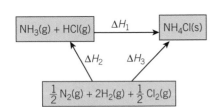

5 The Hess's cycle can be used to determine the value of ΔH_1.

The value of ΔH_1 is given by the expression:

A $\Delta_f H [NH_4Cl] - \Delta_f H [NH_3] + \Delta_f H [HCl]$

B $\Delta_f H [NH_4Cl] - \Delta_f H [NH_3] - \Delta_f H [HCl]$

C $\Delta_f H [NH_3] + \Delta_f H [HCl] - \Delta_f H [NH_4Cl]$

D $\Delta_f H [NH_4Cl] - \frac{1}{2}\Delta_f H [N_2] - 2 \Delta_f H [H_2] - \frac{1}{2}\Delta_f H [Cl_2]$ *(1 mark)*

6 Ethanol, C_2H_5OH, can be used as a liquid fuel. The enthalpy of combustion of ethanol can be determined by burning it in a spirit burner and using the energy released to heat water.

The following results were obtained in an experiment:

Mass of calorimeter	120 g
Mass of calorimeter + water	220 g
Mass of spirit burner at start	43.56 g
Mass of spirit burner at end	42.46 g
Initial temperature of water	20°C
Maximum temperature of water	44°C

a Calculate the energy, in J, transferred to the water in the calorimeter. *(2 marks)*

b i Use your answer to **a** to determine a value for the enthalpy of combustion of ethanol in $kJ\,mol^{-1}$. Give your answer to 3 s.f. (*4 marks*)

   ii The databook value for the standard enthalpy of combustion of ethanol is much more negative than values determined by experiments. Suggest two reasons for this. (*2 marks*)

7 The structure of butanone is shown below:

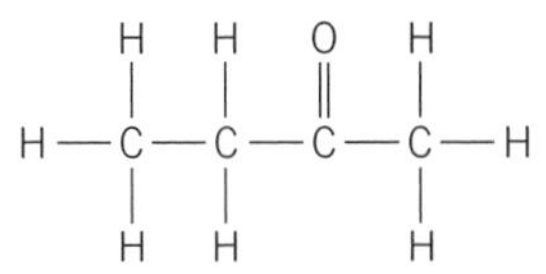

Butanone can undergo complete combustion in a plentiful supply of air:

$$C_4H_8O(l) + 6O_2(g) \rightarrow 4CO_2(g) + 4H_2O(l)$$

a i Use the bond enthalpy data below to calculate a value for the enthalpy change of combustion of butanone.

Bond	Average bond enthalpy / $kJ\,mol^{-1}$
C–C	+347
C–H	+413
C=O	+805
O–H	+464
O=O	+498

(*4 marks*)

   ii The databook value for the enthalpy change of combustion of butanone is significantly different to the value that is calculated using these bond enthalpies. State two reasons for this difference (*2 marks*)

b Butanal also has the molecular formula C_4H_8O. The structure of butanal is shown below:

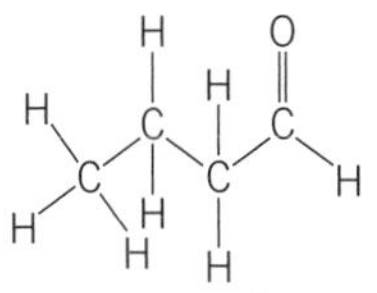

Use ideas about bonds to explain whether the enthalpy change of combustion of butanal is likely to be similar to that of butanone. (*3 marks*)

8 Which of these statements about entropy is true?

A Entropy is a measure of the order of molecules and quanta.

B Entropy is measured in units of $kJ\,mol^{-1}$.

C The total entropy of the system and surroundings always increases if a reaction is feasible.

D Exothermic reactions result in a decrease in total entropy. (*1 mark*)

9 Ammonia can be formed in the Haber process:

$$N_2(g) + 3H_2(g) \rightarrow 2NH_3(g) \quad \Delta H = -92.3\,kJ\,mol^{-1}$$

What can be concluded from this information about the entropy changes at 298 K for this process?

1 The $\Delta_{sys}S$ is positive

2 The $\Delta_{tot}S$ is positive

3 The $\Delta_{surr}S$ is positive

A 1, 2, and 3

B 1 and 2

C 2 and 3

D only 3 *(1 mark)*

10 Which row of the table describes the difference in the properties of a sodium ion compared to a potassium ion?

	Ionic radius of Na^+	Charge density of Na^+	$\Delta_{hyd}H$ of Na^+
A	Greater	Smaller	Greater
B	Smaller	Greater	Greater
C	Greater	Smaller	Smaller
D	Smaller	Greater	Greater

11 Which of these enthalpy changes are **always** exothermic?

1 $\Delta_{sol}H$

2 $\Delta_{hyd}H$

3 $\Delta_{LE}H$

A 1, 2, and 3

B 1 and 2

C 2 and 3

D only 3 *(1 mark)*

12 Sodium sulfate, Na_2SO_4, is soluble in water. A group of students measures the enthalpy change of solution of sodium sulfate. They dissolve a known mass of sodium sulfate in water and obtain a value of $-1.5\,kJ\,mol^{-1}$.

a The standard enthalpy change of solution of sodium sulfate is $-2.5\,kJ\,mol^{-1}$.

Suggest a reason why the value measured by the students is less negative than the databook value. *(1 mark)*

b i Complete this diagram to show the relationship between the enthalpy change of solution, lattice enthalpy and enthalpy change of solution for the ionic compound sodium sulphate.
Label the enthalpy changes and the details of the species involved. *(3 marks)*

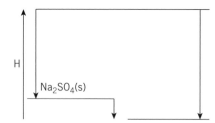

(5 marks)

ii The students find the following data in a databook:

$\Delta_{LE}H[Na_2SO_4] = -1944.0\,kJ\,mol^{-1}$

$\Delta_{hyd}H[Na^+] = -406.0\,kJ\,mol^{-1}$

Use these data, along with the standard enthalpy change of solution of sodium sulfate, to calculate a value for the standard enthalpy of hydration of the sulfate ion. *(4 marks)*

$\Delta_{hyd}H[SO_4^{2-}] = $ _____ $kJ\,mol^{-1}$

13 Many ionic compounds are soluble in water. However, calcium carbonate, $CaCO_3$, is usually regarded as an insoluble compound.

    a Explain why many ionic compounds are soluble in water.   *(3 marks)*

    b An equation can be written to represent the dissolving of calcium carbonate:

$$CaCO_3(s) + aq \rightarrow Ca^{2+}(aq) + CO_3^{2-}(aq) \qquad \Delta H = -13.0\,kJ\,mol^{-1}$$

       i The entropy change for this reaction, $\Delta_{sys}S = -204.8\,J\,K^{-1}\,mol^{-1}$

       Calculate the value for the total entropy change, $\Delta_{tot}S$ at 298 K, and explain how this shows that calcium carbonate is insoluble at this temperature.   *(3 marks)*

      ii Discuss whether the dissolving of calcium could be made feasible by changing the temperature at which the dissolving process was attempted.   *(3 marks)*

5.1 Ionic substances in solution

Specification reference: EL (s)

Many ionic substances dissolve well in water – they are described as being soluble compounds.

- All compounds of Group 1 metals are soluble.
- All compounds containing nitrate ions are soluble.
- All compounds containing ammonium ions are soluble.

Insoluble compounds

Some ionic substances do not dissolve well in water – they are described as insoluble compounds:

- sulfates of barium, calcium, lead, and silver
- halides (chlorides, bromides, and iodides) of silver and lead
- all carbonates (except those of Group 1 ions or ammonium ions)
- hydroxides containing some Group 2 ions, aluminium ions, or *d*-block ions.

Forming precipitates

Certain combinations of ions form insoluble compounds (see above).

If solutions containing these ions are mixed together, a precipitate will form.

 Go further

Almost all substances are soluble to some extent; for example, the solubility of magnesium hydroxide is $2.00 \times 10^{-4}\,mol\,dm^{-3}$

a Convert this into a concentration in $g\,/\,100\,cm^3$ of solution.

b A substance is sometimes described as insoluble if the solubility is less than $0.1\,g\,/\,100\,cm^3$ solution. Should magnesium hydroxide be classified as soluble or insoluble?

Precipitates containing *d*-block ions may have particular colours.

Typical results for combinations of cations and anions are shown below (ppt = precipitate), using solutions with a concentration of around $0.1\,mol\,dm^{-3}$.

Some of the reactions are beyond the scope of the course and have been omitted.

	OH^-	SO_4^{2-}	CO_3^{2-}	Cl^-	Br^-	I^-
Ca^{2+}	White ppt	White ppt	White ppt	Soluble	Soluble	Soluble
Ba^{2+}	White ppt	White ppt	White ppt	Soluble	Soluble	Soluble
Cu^{2+}	Pale blue ppt	Soluble	Green ppt	Soluble	Soluble	Soluble
Fe^{2+}	Green ppt	Soluble	Green ppt	Soluble	Soluble	Soluble
Fe^{3+}	Brown ppt	Soluble	–	Soluble	Soluble	Soluble
Al^{3+}	White ppt	Soluble	–	Soluble	Soluble	Soluble
Pb^{2+}	White ppt	White ppt	White ppt	White ppt	White ppt	Yellow ppt
Zn^{2+}	White ppt	Soluble	White ppt	Soluble	Soluble	Soluble
Ag^+	White ppt	White ppt	White ppt	White ppt	Cream ppt	Yellow ppt

Testing for ions

The formation of precipitates can be used as a test for certain ions:

- Adding barium chloride solution (containing Ba^{2+} ions): white precipitate shows that sulfate ions are present.

- Adding silver nitrate solution (containing Ag^+ ions): a precipitate shows that halide ions are present (white = chloride, cream = bromide, yellow = iodide).

Synoptic link

Precipitation reactions like these are often described using ionic equations. See Topic 1.2, Balanced equations.

Synoptic link

Precipitation reactions involving halide ions (including the solubility of the precipitates in ammonia) are described in Topic 11.3, The p-block: Group 7.

Summary questions

1 Describe how you would carry out a test to show the presence of:
   a bromide ions   (*2 marks*)
   b sulfate ions.   (*2 marks*)

2 What colours are the precipitates formed from the combinations of the following ions:
   a copper(II) ions and hydroxide ions
   b calcium ions and carbonate ions
   c iron(III) ions and hydroxide ions
   d silver ions and iodide ions.   (*2 marks each*)

3 A pale green solution solution containing an ionic salt, A, was investigated using precipitation reactions. When sodium hydroxide solution was added to the solution, a green precipitate was seen. When barium chloride solution was added a white precipitate was formed:
   a identify substance A   (*2 marks*)
   b write ionic equations for the reactions that form the precipitates in the two tests.   (*2 marks*)

5.2 Bonding, structure, and properties

Specification reference: EL (j), EL (l)

Structure and bonding

Types of structure

The types of structure present in a substance depends on the type of bonding present.

Type of bonding	Type of structure	Example
Ionic	Giant ionic	NaCl
Covalent	Simple molecular	CO_2
Covalent	Giant covalent network	SiO_2
Metallic	Giant metallic lattice	Fe

Giant ionic lattice

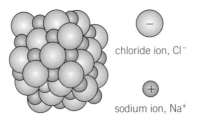

▲ **Figure 1** *Structure of a sodium chloride lattice*

The structure of a giant ionic lattice

There is a regular repeating pattern of positive and negatively charged ions in all three dimensions. The attraction between these oppositely charged ions outweighs the repulsion between ions with the same charge, because the oppositely charged ions are closer.

The sodium chloride structure

You can see in the diagram above that chloride, Cl^- ions are larger than sodium ions, Na^+. The ions are arranged so that 6 Cl^- ions surround 1 Na^+ ion in a 3-dimensional arrangement.

The ratio of Cl^- to Na^+ is 1:1, so 6 Na^+ also surround 1 Cl^-.

Characteristic properties

- High melting point, because there are strong electrostatic attractions between ions
- Often soluble in water
- Conduct electricity when molten or in solution, because the charged ions are able to move in response to a voltage.

Simple molecular structure

There are strong covalent bonds within the molecules, but only weak intermolecular bonds between the molecules.

Characteristic properties

- Low melting point
- Usually insoluble in water
- Do not conduct.

Synoptic link

A description of the electrostatic forces in ionic and covalent bonds is found in Topic 3.1, Chemical bonding.

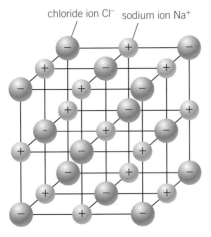

chloride ion Cl^- sodium ion Na^+

▲ **Figure 2** *An alternative way of representing the lattice to show the 3-dimensional arrangement of ions*

Go further

The 6 Cl^- ions will arrange themselves to be as far apart as possible. Suggest the name given to the 3-dimensional arrangement that they will take up.

Synoptic link

Intermolecular bonds and the properties of molecular substances are explained in Topics 5.3 and 5.4.

Giant covalent network

Some covalent substances have a giant network structure. There are strong covalent bonds between the atoms in the network.

Characteristic properties

- High melting point, because all the bonds in the structure are strong covalent bonds

- Insoluble in water

- Do not conduct electricity (except for graphite).

▲ **Figure 3** *Diamond has a giant covalent network structure*

Common misconception: Molecules or giant structures?

Silicon dioxide, SiO_2 has a giant structure, but does not consist of separate SiO_2 molecules. The giant structure simply contains Si and O atoms in the ratio 1:2, all strongly bonded together in a 3-dimensional lattice.

Giant metallic lattice

All metals have a giant metallic lattice structure. There is a strong electrostatic attraction between the positive metal ions and the delocalised electrons between the ions.

Characteristic properties

- High melting point, because there is strong attraction between ions and electrons

- Insoluble in water

- Conduct electricity when solid or molten because the electrons are free to move in response to a voltage.

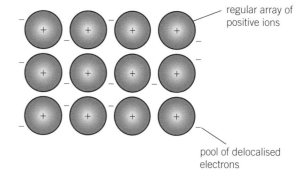

regular array of positive ions

pool of delocalised electrons

▲ **Figure 4** *A giant metallic lattice structure*

➕ Go further

Group 1 metals have relatively low melting points. Use ideas about the charge and size of Group 1 ions to explain this.

Key term

Delocalised electrons: Electrons that are not associated with a particular atom or bond, but are free to move over several atoms.

Summary questions

1 Describe the structure and bonding in the metal calcium. *(4 marks)*

2 Calcium oxide, $CaCl_2$ has a giant ionic structure and sulfur dichloride, SCl_2 has a simple molecular structure. Describe the differences in properties between these substances. *(3 marks)*

3 Silicon dioxide is a high melting point solid, but carbon dioxide is a gas at room temperature. Use ideas about structure and bonding to explain this difference in properties. *(5 marks)*

5.3 Bonds between molecules: temporary and permanent dipoles

Specification reference: OZ (a), OZ (b)

Bonds between molecules

In any liquid or solid, there are bonds between molecules. These are called intermolecular bonds. One type of intermolecular bond is a dipole–dipole bond. Dipole–dipole bonds can be:

- permanent dipole–permanent dipole bonds
- instantaneous dipole–induced dipole bonds.

Electronegativity and dipoles

Permanent dipoles can arise in molecules, because some molecules contain polar bonds:

- Covalent bonds can be polar if the two atoms in the bond have different electronegativities.
- The charges in a polar bond are shown using δ^+ and δ^- symbols.

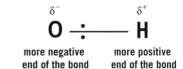

more negative end of the bond — more positive end of the bond

because O has greater share of electrons

▲ **Figure 1** *A polar O—H bond*

Electronegativity values

Electronegativity is related to the position of an atom in the periodic table.

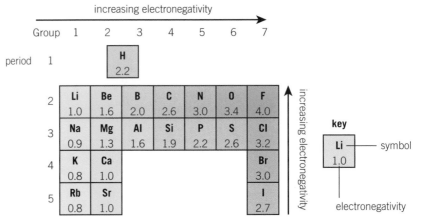

▲ **Figure 2** *Trends in electronegativity in the periodic table*

In a bond between two atoms, the more electronegative atom has a δ^- charge and the less electronegative has a δ^+ charge.

Dipoles in molecules

Molecules that contain polar bonds may have a permanent dipole (also called an overall dipole). They are also sometimes called polar molecules.

 Worked example: Overall dipole of CCl$_4$

The molecule tetrachloromethane, CCl$_4$, has a tetrahedral structure. Does it have an overall (permanent) dipole?

Step 1: Decide whether there are any polar bonds in the molecule:
C—Cl is polar because Cl is more electronegative than C

Step 2: If there are any polar bonds, draw out the 3-dimensional shape of the molecule, marking on the partial charges:

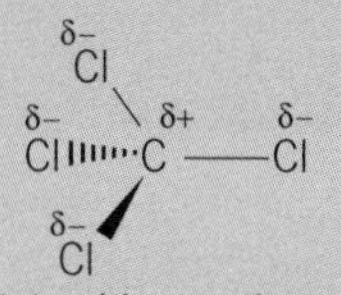

$$
\begin{array}{c}
\overset{\delta-}{Cl} \\
\overset{\delta-}{Cl} \text{\tiny\textbackslash}\ \overset{\delta+}{C} — \overset{\delta-}{Cl} \\
\overset{\delta-}{Cl}
\end{array}
$$

tetrachloromethane

Step 3: Decide whether the charges are arranged symmetrically – in other words, whether the centre of positive charge is in the same place as the centre of negative charge:

The charges are arranged symmetrically, so there is no overall dipole in a CCl$_4$ molecule

Permanent dipole–permanent dipole bonds

If two neighbouring molecules both have a permanent dipole, then there will be an electrostatic attraction between the charges of these dipoles. This is called a permanent dipole–permanent dipole bond.

Instantaneous dipole–induced dipole bonds

If two neighbouring molecules do not have a permanent dipole, there will still be an attraction between the molecules. This is called an instantaneous dipole–induced dipole bond. Instantaneous dipole–induced dipole attraction occurs between molecules, even if permanent dipoles are also present.

Explaining how these bonds arise

- Electrons in a molecule are in continuous random motion.
- At a particular instant, the electrons may be distributed unevenly. This creates an instantaneous dipole.
- The dipole induces a dipole on a neighbouring molecule, creating an induced dipole.
- There is an electrostatic attraction between the two dipoles.

Because the electron distribution in a molecule is constantly changing, instantaneous dipole–induced dipole bonds are continually breaking and reforming.

Factors affecting strength of instantaneous dipole–induced dipole bonds

- Number of electrons in the molecule: more electrons mean more chance of an instantaneous dipole arising
- Distance between the molecules – closer packing means a greater electrostatic attraction.

Summary questions

1 Show the partial charges that exist on the atoms in these bonds:
   a C–Cl,   **b** Cl–F,   **c** H–N,   **d** Cl–S   *(4 marks)*

2 CCl$_4$ does not have an overall dipole:
   a State the type of intermolecular bonds between two molecules of CCl$_4$.   *(1 mark)*
   b Explain how these bonds arise.   *(5 marks)*

3 The molecule CCl$_2$F$_2$ contains polar bonds. Discuss whether the molecule has an overall dipole.   *(4 marks)*

$$
\overset{\delta+}{\cdots H} — \overset{\delta-}{Br} \cdots \overset{\delta+}{H} — \overset{\delta-}{Br} \cdots \overset{\delta+}{H} — \overset{\delta-}{Br}
$$

▲ **Figure 3** *Permanent dipole–permanent dipole bonds between HBr molecules*

Common misconception: Dipole–dipole attractions

If a molecule has an overall dipole, then there will be two types of intermolecular bonds: permanent dipole–permanent dipole bonds and instantaneous dipole–induced dipole bonds. There may even be hydrogen bonds as well (see Topic 5.4, Bonds between molecules).

Synoptic link

You need to be able to apply these ideas to explain the patterns in boiling point of alkanes (Topic 12.1, Alkanes) and of the halogens (Topic 11.3, The p-block: Group 7).

Longer chains increase the number of instantaneous dipole-induced dipole bonds

octane: T$_b$: 399 K

pentane: T$_b$: 309 K

Increased branching decreases the surface area in contact and decreases the strength of instantaneous dipole-induced dipole bonds

pentane: T$_b$: 309K

methyl butane: T$_b$: 301K

Molecules with more electrons have more chance of forming instantaneous dipoles, and the strength of instantaneous-dipole induced dipole bonds increases

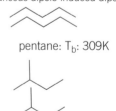

Cl—Cl	Br—Br	I—I
34 electrons	70 electrons	106 electrons
T$_b$: 238K	T$_b$: 332 K	T$_b$: 457 K

▲ **Figure 4** *The factors that affect instantaneous dipole-induced dipole bonds can be used to explain the boiling points of alkanes and halogen molecules*

5.4 Bonds between molecules: hydrogen bonding

Specification reference: OZ (c), OZ (d)

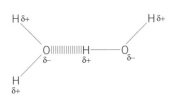

▲ **Figure 1** *In a collection of water molecules there will be 2 hydrogen bonds per molecule*

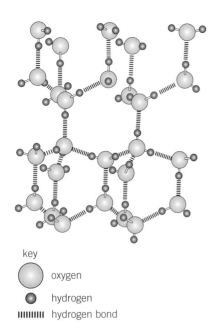

key

○ oxygen

● hydrogen

||||||||| hydrogen bond

▲ **Figure 2** *Ice has a very open structure*

Hydrogen bonds

In general, hydrogen bonds are the strongest type of intermolecular bond.

Requirements for a hydrogen bond between two molecules

A δ^+ H atom in one molecule (e.g. a H atom bonded to an electronegative atom such as O, N, or F). A small electronegative atom (O, N, or F) in the other molecule. A lone pair on the electronegative atom.

Arrangement of atoms in a hydrogen bond

The lone pair points directly at the δ^+ H atom. The arrangement of atoms around the δ^+ H is linear, so the bond angle is 180°.

Hydrogen bonds in water and ice

Liquid water

Water molecules contain an O atom with 2 lone pairs. There are 2 δ^+ H atoms in each water molecule. In a collection of several water molecules, each water molecule can, on average, form up to *two* hydrogen bonds with neighbouring molecules. Liquid water has a higher boiling point than other small molecules with similar M_r values, such as CH_4 or NH_3.

Ice

The hydrogen bonds formed when water freezes give ice a regular structure. The hydrogen bonds and covalent bonds around each O atom are arranged tetrahedrally. This arrangement of bonds around the O atom gives ice a very open structure. As a result, it has a lower density than water. So ice floats on water.

Comparing boiling points of molecules

Ideas about intermolecular bonds can be used to compare the boiling points of two substances. The stronger the intermolecular bonds, the higher the boiling point.

> **Model answer: Explain why propan-1-ol has a higher boiling point than propanone**
>
> The strongest type of intermolecular bond in propan-1-ol is hydrogen bonding; in propanone, it is permanent dipole–permanent dipole. Hydrogen bonding is a stronger type of force than permanent dipole–permanent dipole bonding. So more energy is needed to separate molecules of propan-1-ol than molecules of propanone.

Summary questions

1 A hydrogen bond can form between two molecules of ammonia, NH_3. Explain why. *(3 marks)*

2 Draw a diagram showing how a hydrogen bond can form between two molecules of ethanol. Show partial charges and relevant lone pairs. *(4 marks)*

3 The boiling points of water, hydrogen fluoride, and methane are: 373 K, 293 K, and 109 K. Explain the differences in boiling points of these molecules. *(4 marks)*

Proteins

Proteins are natural polymers, made from many amino acids bonded together by peptide links.

Levels of structure

Most proteins are large molecules with several levels of structure.

Primary structure

The primary structure is the **sequence of amino acids** in the protein chain. It is maintained by the covalent bonds in the peptide links that form when the amino group of one amino acid condenses with the carboxylic acid group in another amino acid.

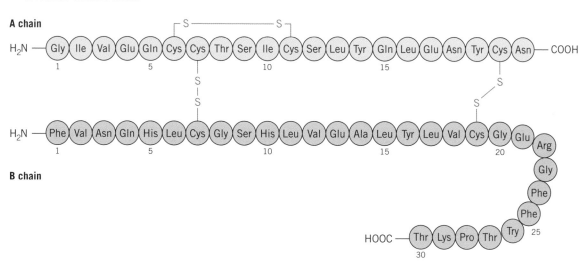

▲ **Figure 1** *The sequence of amino acids in the protein insulin*

Secondary structure

Secondary structure is the folding of the protein chain **into three-dimensional features (alpha helix, or beta pleated sheet).** The secondary structure is maintained by hydrogen bonds between the atoms in adjacent peptide links

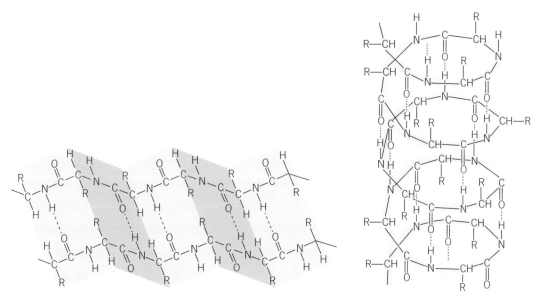

▲ **Figure 2** *Secondary structure within a protein is determined by hydrogen bonds between peptide links*

Go further

Regularly repeating primary sequence may allow polypeptide chains to pack together more closely; this means that the intermolecular bonds (strictly intramolecular bonds) are likely to be stronger; so more force is needed to allow the chains to slide past each other.

Tertiary structure

Further folding can occur, which causes the protein to take up a specific, complex three–dimensional shape. The folding happens because of the formation of bonds such as hydrogen bonds, ionic bonds, or instantaneous dipole–induced dipole bonds between the side groups of amino acids. One particular side group, on a cysteine amino acid, contains an –SH group, and these can form a covalent bond called a disulfide bridge.

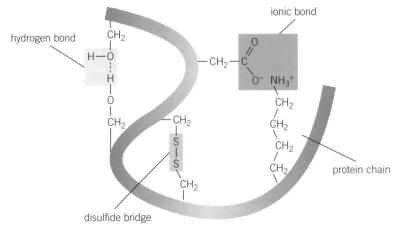

▲ **Figure 3** *The further folding to form a tertiary structure of a protein is determined by bonds between amino acid side groups*

Levels of structure and protein properties

The amino acid sequence (primary structure) determines the pattern of bonds that form with the protein and hence the secondary and tertiary structure. This in turn will determine the properties of a protein and the function of a protein in a biological system.

Enzymes always have a complex tertiary structure which includes a cleft called the **active site**. Small changes to the primary structure can make a big difference to the shape of the active site and its ability to bind to a substrate.

Proteins with a regularly repeating primary sequence will form extensive regions of regular secondary structure and may have a structural role in an organism such as in muscle fibres or hairs.

Summary questions

1 Describe the differences between the secondary and tertiary structure of a protein. *(4 marks)*

2 Draw a diagram to show how a hydrogen bond can form between two neighbouring peptide groups in a region of secondary structure. Indicate relevant lone pairs and partial charges. *(3 marks)*

3 Changes in the primary structure of an enzyme can result in it becoming inactive. Explain why. *(3 marks)*

5.6 Molecular recognition

Specification reference: PL (e)

Pharmacophores and receptor sites

Molecules that act as medicines possess structural features known as pharmacophores. These enable the molecule to bonds to a receptor site (or active site) in a target organism.

Some drug molecules act as **inhibitors** of enzymes.

Identifying pharmacophores

Chemists identify pharmacophores by examining the structures of several molecules that have similar biological actions. These molecules will have often have a structural feature in common: this will be the part of the molecule that binds to the receptor site and so it will be the pharmacophore.

Modifying the pharmacophore

Once chemists have identified the pharmacophore in a drug molecule, they will try and improve the properties of the drug (to make it more effective or have fewer side effects) by modifying the pharmacophore. They do this by changing the groups of atoms that are attached to the pharmacophore.

Interactions with receptor sites

Pharmacophores interact with receptor sites in a similar way to that in which a substrate binds to the active site of an enzyme.

The size and three-dimensional shape of a pharmacophore will be complementary to the shape of a particular receptor site. So pharmacophores can fit into, and bond to, that receptor site.

The functional groups on the pharmacophore will be oriented in a particular way that enables them to bond to groups in the receptor site by forming ionic bonds, hydrogen bonds, instantaneous dipole–induced dipole bonds, etc.

Many receptor sites contain chiral carbon atoms. If the pharmacophore is also chiral then only one enantiomer of the pharmacophore will have groups in the correct three-dimensional arrangement to bond to the receptor site.

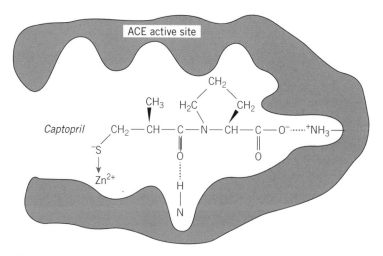

▲ **Figure 2** The pharmacophore in the drug captopril, binding to the active site of the enzyme ACE

Synoptic link

Active sites in enzymes are described in Topic 10.8, Enzymes.

Key terms

Pharmacophore: The part of a molecule that is responsible for its pharmacological activity.

Receptor site: These are found within certain proteins on the surface of cells. They recognise and bond to specific molecules in a similar way to the way that active sites bind to substrate molecules.

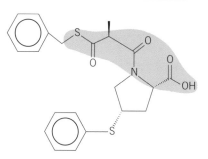

▲ **Figure 1** The drug zofenopril, showing the pharmacophore that is also found in other similar drugs

Key term

Inhibitor: A molecule that competes with a substrate molecule for the active site of an enzyme and prevents the substrate from binding.

Synoptic link

The action of enzyme inhibitors is described in Topic 10.8, Enzymes.

Go further

Examples of properties that might be altered by modifying a pharmocophore can include:

its solubility in water (blood) or fats (cell membranes)

its ability to bind to other, similar receptor sites

how quickly it is broken down in a living organism.

Question: Suggest how a pharmacophore could be modified to:

a make it more water soluble   **b** make it more soluble in fats.

Revision tip

You may be asked to identify the pharmacophore present in a drug molecule. Compare the molecule with the structure of other molecules that have the same biological effect, and find the largest part of the structure that all these molecules have in common.

 Worked example: Enzyme inhibitors

The drug molecule captopril acts by inhibiting the enzyme ACE (see Figure 2 above). Explain how it does this.

- Captopril is likely to have a similar three-dimensional structure to that of the substrate of ACE.
- So it binds to the active site, competing with the substrate.
- This prevents substrate molecules bonding to the active site, so the enzyme becomes inactive.

Summary questions

1 a Explain what is meant by the term 'pharmacophore'.   (*1 mark*)
   b Describe one way in which chemists modify the structure of a pharmacophore.   (*2 marks*)

2 The two molecules below are both a type of penicillin. Suggest the structure of the pharmacophore in penicillin drugs.   (*1 mark*)

3 The drug thalidomide exists as two enantiomers. These enantiomers have different biological effects.
   a Explain why thalidomide exists as two enantiomers.   (*2 marks*)

   b Use ideas about receptor sites to explain why these enantiomers have very different biological effects.   (*2 marks*)

5.7 Bonding dyes to fibres

Specification reference: CD (a)

Colourfast dyes

Dyes' molecules possess a chromophore and are therefore coloured. However in order for them to be effective as dyes, they must also be able to attach strongly to fabrics or other solid substances. Dyes with this property are described as **colourfast**; it is difficult for the dye molecule to detach from the fibre when it is washed.

Dye–fibre interactions

Dyes can bond to fibres by making use of a range of interactions:

- intermolecular bonds, especially hydrogen bonding and instantaneous dipole–induced dipole
- ionic bonds
- covalent bonds, including dative covalent bonds to metal ions (mordants).

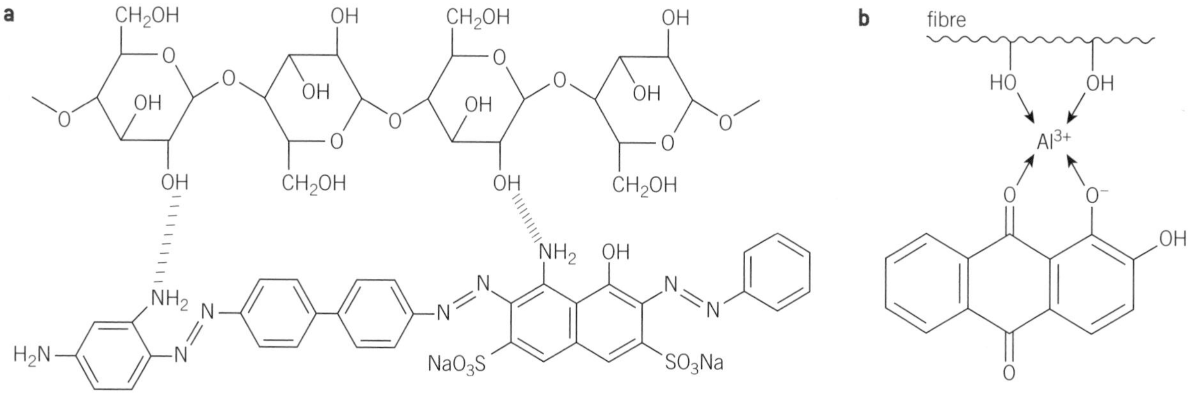

Revision tip

You can use ideas about the strength of bonds broken and formed to explain why dyes do not easily wash out of fibres. Think about the bonds present between the dye and the fibre and the bonds that would be possible between the dye and water molecules if the dye dissolved. This is similar to the way in which you would discuss why a solute may be insoluble in a particular solvent.

▲ **Figure 1 a** *OH groups on a dye molecule allow it to form hydrogen bonds to a cellulose fibre* **b** *an aluminium ion acts as a mordant by forming dative covalent bonds to both a dye molecule and a fibre*

Modifying the dye molecule

Certain groups are introduced into dye molecules in order to enable bonding to fibres:

- Acidic groups that ionise in aqueous solutions – the sulfonic acid group SO_3H is the most common acidic group.

- Fibre-reactive groups that can react in a condensation reaction with groups on the fibre.

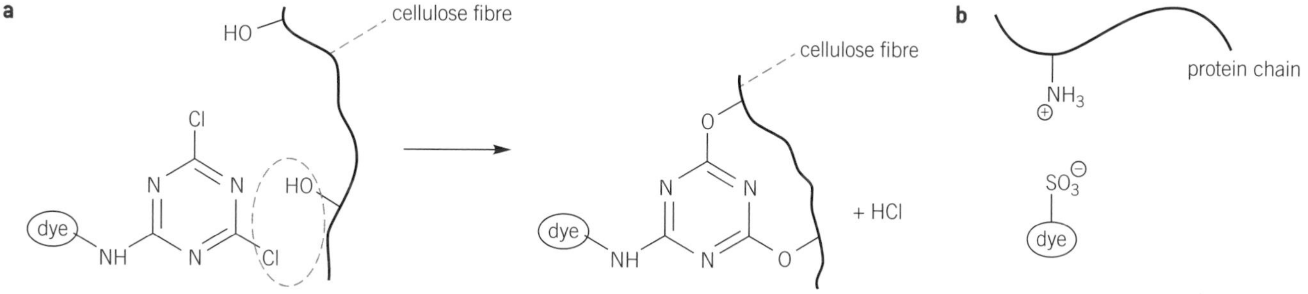

Synoptic link

The structure and synthesis of dyes is described in Topic 12.5, Reactions of arenes.

▲ **Figure 2 a** *a fibre-reactive dye forming a covalent bond to the amine group on a fibre* **b** *an acidic dye forming ionic bonds to a positively charged fibre*

Structure and properties

Explaining the properties of dyes

A colourfast dye stays attached to the dye even when washed in warm water. You can use ideas about intermolecular bonding to explain why dyes are colourfast.

Model answer: Colourfast dyes

Disperse red is a non-polar dye and is used to dye polyester fibres. Explain why disperse red is a colourfast dye when in contact with water.

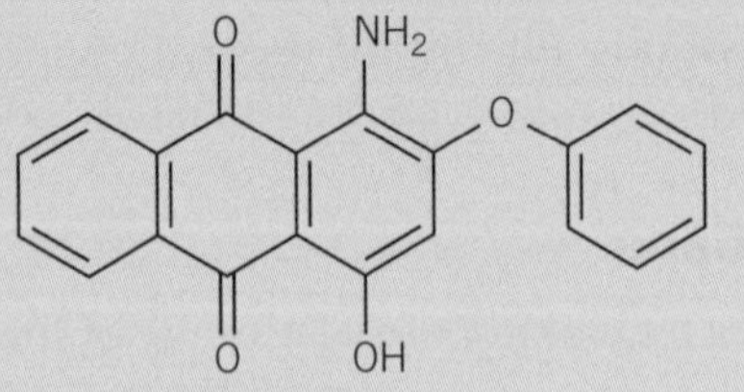

▲ Figure 3 *Disperse red*

- There are (strong) instantaneous dipole–induced dipole bonds between the dye molecule and the fibre.
- There are hydrogen bonds between water molecules.
- Some (weak) instantaneous dipole–induced dipole and a few hydrogen bonds can form between the dye molecule and water.
- The bonds that would need to break are stronger than the bonds that could form.
- The energy released from bond formation is not sufficient to compensate for the energy required to break bonds. So the dye will remain bonded to the fabric.

State the bonds present between the dye molecule and the fibres.

State the bonds present in the water.

State the bonds that can form between water and the dye molecule.

Compare the strength of the bonds described above.

Comment on the energy changes, and relate to the colourfast property of the dye.

Synoptic link

Intermolecular bonds, such as hydrogen bonds and instantaneous dipole–induced dipole bonds, are described in Topic 5.3, Bonds between molecules: temporary and permanent dipoles.

Summary questions

1 List four types of bond that are used to bond dyes to fibres. (*4 marks*)

2 The dye acid blue contains a sulfonic acid group, SO_3H, and is used to dye nylon. The ends of nylon chains often contain basic NH_2 groups. Suggest how acid blue bonds to nylon. (*3 marks*)

3 Cotton (cellulose) can be dyed by using the colourfast dye direct red. Use ideas about the breaking and forming of bonds to explain why direct red is colourfast in water when it is used to dye cotton. (*6 marks*)

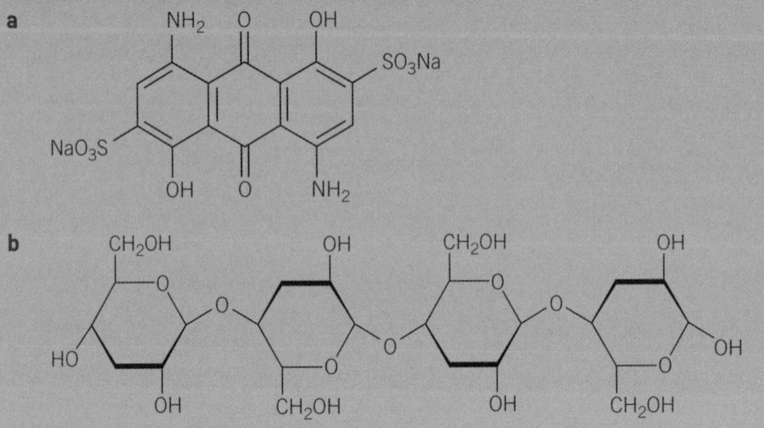

▲ Figure 4 *The structures of a direct red and b cellulose*

58

Experimental techniques

Specification reference: EL (s), EI (w)

Analysing an unknown salt X

You need to be able to describe how to carry out tests to identify a range of possible cations and anions present in a sample of an unknown salt, X.

Testing for anions

You can use simple test tube reactions to test for the presence of some common negative ions in an unknown compound X.

Carbonate

Carbonates react with acid to form carbon dioxide gas:

- In a test tube, add dilute acid to a solid or solution of X.
- If bubbles of gas are formed, then the solid probably contains a carbonate (or hydrogen carbonate) ion.
- To check that the gas is carbon dioxide, bubble the gas through limewater (saturated calcium hydroxide solution). A milky-white precipitate will be formed.

Sulfate

Barium sulfate is insoluble:

- Add barium chloride (or barium nitrate) solution to a solution of X.
- A white precipitate indicates the presence of sulfate ions.

Halide ions

Silver halides are insoluble and have noticeably different colours:

- Add silver nitrate solution (acidified with a few drops of nitric acid) to a solution of X.
- A precipitate indicates the presence of halide ions.
- Chloride ions produce a white precipitate, bromide ions produce a cream precipitate, and iodide ions a yellow precipitate.

Testing for metal cations

Precipitation reactions

Some metal cations produce coloured precipitate with sodium hydroxide, which can be used to identify the cation. Add sodium hydroxide solution to a solution of X. Observe the colour of any precipitate.

Flame tests

A piece of nichrome wire is cleaned in concentrated hydrochloric acid and dipped into a solution of X or a powdered sample of X. The end of the wire is heated in a blue Bunsen flame. Different metal ions cause different colours to be observed in the flame.

> **Revision tip**
>
> **Sequence of tests for anions**
>
> When testing an unknown solution for a range of anions, the tests should be done in a particular order:
>
> 1 Test for carbonate ions using nitric acid as the acid.
>
> 2 Test for sulfate ions; filter off any precipitate formed.
>
> 3 Test for halide ions.

Cation present	Observation
Cu^{2+}	Blue precipitate
Fe^{2+}	Green precipitate
Fe^{3+}	Brown precipitate
Ca^{2+}, Ba^{2+}, Al^{3+}, Pb^{2+}, Zn^{2+}	White precipitate

> **Revision tip**
>
> If no precipitate is observed in this test, then X probably contains a Group 1 ion or an ammonium ion.

> **Synoptic link**
>
> You need to be able to use ideas about electron energy levels to describe why metal ions emit visible light. See Topic 6.1, Light and electrons.

Chapter 5 Practice questions

1 Which combination of ions will produce a precipitate?

    1 Ba^{2+} ions and Cl^- ions

    2 Fe^{2+} ions and OH^- ions

    3 Ag^+ ions and I^- ions

    A 1, 2, and 3

    B Only 1 and 2

    C Only 2 and 3

    D Only 3   *(1 mark)*

2 Which of these options is the best description of the likely properties of the compound boron trifluoride, BF_3?

    A High melting point, insoluble in water, conducts electricity when molten.

    B Low melting point, insoluble in water, does not conduct electricity when molten.

    C Low melting point, soluble in water, conducts electricity when molten.

    D High melting point, insoluble in water, does not conduct electricity when molten.   *(1 mark)*

3 Which of these molecules has an overall dipole?

    1 CCl_2H_2

    2 CCl_3F

    3 CCl_4

    A 1, 2, and 3

    B Only 1 and 2

    C Only 2 and 3

    D Only 3   *(1 mark)*

4 Ice has a lower density than water. The best explanation of this is:

    A Hydrogen bonds are longer in ice than in liquid water.

    B Ice contains pockets of air that are trapped by the hydrogen bonds in the structure.

    C The tetrahedral arrangement of bonds and hydrogen bonds around each O atom holds the molecules in an open structure.

    D The hydrogen bonds cause the covalent bonds in the water molecules to lengthen so the molecules take up more space.   *(1 mark)*

5 Propanone, butane, propan-1-ol, and ethane-1,2-diol all have similar molecular masses. The most likely order of the boiling points of these molecules, from low to high, is:

    A propanone, butane, ethane-1,2-diol, propan-1-ol

    B ethane-1,2-diol, propanone, propanol, butane

    C propanone, butane, propanol, ethane-1,2-diol

    D butane, propanone, propan-1-ol, ethane-1,2-diol   *(1 mark)*

6 A solution of an ionic salt, X is investigated using a series of tests. No precipitate is observed when barium chloride is added, but a cream precipitate is seen when acidified silver nitrate is added.

Sodium hydroxide solution was added to a second sample and a brown precipitate was seen. The most likely identity of X is:

A Iron(III) bromide

B Iron(II) sulfate

C Copper(II) chloride

D Iron(III) sulfate   (*1 mark*)

7 The concentration of carbon dioxide in the troposphere is 0.4%, whereas the concentration of carbon monoxide is 0.1 ppm. How many times greater is the carbon dioxide concentration than the carbon monoxide concentration?

A 400

B 40 000

C 3.9

D 4 000 000   (*1 mark*)

8 During the second half of the twentieth century, a class of synthetic molecules known as CFCs (chlorofluorocarbons) were used in large quantities in several different applications.

These molecules are relatively volatile and the widespread use of CFCs resulted in large amounts of these molecules being released into the atmosphere.

a One such CFC molecule is CCl_3F. The structure of this molecule is shown below:

The C–Cl and C–F bonds in this molecule are polar and the molecule itself has an overall dipole

i State the meaning of the word polar when used to describe bonds   (*1 mark*)

ii Explain why the bonds in this CFC molecule are polar   (*2 marks*)

iii Explain why CCl_3F has an overall dipole   (*2 marks*)

iv Name the strongest type of intermolecular bond that exists between two molecules of CCl_3F.   (*1 mark*)

b A related molecule, CCl_4 has a similar structure to CCl_3F, but a Cl atom replaces the F atom. The strongest intermolecular bonds between two molecules of CCl_4 are instantaneous dipole–induced dipole bonds.

i Describe how these bonds arise.   (*4 marks*)

ii CCl_4 has a higher boiling point than CCl_3F. Discuss the factors that account for this difference   (*5 marks*)

9 Which of these statements about the secondary structure of proteins is correct?

A It is maintained by covalent bonds within peptide groups.

B It is maintained by hydrogen bonds between amino acid side chains.

C It consists of regions of the protein that have specific three-dimensional shapes.

D It has an important role in controlling the specificity of enzymes.   (*1 mark*)

10 The tertiary structure of a protein is maintained by several types of bond. Which type of bond is most likely to be disrupted when the pH of a protein is altered.

    A Covalent bonds   **C** Instantaneous dipole–induced dipole

    B Ionic bonds   **D** Hydrogen bonds   *(1 mark)*

11 Related drug molecules often contain a specific pharmacophore. Which of these statements about pharmacophores is correct?

    1 The pharmacophore is the smallest section of a structure that is common to all the drug molecules.

    2 All pharmacophores are chiral molecules.

    3 If the structure of a pharmacophore is modified it may become even more effective.

    A 1,2, and 3   **B** 1 and 2   **C** 2 and 3   **D** only 3   *(1 mark)*

12 Dyes can bond to fibres by means of a range of intermolecular bonds.

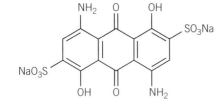

    a Acid Blue dyes bond to wool fibres by ionic bonds.

        i Name the group of atoms in this structure responsible for the formation of ionic bonds to wool.   *(1 mark)*

        ii Identify a group of atoms present in a protein structure that could form ionic bonds to this group.   *(1 mark)*

    b Disperse dyes form instantaneous dipole–induced dipole bonds with polymer chains.

        i Describe how instantaneous dipole–induced dipole bonds form between two molecules.   *(3 marks)*

        ii Suggest features in a dye molecule that will encourage the formation of relatively strong instantaneous dipole–induced dipole bonds.   *(2 marks)*

    c Direct dyes often have hydroxyl groups which can hydrogen bond to alcohol or amine groups in fibres. Draw a diagram to show how a hydroxyl group can hydrogen bond to a primary amine group. Indicate partial charges and relevant lone pairs.   *(4 marks)*

13 Dextroamphetamine is a drug used to treat disorders of the nervous system. The structure of dextroamphetamine is shown. It is thought that dextroamphetamine forms hydrogen bonds to receptor sites on the surface of nerve cells.

    a **i** Name the functional group in the dextroamphetamine molecule involved in binding to the receptor site.   *(1 mark)*

        ii Explain how this group of atoms is able to form hydrogen bonds to the receptor site.   *(2 marks)*

    b Dextroamphetamine exists as a pair of stereoisomers.

        i Explain why it can exist as stereoisomers.   *(2 marks)*

        ii Draw suitable diagrams to show how the structures of these stereoisomers are related.   *(2 marks)*

6.1 Light and electrons

Specification reference: EL (w), EL (v), OZ (u)

Bohr's theory and wave-particle duality

Bohr model of the atom

Electrons in an atom occupy shells, or energy levels. Electrons in an energy level have a specific amount of energy; the energy of electrons is said to be quantised.

Atomic emission spectra

Appearance of an emission spectrum

An emission spectrum is seen when atoms are heated up to a high temperature. It consists of coloured lines on a black background:

- The lines become closer at higher frequencies.
- There are several series of lines (although some of these may fall outside the visible part of the spectrum).

Explaining the formation of an emission spectrum

- Electrons in the ground state absorb energy.
- This energy causes electrons to be excited to a higher electron energy level.
- Electrons then drop back to lower energy levels. The energy lost (ΔE) is emitted as a photon of light.
- The light emitted is seen as a specific line in the spectrum.
- The frequency of the photon is related to the energy lost by the equation: $\Delta E = h\nu$.
- Different energy gaps produce photons of different frequencies.

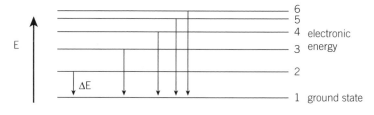

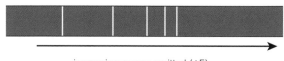

increasing energy emitted (ΔE)
increasing frequency of light emitted

▲ **Figure 1** *An emission spectrum*

Flame colours

Different metal ions produce different flame colours (see Table 1).

 Go further

You can use the diagrams on this page showing electron transitions to help you explain the key features of atomic spectra:

1. Lines get closer together at high frequency (hint: which transitions produce the highest frequency lines in these diagrams?).

2. Several series of lines are seen (hint: these diagrams show one series of lines, starting or finishing at $n = 1$).

Construct explanations of these two patterns, using the given hints.

Synoptic link

The lines in atomic spectra provided evidence for the existence of electron shells and so helped the development of the atomic model. See Topic 2.1, A simple model of the atom.

Key term

Quantised: Energy that can take only particular values (known as quanta).

Synoptic link

You may need to recall how these energy levels can be split into sub-shells with slightly different energies. See Topic 2.3, Shells, sub-shells, and orbitals.

Key terms

Ground state: The lowest energy level that an electron can occupy.

Photons: Quanta of energy in the form of electromagnetic radiation.

▼ **Table 1**

Ion	Colour
Li^+	Bright red
Na^+	Yellow
K^+	Lilac
Ca^{2+}	Orange-red
Ba^{2+}	Pale green
Cu^{2+}	Blue-green

Key term

Flame colour: The light emitted by metal ions when a vaporised metal salt is heated up in a flame.

Synoptic link

Flame colours can be used as part of a practical investigation to identify the ions present in a compound. See Chapter 5, Experimental techniques.

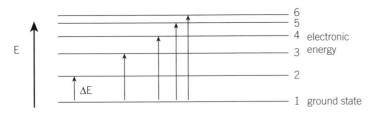

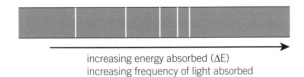

increasing energy absorbed (ΔE)
increasing frequency of light absorbed

▲ **Figure 2** *The formation of an atomic absorption spectrum*

Atomic absorption spectra

Appearance of an absorption spectrum

If white light is passed through a sample containing vaporised atoms, an atomic absorption spectrum is seen.

Formation of an absorption spectrum

- Electrons in the ground state absorb photons of light
- This energy from these photons causes electrons to be excited to a higher electron energy level
- The frequency of the photon is related to the energy gained by the equation: $\Delta E = h\nu$
- Light of this frequency does not pass through the sample, so a black line is seen in the spectrum.

Similarities and differences between the two types of spectrum

Similarities

- For a given element, lines appear at the same frequency
- Lines converge at a higher frequency
- Several series of lines are seen.

Differences

- Emission spectrum: coloured lines on a black background
- Absorption spectrum: black lines on a coloured background.

Calculations using data from lines in the spectrum

You need to be able to link energy (of the photon absorbed or emitted), frequency and wavelength. The key equations are:

$$\Delta E = h\nu \text{ and } c = \nu\lambda.$$

> **Revision tip**
> You will need to learn both of these equations. $\Delta E = h\nu$ is also useful when explaining the formation of the lines in atomic spectra.

ν = frequency in Hz

h = Planck constant, $6.63 \times 10^{-34}\,\text{J\,Hz}^{-1}$

c = speed of light, $3 \times 10^8\,\text{m\,s}^{-1}$

λ = wavelength in m.

> **Revision tip**
> When you are doing these calculations, remember that ΔE refers to an electron transition in a single atom, so the value of ΔE, in J, will be very small.

Summary questions

1 State the flame colour associated with:
   a sodium ions
   b barium ions. (*2 marks*)

2 Discuss the difference between an atomic emission spectrum and an atomic absorption spectrum. You should include comments about the appearance of the spectra and the processes in atoms that produce the spectra.
   (*4 marks*)

3 The atomic emission spectrum of a hydrogen atom consists of several series of lines. Explain how these series of lines are formed. (*5 marks*)

> 🖩 **Worked example: Calculations using the wavelength of light**
>
> The emission spectrum of sodium contains a line with λ = 590 nm (1 nm = 10^{-9} m). Calculate the change in energy of the electronic transition that causes this line in the spectrum.
>
> **Step 1:** Convert the wavelength to m: 590 nm = 590×10^{-9} = 5.90×10^{-7} m.
>
> **Step 2:** Calculate the frequency of this light: $c = \nu\lambda$, so $\nu = c/\lambda = \frac{3.00 \times 10^8}{5.90 \times 10^{-7}}$
> = 5.085×10^{14} Hz
>
> **Step 3:** Calculate the energy change that produces light of this frequency: $\Delta E = h\nu$,
> so $\Delta E = 6.63 \times 10^{-34} \times 5.085 \times 10^{14} = 3.37 \times 10^{-19}$ J.

6.2 What happens when radiation interacts with matter?

Specification reference: EL (v), OZ (s), OZ (t), OZ (u)

The electromagnetic spectrum

Visible light, ultraviolet radiation, and infrared radiation are all examples of electromagnetic radiation.

	radiofrequency	micro-wave	infrared	visible	UV	X-rays	γ-rays

Frequency (Hz) 10^5 10^6 10^7 10^8 10^9 10^{10} 10^{11} 10^{12} 10^{13} 10^{14} 10^{15} 10^{16} 10^{17} 10^{18} 10^{19} 10^{20}

Wavelength (m) 10^3 1 10^{-3} 10^{-6} 10^{-9}

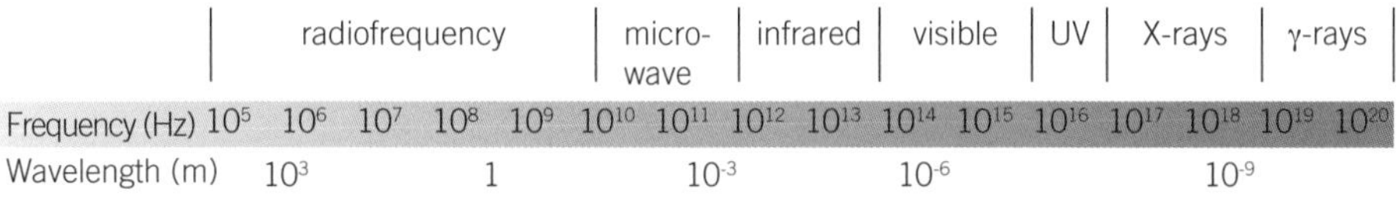

▲ **Figure 1** *The electromagnetic spectrum, showing the wavelengths and frequencies of the different types of radiation*

Radiation from the Earth and the Sun

The surface of the Earth and the surface of the Sun have very different temperatures. They both emit electromagnetic radiation, but of very different types.

The effect of radiation on matter

Electromagnetic radiation can interact with matter, transferring energy to the atoms or bonds present. The effect that this has on matter depends on the type of radiation involved.

Type of radiation	Effect on matter
Infrared	Bonds vibrate with greater energy
Visible	Electrons excited to higher energy levels; some bonds can break
Ultraviolet	Electrons excited to higher energy levels and bonds break

Calculations to do with bond breaking

You can use information about the bond enthalpy of covalent bonds to calculate the minimum energy needed to cause a bond to break.

 Worked example: Calculating the frequency needed to cause photodissociation

Electromagnetic radiation can cause the photodissociation of bromomethane, CH_3Br in the Earth's atmosphere. The C—Br bond has a bond enthalpy of 276 kJ mol⁻¹. Calculate the minimum frequency of radiation needed to break the C—Br bond.

Step 1: Convert the bond enthalpy to J: 276 kJ mol⁻¹ = 276 000 J mol⁻¹

Step 2: This is the energy needed for 1 mole of bonds (= N_A bonds, where N_A = Avogadro constant). Calculate the energy needed to break 1 bond:
$$N_A = 6.02 \times 10^{23} \text{ mol}^{-1}.$$
So energy needed to break 1 bond = 276 000/6.02 × 10²³ = 4.585 × 10⁻¹⁹ J.

Step 3: Calculate the frequency of a photon with this energy, using E = h ν:
$$\nu = \frac{E}{h} = \frac{4.585 \times 10^{-19}}{6.63 \times 10^{-34}} = 6.92 \times 10^{14} \, Hz$$

Key term

Electromagnetic spectrum: The range of wavelengths or frequencies over which electromagnetic radiation extends.

Revision tip

You need to know how the wavelength, frequency (and hence energy) change as the type of radiation changes from infrared to ultraviolet. You do not need to know the actual values.

Revision tip

You may need to do this calculation in reverse and calculate the bond enthalpy of a bond that can be broken by radiation of a particular frequency.

Key term

Photodissociation: Bond breaking caused by visible light or ultraviolet radiation.

Summary questions

1 Name the principal type of electromagnetic radiation emitted by:
   a the surface of the Earth
   b the Sun   (2 marks)

2 Visible light and infrared radiation differ from each other in terms of:
   a the properties of the photons, and
   b the effect they have on matter. Describe these differences. (5 marks)

3 The minimum frequency of radiation needed to break one C—H bond is 1.09 × 10¹⁵ Hz. Calculate the bond enthalpy of a C—H bond, in kJ mol⁻¹   (3 marks)

6.3 Radiation and radicals

Specification reference: OZ (o). OZ (p), OZ (q)

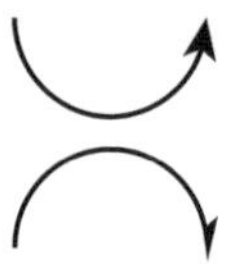

▲ **Figure 1** *A full curly arrow (top) and a half curly arrow (bottom)*

Synoptic link

You need to be able to use full curly arrows to show the movement of a pair of electrons in organic mechanisms. See Topic 12.2, Alkenes and 13.2, Haloalkanes.

Key terms

Heterolytic fission: A type of covalent bond breaking, in which both electrons from the shared pair go to just one atom.

Homolytic fission: A type of covalent bond breaking, in which one electron from the shared pair goes to each atom.

Radical: A species with one (or more) unpaired electrons.

Key term

Radical chain reaction: A reaction in which new radicals are formed at the end of one step, which continue. These radicals then continue the process.

Revision tip

Dots are not always included in the formulae of radicals. You may need to deduce whether a species is a radical by counting the total number of outer shell electrons: if there is an odd number, then the species must be a radical. Some species (e.g. O atoms) have two unpaired electrons and are called biradicals. Dots are not usually used for biradicals.

Bond fission

Covalent bonds can break in two different ways, producing either ions or radicals:

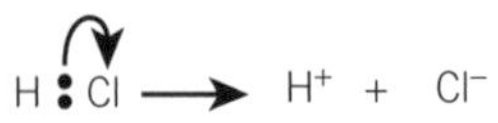

$$H\!:\!Cl \longrightarrow H^+ + Cl^-$$

▲ **Figure 2** *Heterolytic fission, producing ions*

▲ **Figure 3** *Heterolytic fission, producing radicals*

Curly arrows

These are used to show the movement of electrons.

- A half curly arrow shows the movement of one electron.
- A full curly arrow shows the movement of a pair of electrons.

Radicals and chain reactions

Radicals are highly reactive and undergo chain reactions. These reactions are often very fast, and may occur in the presence of light. You can often recognise radicals in equations, because they are written with a dot to indicate the unpaired electron:

$$Cl\bullet \qquad NO\bullet \qquad CH_3\bullet$$

Mechanism of a chain reaction

Radical chain reactions happen in three stages:

1 Initiation: radicals are formed from a stable molecule.

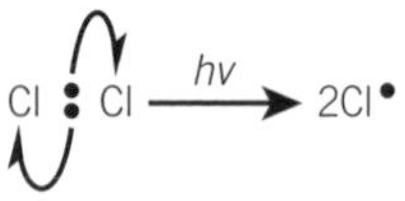

$$Cl\!:\!Cl \xrightarrow{h\nu} 2Cl\bullet$$

2 Propagation: a radical reacts and the process forms a new radical.

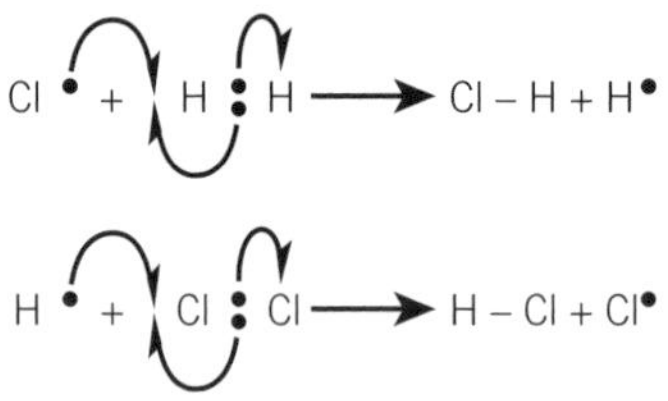

$$Cl\bullet + H\!:\!H \longrightarrow Cl-H + H\bullet$$

$$H\bullet + Cl\!:\!Cl \longrightarrow H-Cl + Cl\bullet$$

Propagation steps often occur in pairs; the radical formed by the 1st step reacts again in the 2nd step.

3 Termination: two radicals collide to form a stable molecule.

$$H\bullet + H\bullet \longrightarrow H-H$$

Examples of radical chain reactions

Reactions of alkanes with halogens

A halogen atom can substitute for a hydrogen atom in an alkane chain. The mechanism of this reaction is radical substitution.

1. Initiation: Homolytic fission of a halogen molecule in the presence of UV light:

$$Cl_2 + h\nu \rightarrow 2Cl\bullet$$

2. Propagation steps: A methyl radical is formed, which then reacts to reform the $Cl\bullet$.
   The $Cl\bullet$ can be thought of as a catalyst in these stages:

$$CH_4 + Cl\bullet \rightarrow CH_3\bullet + HCl$$

$$CH_3\bullet + Cl_2 \rightarrow CH_3Cl + Cl\bullet$$

3. Termination: $Cl\bullet$ or $CH_3\bullet$ radicals collide:

$$Cl\bullet + Cl\bullet \rightarrow Cl_2$$

$$CH_3\bullet + Cl\bullet \rightarrow CH_3Cl$$

The depletion of ozone by Cl atoms

Ozone, O_3 is present in the stratosphere. Halogen atoms from synthetic compounds such as haloalkanes act as catalysts for the breakdown of ozone. (In the mechanism below, the dots showing unpaired electrons have not been included.):

1. Chloroalkanes reach the stratosphere and photodissociate, forming Cl:

$$CH_3Cl + h\nu \rightarrow CH_3\bullet + Cl\bullet$$

2. The chlorine reacts with ozone in a catalytic cycle, involving O atoms, which are also present in the stratosphere:

$$Cl\bullet + O_3 \rightarrow ClO\bullet + O_2$$

$$ClO\bullet + O \rightarrow Cl\bullet + O_2$$

3. Chlorine atoms are removed from the cycle by termination reactions:

$$Cl\bullet + Cl \rightarrow Cl_2$$

 Go further

The overall effect of a 2-stage process can be deduced by adding together the reactants and products of both stages and then cancelling out any species that appear on both sides. Write overall equations for the propagation steps in:

a   the reaction of methane with chlorine

b   the depletion of ozone by chlorine radicals.

Photodissociation of other haloalkanes

Iodoalkanes and bromoalkanes photodissociate more easily than chloroalkanes, because C−I and C−Br bonds are weaker than C−Cl bonds. This means that they can be broken down by the lower frequency radiation in the troposphere and do not reach the stratosphere. Fluoroalkanes do not photodissociate in the stratosphere, because the C−F bond is too strong to be broken by the ultraviolet radiation present.

The importance of ozone

Ozone in the stratosphere

Ozone absorbs high-energy ultraviolet radiation from the Sun. It prevents this radiation reaching the surface of the Earth. High-energy ultraviolet radiation can cause health problems, including skin cancer and eye cataracts.

Ozone in the troposphere

Ground level ozone is a secondary pollutant, present as a component of photochemical smog.

Revision tip

The presence of ultraviolet radiation (or visible light) in a reaction is sometimes shown using the symbol 'hv' in the equation.

Synoptic link

You may need to be able to compare radical substitution with other substitution reactions, such as nucleophilic substitution in Topic 13.2, Haloalkanes.

Revision tip

Ozone can also be depleted by the action of other radicals present in the stratosphere, such as NO and OH. You should be able to deduce the equations for depletion processes involving these radicals.

Synoptic link

You also need to be able to link the trend in bond enthalpy of C-halogen bonds to the reactivity of haloalkanes with nucleophiles. This is covered in Topic 13.2, Haloalkanes.

Key terms

Troposphere: The layer of the atmosphere immediately above the Earth's surface.

Stratosphere: The layer of the atmosphere above the troposphere.

Synoptic link

Photochemical smog is formed from a reaction between hydrocarbons and nitrogen oxides. The formation of these pollutants is described in Topic 15.1, Atmospheric pollutants. Photochemical smog causes breathing problems and the corrosion of plastics, rubber, and textiles.

The formation and destruction of ozone

Ozone formation

Ozone is formed naturally in the stratosphere:

1 Oxygen molecules photodissociate into oxygen atoms:

$$O_2 + h\nu \rightarrow 2O$$

2 Ozone is formed when an oxygen atom combines with an oxygen molecule:

$$O_2 + O \rightarrow O_3$$

Ozone destruction

Ozone is destroyed naturally when it absorbs high energy ultraviolet radiation:

$$O_3 + h\nu \rightarrow O_2 + O$$

Summary questions

1 State two differences between homolytic and heterolytic bond fission. *(2 marks)*

2 Ozone molecules in the stratosphere absorb high energy ultraviolet light:
   a describe the chemical process that occurs when this happens
   b explain why this process is important for human health at the surface of the Earth. *(3 marks)*

3 NO radicals can deplete ozone in a similar way to the action of Cl radicals. Write equations for a catalytic cycle involving NO and ozone molecules. *(2 marks)*

6.4 Infrared spectroscopy

Specification reference: WM (j)

The effect of infrared radiation on molecules

Each type of covalent bond in a molecule vibrates at a specific frequency. When a bond absorbs infrared radiation of the correct frequency, it vibrates with greater energy. So molecules may absorb infrared radiation at several different frequencies, depending on the bonds that are present.

The appearance of an infrared spectrum

An infrared spectrum shows the % transmittance of infrared over a range of wavenumbers.

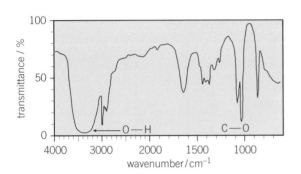

$CH_3—CH_2—OH$

▲ **Figure 1** *The infrared spectrum of ethanol*

- Absorption of infrared radiation is shown by the downward pointing troughs – called peaks – in the spectrum.
- The strength and shape of the peak may also be helpful when interpreting the spectrum – peaks may be strong or broad.

Fingerprint region

The peaks above 1500 cm⁻¹ help you to identify the types of bond in the molecule. The pattern of peaks below 1500 cm⁻¹ is very complex and difficult to analyse. This fingerprint region is unique to a specific molecule and so, by comparing it to a computer database, the exact identity of a molecule can be found.

Summary questions

1 Describe what happens in a molecule when it absorbs infrared radiation. *(2 marks)*

2 The infrared spectrum of ethanoic acid contains a broad peak between 2 500 and 3 200 cm⁻¹ and a sharp peak at 1 715 cm⁻¹. Explain why ethanoic acid has peaks at these wavenumbers. *(4 marks)*

3 A C–H bond vibrates at a frequency of 9.04×10^{13} Hz. Calculate the wavenumber of infrared radiation that will be absorbed by this C–H bond. *(3 marks)*

Revision tip

When describing or interpreting a spectrum, you refer to the bond, the location (the functional group containing the bond), and wavenumber range.

Revision tip

In exam answers, you should always discuss regions 3200–3600 cm⁻¹ (where O–H bonds show up) and 1700–1750 cm⁻¹ (where most C=O bonds show up). The absence of a peak may be as important as its presence.

Common misconception: C–H bonds

C–H bonds are present in all molecules, so infrared spectra of organic molecules generally have a peak or set of peaks at around 3000 cm⁻¹. However, they are not often useful in helping to identify functional groups in molecules, except in some cases, for example alkenes or arenes.

Key term

Fingerprint region: The region in an infrared spectrum below 1500 cm⁻¹.

6.5 Mass spectrometry

Specification reference: EL(x), WM(i)

Key terms

Mass spectrometry: Use of a mass spectrometer to identify the atoms or molecules in a sample.

Mass spectrum: A graph of abundance against mass/charge (m/z) ratio for the ions detected by the mass spectrometer.

Synoptic link

You need to be able to calculate the relative atomic mass of an element from the % abundance values in a mass spectrum. See Topic 1.1, Amount of substance.

Revision tip

If you are giving the formula of a fragment that causes a peak in the mass spectrum, you must include a + charge.

A mass spectrometer is an instrument that measures the masses and abundances of positive (1+) ions produced from atoms and molecules.

You need to know how to interpret two types of mass spectra.

- the mass spectrum produced from a sample of an element
- the mass spectrum produced from a sample of a molecule.

Mass spectra of elements

A sample of an element may contain several isotopes, with different abundances.

Deductions from the mass spectrum of an element

From the mass spectrum of an element you can deduce:

- the number of isotopes in the sample from the number of peaks in the spectrum
- the relative isotopic masses of each isotope from the m/z values
- the % abundance of each isotope from the height of the peaks.

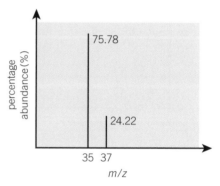

 Go further

The mass spectrum of chlorine shows that there are two isotopes present in this sample and the relative abundances of these isotopes. However, samples of the element chlorine contain **molecules**, each containing 2 chlorine atoms.

A mass spectrum of Cl_2 has three peaks.

a   Use information from the mass spectrum of Cl atoms to explain why there are 3 peaks, and suggest the m/z value for each peak.

b   Calculate the % abundance of each of the 3 peaks to 3 s.f.

Mass spectra of molecules

The mass spectrum produced from a molecule consists of a number of peaks, with a wide range of masses (m/z values).

The peak with the largest mass is called the molecular ion, or M^+ ion.

Most of these peaks are caused by fragmentation:

$$C_2H_6O^+ \quad \rightarrow \quad C_2H_5^+ \quad + \quad OH$$
M^+ ion $\qquad\qquad$ fragment ion $\qquad$ uncharged fragment (not detected)

Common misconception: Charges on fragments

The fragments formed in a mass spectrum do not have the charges that you would normally expect from their chemical properties. OH often exists as OH^- ions in chemical reactions, but in a mass spectrometer it may be formed as OH^+ or just as an uncharged OH group.

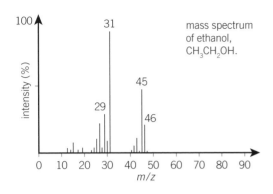

mass spectrum of ethanol, CH_3CH_2OH.

Interpreting the mass spectrum of a molecule

From the mass spectrum you can make various deductions:

- the relative molecular mass of the molecule (from the *m/z* value of the molecular ion)
- the molecular formula of the fragment ions, and of the other fragment that has been lost in order to form the fragment.

Isotope peaks

You may notice on the mass spectrum that there is a tiny peak at 47, just to the right of the peak caused by the M⁺ ion.

This is caused by the presence of ¹³C isotopes in the ion.

About 1% of C atoms are ¹³C, rather than ¹²C, so a small fraction of ethanol molecules will contain one ¹³C atom. The molecular mass of an ethanol molecule with one ¹³C is 47, not 46.

The peak at 47 is known as the M+1 peak.

Summary questions

1 A sample of an element contains several isotopes. State 3 pieces of information about the isotopes that can be deduced from the mass spectrum of the element. *(3 marks)*

2 A mass spectrum of the organic molecule benzene (C_6H_6) includes a peaks at *m/z* = 78 and a much smaller peak at *m/z* = 79. Explain the origin of these peaks. *(2 marks)*

6.6 Further mass spectrometry

Specification reference: PL(r)

Synoptic link

The terms m/z and M⁺ peak were explained in Topic 6.5, Mass spectrometry.

Deducing molecular formulae and structures

The m/z value of the M⁺ peak gives you the relative molecular mass of a molecule. However it is not possible to deduce the molecular formula without further information, which can be provided by the technique of high-resolution mass spectrometry.

The fragmentation pattern obtained by low-resolution mass spectrometry can then be used to deduce the likely structure of a molecule.

High resolution mass spectrometry

High resolution mass spectrometry measures the mass of the M⁺ peak to four decimal places.

With this level of precision, it turns out that the relative mass of individual atoms are not whole numbers ($^{14}N = 14.0031$, $^{1}H = 1.0078$, $^{16}O = 15.9949$). ^{12}C however, is defined as exactly 12.0000.

This means that each molecular formula has a unique relative mass, to 4 d.p., e.g.

$$C_2H_6O = 46.0417$$

The mass of the M⁺ ion can be compared to a database and the molecular formula can be deduced.

Using the fragmentation pattern

A combination of the mass of the M⁺ ion and other information can tell you the molecular formula of the molecule.

However, there may be several isomers with the same molecular formula. The fragmentation pattern helps you to deduce which isomer is actually present.

From a knowledge of the molecular formula of the molecule and the mass of the fragment ions, you can deduce:

- The molecular formula of the fragment ions.

- The molecular formula of the other fragment that has been lost in order to form the fragment.

It may then be possible to deduce the structure of the molecule.

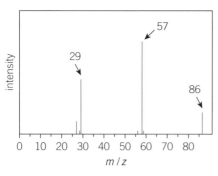

▲ **Figure 1** *The low resolution mass spectrum of the ketone $C_5H_{10}O$*

Revision tip

Remember that if a fragment is shown in the mass spectrum then you must write its formula with a positive charge. The groups of atoms that are lost from these fragment ions are uncharged and so do not show up in the spectrum.

 Worked example: Interpreting a mass spectrum

What information can be deduced from the mass spectrum of the ketone with molecular formula $C_5H_{10}O$, shown in Figure 1 above?

Step 1: Identify the molecular ion peak and use it to find the M_r value of the molecule: the M⁺ peak has the largest m/z value, i.e. 86. The M_r value is therefore 86, which is consistent with the molecular formula given.

Step 2: Identify the fragments that give rise to other peaks in the spectrum, using the difference between the M⁺ peak and the fragment peak, or by comparing the fragment peak mass with a list of common masses.
Peak at 57: 86 − 57 = 29, therefore a C_2H_5 *group* has been **lost**. The other alternative is CHO, but this group of atoms is only found in aldehydes, not ketones. The *fragment* is $C_3H_5O^+$.
Peak at 29: 29 suggests the presence of a $C_2H_5^+$ *fragment*. This suggests a *group* with m/z = 57 has been **lost,** which is likely to be C_3H_5O.
Small peak at 28: There is also a small peak at 28, which could be due to a CO^+ *fragment*. 86 − 28 = 58, which is 2 × 29, so maybe this is formed by the **loss** of two C_2H_5 *groups.*

Step 3: Use the information to suggest a structure for the molecule: the molecule is a ketone and also contains at least one C_2H_5 group. There do not seem to be any fragments suggesting larger alkyl groups (e.g C_3H_7) and so the most likely structure is $CH_3CH_2COCH_2CH_3$, pentan-3-one.

▼ **Table 1** *Common masses of species*

mass	possible formula
15	CH_3
17	OH
18	CO
29	C_2H_5 or CHO
43	C_3H_7 or CH_3CO
45	C_2H_5O or $COOH$
77	C_6H_5

Common misconception: Charges on fragments

Remember that the fragments that are detected as peaks in the mass spectrum will have a positive charge, even if this is not the usual charge for this group of atoms. So a fragment with a *m/z* value of 17 will be OH^+ not OH^-.

Summary questions

1 Explain how high-resolution mass spectrometry enables chemists to find the molecular formula of a molecule. (*4 marks*)

2 The molecule butanone, C_4H_8O, has a structure $CH_3CH_2COCH_3$. The mass spectrum of butanone has fragment peaks at *m/z* = 57, 43, and 29.
   a Explain how each of these fragments has formed from the molecular ion. (*3 marks*)
   b Give the formula of the species that causes each of these peaks. (*3 marks*)

3 The molecules propyl methanoate ($HCOOC_3H_7$) and ethyl ethanoate ($CH_3COOC_2H_5$) are isomers. Discuss the value of high- and low-resolution mass spectrometry to enable a chemist to distinguish between these molecules. (*6 marks*)

6.7 Nuclear magnetic resonance (NMR) spectroscopy

Specification reference: PL (s)

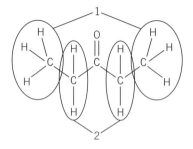

▲ **Figure 1** *Pentan-3-one has two different proton environments*

Key terms

Chemical shift (δ): A measure of the difference between the frequency absorbed by the protons in a peak and that absorbed by the protons in a reference compound (tetramethylsilane, TMS).

Multiplet: A peak that is split into several peaks, each with a very slightly different chemical shift.

Nuclear magnetic resonance spectroscopy is based on the fact that certain nuclei, such as ^1H and ^{13}C, have magnetic properties and can occupy high energy or low energy states when they are placed in a magnetic field.

The energy gap between these states corresponds to the absorption or emission of a photon of radio frequency electromagnetic radiation.

Proton (^1H) NMR spectroscopy

This provides information about the **chemical environment** of each type of proton in a molecule.

Chemical environment

The chemical environment of a proton is determined by factors such as how close it is in space to polar atoms, such as the O atoms in a $C=O$ bond.

Proton (^1H) NMR spectra

Proton NMR spectra can either be low-resolution (low-res) or high-resolution (high-res).

Low-resolution NMR spectrum

In a low-resolution spectrum you will see a number of separate peaks at different chemical shifts.

From a low-resolution NMR spectrum you can deduce three pieces of information:

- The number of proton environments from the number of peaks.
- The relative number of protons in each environment from the height of the peaks (or strictly the area of the peaks) (although this is often indicated by a label on the peak such as '3H (or just 3)' etc.
- Some information about the environment that produces each peak, from the value of the chemical shift.

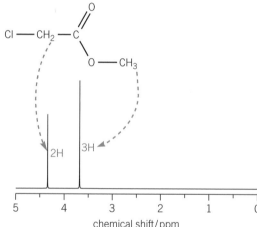

▲ **Figure 2** *A low-resolution NMR spectrum of methyl chloroethanoate. There are two peaks showing that there are two different proton environments in this molecule. The ratio of protons in each environment is 2:3*

^1H NMR chemical shifts relative to TMS

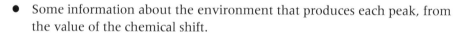

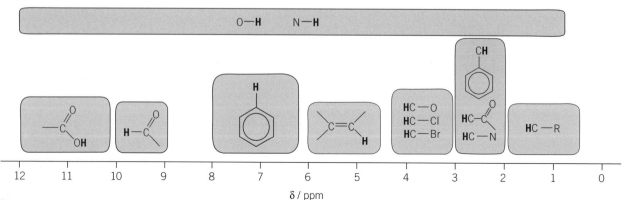

▲ **Figure 3** *The ^1H chemical shift data included in the data sheet*

High-resolution proton NMR spectra

In a high-resolution NMR spectrum, many of the peaks are shown to be split to form **multiplets**, such as doublets, triplets, or quartets. The pattern of this splitting provides more information about the groups of atoms that are adjacent to each proton environment.

Splitting patterns and the *n* + 1 rule

To predict the pattern of splitting observed for the peak caused by a proton environment, you use the *n* + 1 rule:

"The number of peaks in the multiplet caused by a proton environment equals n + 1, where n is the total number of equivalent protons on adjacent carbon atom(s)."

Go further: Coupling

Splitting arises because the magnetic field experienced by the protons in one environment can be affected by the alignment of the protons in an adjacent environment. Each proton can be aligned with the field ($\uparrow$) or against the field ($\downarrow$). So if there are two protons in a second, adjacent, environment these can be aligned in the following patterns:

$\uparrow\uparrow$ $\uparrow\downarrow$ $\downarrow\uparrow$ $\downarrow\downarrow$

Since $\uparrow\downarrow$ and $\downarrow\uparrow$ have the same effect, this means that there will be three different patterns of alignment, in the ratio 1: 2: 1. This causes the peak from the first environment to be split into a triplet, with the ratio of heights following the same 1: 2: 1 pattern.

Question: Use the same ideas of magnetic alignments to: **a** show that 3 protons in an adjacent environment will split a peak into a quartet; **b** predict the ratios of the peaks in the quartet.

Common misconception: OH and NH groups and splitting patterns

At this level, you only need to consider splitting caused by protons on adjacent C atoms.

You need to remember that the protons on OH, NH, or NH_2 groups do not cause splitting.

So even if there is a H attached to an O that is adjacent to a CH proton environment the peak due to this environment is not split.

Peaks due to OH or NH_2 groups are never split.

Carbon-13 (^{13}C) NMR spectroscopy

About 1% of the C atoms in a sample of an organic compound are in the form of a ^{13}C isotope. The nucleus of ^{13}C has magnetic properties, so a ^{13}C NMR spectrum can be obtained from every organic compound.

Revision tip

The peaks in the ^{13}C spectra that you will interpret will not be split into multiplets. Because only 1 in 100 atoms of C are ^{13}C isotopes it will be very unlikely that there will be two ^{13}C atoms adjacent to each other.

The presence of ^{1}H atoms does not cause splitting either because the spectra are obtained using a 'proton decoupling' technique.

Revision tip

You can use the data sheet to find the information about the likely chemical shifts of various common types of environment. Treat this information cautiously, as the actual chemical shift varies quite a lot depending on other factors. In general it is best to use chemical shift values to confirm the structure that you have deduced using other data, rather than starting with the chemical shift data.

Revision tip

You should make it clear how you have applied the *n*+1 rule in your answer. For example a triplet peak is caused by the presence of 2 protons on the carbon atom adjacent to the proton environment responsible for this peak.

You will not be expected to interpret multiplets more complex than a quartet (a peak split into 4).

Revision tip

You may be asked to predict the ^{1}H or ^{13}C spectrum of a compound. You should include information about the number of peaks and their likely chemical shifts. For ^{1}H spectra you should also predict the relative number of protons responsible for each peak and any splitting that might be seen for each peak.

Revision tip

Certain patterns of splitting are good evidence for particular groups of atoms. If there is a quartet and a triplet in the high-res ^{1}H NMR spectrum of a molecule, then this is good evidence for the presence of an ethyl group ($-CH_2-CH_3$).

The two protons in the CH_2 group split the peak for the adjacent CH_3 group into a triplet, and the three protons in the CH_3 group split the peak for the CH_2 group into a quartet.

Revision tip

The heights of the peaks in a
^{13}C spectrum do not give any
information about the number of C
atoms in each environment.

Interpreting a ^{13}C spectrum

From a ^{13}C NMR spectrum you can deduce two pieces of information:

- The number of different environments of C atoms in the molecule – from the number of peaks.

- Some information about these environments, such as the group that the C is part of – from the chemical shift of each of these peaks.

^{13}C NMR chemical shifts relative to TMS

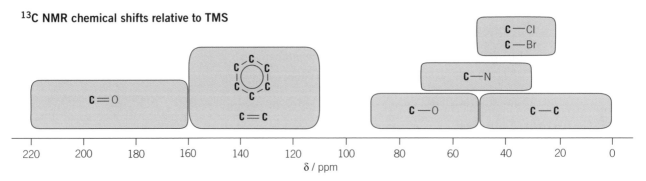

▲ **Figure 4** *Chemical shifts of different 13C environments as shown in the OCR data sheet*

🖩 Worked example: Interpreting a ^{13}C NMR spectrum

The ^{13}C spectrum of a molecule with the molecular formula C_3H_6O is shown below:

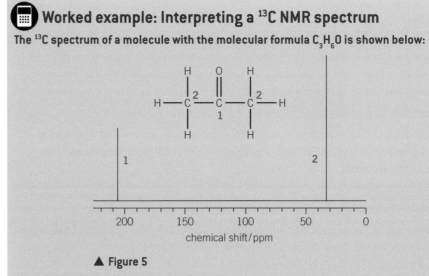

▲ Figure 5

Deduce the structure of the molecule:

Step 1: Work out the number of different carbon atom environments: there are two peaks, so there are two environments. There are three C atoms in the molecule, so two of these C atoms must be in an identical environment.

Step 2: Use the approximate ranges of chemical shifts to get an idea about the atoms that these C atoms are bonded to: $\delta = 207$ suggests C=O; $\delta = 32$ suggests C—C.

Step 3: Consider the possible structures of molecules with the formula C_3H_6O and explain which one matches the ^{13}C evidence most closely: Possible structures are propanone (CH_3COCH_3) or propanal (CH_3CH_2CHO). Propanal has three C environments, propanone has two C environments, so the molecule must be propanone. There are C=O and C—C groups in propanone, so this helps to confirm the choice.

Interpreting NMR spectra of aromatic compounds

The ^{1}H and ^{13}C nuclei in the molecule benzene are all in identical environments, so only one peak will be observed in the low-resolution ^{1}H or ^{13}C NMR spectrum of benzene.

If there are other groups substituted onto the benzene ring, then there may be several similar (but non-identical) environments.

a 2H environments and   **b** 3H and 4C environments
4C environments

▲ **Figure 6** *The different environments of H and C atoms in two substituted benzene compounds*

Summary questions

1   Predict the number of peaks in **a** the low-resolution ^{1}H NMR spectrum and **b** the ^{13}C NMR spectrum for the molecule shown below: (*2 marks*)

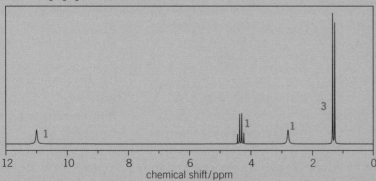

2   Explain why the molecule ethanal (CH_3CHO) produces a high-res ^{1}H spectrum consisting of a quartet and a doublet group.   (*2 marks*)

3   The high-resolution ^{1}H NMR spectrum of a molecule with molecular formula $C_3H_6O_3$ is shown below.

chemical shift/ppm

a   Deduce the structure of this molecule.   (*5 marks*)
b   Predict the appearance of the ^{13}C spectrum of the molecule.   (*4 marks*)

6.8 Using combined spectroscopic techniques to deduce the structure of organic molecules

Specification reference: PL (t)

Information from different spectroscopic techniques

Different spectroscopic techniques provide different pieces of information that enable chemists to deduce the structure of an unknown compound. The empirical formula can be worked out from the % mass data. The M^+ peak from the mass spectrum gives the M_r so the molecular formula can be worked out from the empirical formula. (Alternatively the molecular formula can be worked out directly from the M^+ peak in the high-resolution mass spectrum.)

▼ **Table 1** *Information from different spectroscopic techniques*

Technique	Information	How to deduce the information
Mass spectrometry	the relative molecular mass	from the M^+ ion
Mass spectrometry	the groups of atoms present in the molecule	from the fragmentation pattern
High-resolution mass spectrometry	the molecular formula	from the mass of the M^+ peak (to 4 d.p.)
Low-res ^{1}H spectroscopy	the number of proton environments	from the number of peaks
Low-res ^{1}H spectroscopy	relative number of the protons in each environment	from the height of the peaks (or of the integrated trace)
Low-res ^{1}H spectroscopy	the types of proton environments	from the chemical shift of each peak
High-res ^{1}H spectroscopy	the number of H atoms on C atoms adjacent to each proton environment	from the splitting pattern of each peak
^{13}C spectroscopy	the number of ^{13}C environments	from the number of peaks
^{13}C spectroscopy	the types of ^{13}C environments	from the chemical shift of each peak
Infrared spectroscopy	the functional groups present in the molecule	from the wavenumbers of the peaks

The molecular formula can also be deduced from percentage by mass data. This enables you to work out the empirical formula. You can then use the relative molecular mass to find the molecular formula.

Using combined spectroscopic techniques to identify a molecule

An unknown molecule has the empirical formula C_3H_4O. The M^+ peak in the mass spectrum occurs at an *m/z* value of 112. The infrared spectrum shows a peak at 1725 cm^{-1}, but no peak above 3100 cm^{-1} or in the region 1600–1725 cm^{-1}. The ^{1}H NMR spectrum shows just one peak, at δ = 2.7 ppm. In the high-res spectrum this peak is split into a triplet. The ^{13}C NMR spectrum has two peaks, at δ = 36 and 210 ppm. You can use this information to deduce the structure of the molecule.

Step 1: Find the molecular formula of the molecule: The M_r of a C_3H_4O formula would be 56. The M_r of the actual molecule, from the M$^+$ peak, is 112. So there are two C_3H_4O units in the molecular formula, which is therefore $C_6H_8O_2$.

Step 2: Deduce the functional groups present in the molecule: There are no O–H bonds present (no peak above 3100) or C=C bonds (no peak between 1600–1700 cm^{-1}). However the peak at 1725 cm^{-1} indicates a C=O in either an aldehyde or a ketone; it can't be a carboxylic acid because there are no O–H bonds present. There must be two identical C=O groups, because there is only one peak.

Step 3: Deduce the number of ^{1}H and ^{13}C environments: There is just one peak in the ^{1}H NMR spectrum so all the ^{1}H nuclei must be in the same environment. There are two peaks in the ^{13}C spectrum so there are just two different ^{13}C environments. This strongly suggests that the molecule must be quite symmetrical.

Step 4: Use the splitting of the ^{1}H peaks to suggest some structural features present in the molecule: The presence of triplets suggest that there are CH$_2$ groups adjacent to each ^{1}H environment. Because the four ^{1}H environments are identical, there must be two CH$_2$–CH$_2$ groups present.

Step 5: Put the information together to suggest a structure: two ketone groups (C=O) and two CH$_2$–CH$_2$ groups. The only structure which works is Figure 1.

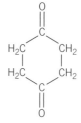

▲ Figure 1

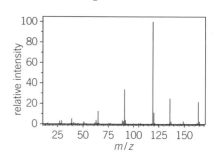

▲ Figure 2

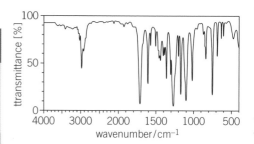

▲ Figure 3

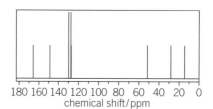

▲ Figure 4

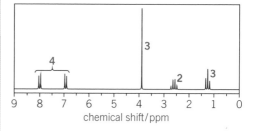

▲ Figure 5

Summary questions

1 Explain what information can be obtained from a combination of percentage by mass data and the M$^+$ ion peak from mass spectrometry. *(3 marks)*

2 Infrared spectroscopy and mass spectrometry can be used to distinguish between pairs of molecules. For each pair of molecules, state which method would allow these pairs of molecules to be distinguished most easily. For your chosen method, describe **one** way in which the results would be different for the two molecules.

   a

   i   **ii**

   b i CH$_2$=CH–CH$_2$OH and **ii** CH$_3$CH$_2$CHO   *(6 marks)*

3 An unknown compound has the following percentage composition by mass: C, 73.17%   H, 7.32%   O, 19.51%
   The mass spectrum, infrared spectrum, carbon-13, and proton NMR spectra are shown in Figures 2–5. Analyse this information to suggest the structural formula of the compound.   *(6 marks)*

6.9 Coloured organic molecules

Specification reference: CD (m)

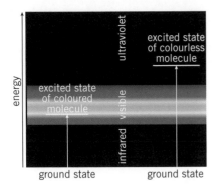

▲ **Figure 1** *The energy gap between energy levels in a molecule determines whether a molecule is coloured or colourless*

Electronic transitions and colour

When electromagnetic radiation from the visible or ultraviolet part of the spectrum is absorbed by molecules, electrons are excited from a lower energy level (the ground state) to a higher one (an excited state).

The frequency of radiation absorbed is related to the energy gap between these energy levels by the equation $\Delta E = h\nu$.

If the energy gap corresponds to a frequency of radiation in the range of visible light, the molecule will absorb some of the colours of visible light. The **complementary colour** will be transmitted or reflected and the molecule will appear coloured.

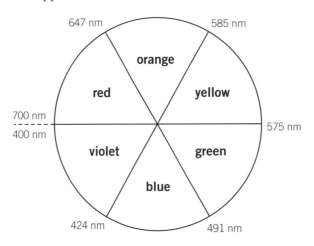

▲ **Figure 2** *The colour wheel*

Chromophores

The part of the molecule responsible for the absorption of visible light (or uv radiation) is known as the **chromophore**.

Many chromophores consist of a conjugated system in which there is extensive delocalisation of electrons. This occurs when there are alternating double and single bonds in a molecule.

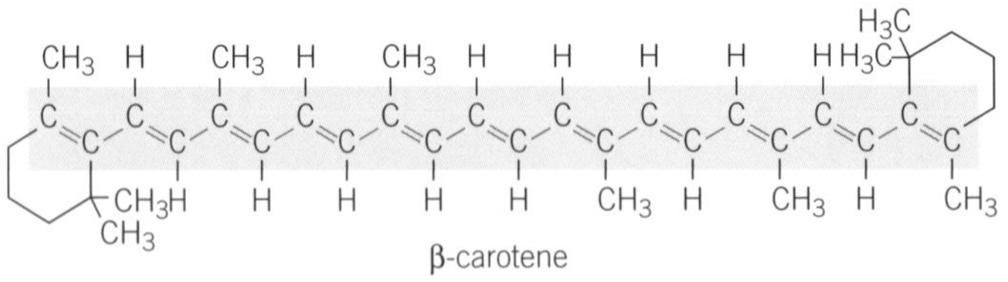

β-carotene

▲ **Figure 3** *The chromophore (shaded) in β-carotene consists of a conjugated system of alternating double and single bonds*

The relationship between chromophore and colour

The more extensive the delocalised system in a chromophore, the smaller the energy gap between energy levels.

Because a smaller energy gap corresponds to a lower frequency of light, molecules with extensive delocalised systems, such as azo dyes, absorb in the visible part of the spectrum.

Molecules with less extensive delocalisation, such as benzene, have a larger gap between energy levels and absorb a higher frequency of radiation. So benzene absorbs in the ultraviolet part of the spectrum.

Modifying the chromophore

Chemists modify the chromophore in a coloured molecule by adding groups that extend the delocalised system. This reduces the gap between energy levels and results in the molecule absorbing a lower frequency of radiation.

Synoptic link

Topic 12.5, Reactions of arenes, explains how the chromophore is modified in azo dye molecules.

Summary questions

1 Dye molecules are organic molecules that possess a chromophore:
   a What is meant by the term 'chromophore'?
   b What structural feature is usually present in the chromophores of organic molecules?   (2 marks)

2 The structure of the dye molecule disperse red 60 is shown below:

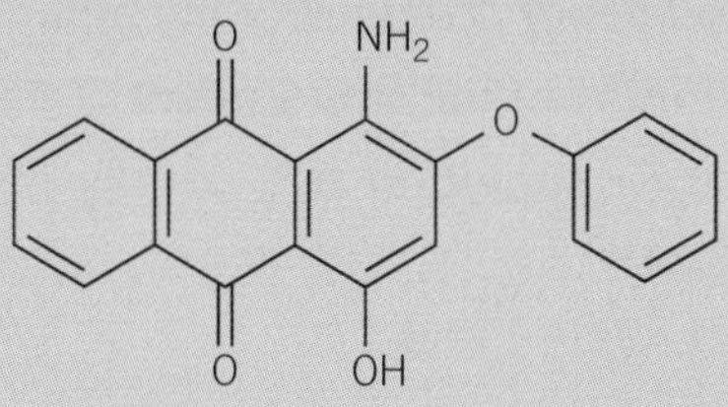

   Copy the structure and circle the chromophore in this molecule.

3 The structures of two molecules with different chromophores are shown below:

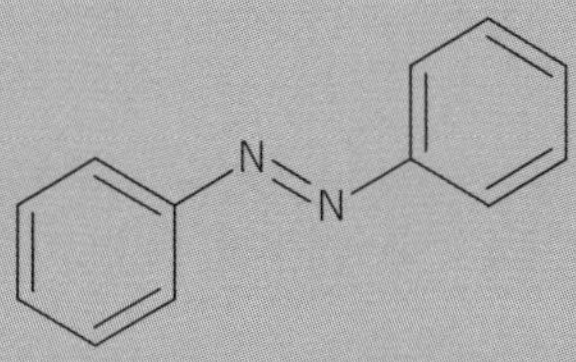

   a is orange, while b is colourless. Explain the difference in appearance of these two molecules.   (5 marks)

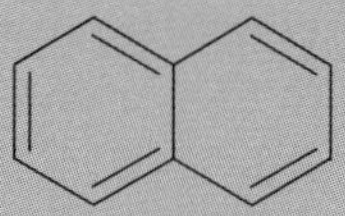

Key terms

Chromophore: The part of an organic molecule responsible for absorbing visible light or ultraviolet radiation.

Conjugated system: A region of the molecule in which there is extensive delocalisation of electrons.

1 Visible light, ultraviolet radiation, and infrared radiation are all types of electromagnetic radiation.
   What is the correct order for the wavelengths of these types of radiation, starting with the shortest:

   A visible light, ultraviolet, infrared

   B infrared, visible light, ultraviolet

   C ultraviolet, infrared, visible light

   D ultraviolet, visible light, infrared.   (*1 mark*)

2 Which of the following statements about ozone is true?

   1 It is broken down into oxygen molecules and oxygen atoms when it absorbs high-frequency ultraviolet radiation.

   2 It is formed naturally in the stratosphere.

   3 It is formed in the troposphere by reactions involving sunlight and other pollutant gases.

   A 1, 2, and 3   **B** Only 1 and 2

   C Only 2 and 3   **D** Only 3   (*1 mark*)

3 Sodium ions can be identified using a flame test. Sodium ions produce an intense yellow colour when introduced into a Bunsen flame. The best explanation for this is:

   A A photon of yellow light is released when electrons drop from higher to lower energy levels.

   B Some frequencies of white light are absorbed when electrons are excited to higher energy levels, leaving only yellow light able to pass through the sodium ions.

   C Sodium ions reflect yellow light from the Bunsen flame.

   D Sodium ions provide energy to the molecules in the flame, causing their electrons to release photons of yellow light.   (*1 mark*)

4 Which of the following statements about the fingerprint region of the spectrum are true?

   1 It shows the absorptions of infrared radiation above 1500 cm^{-1}.

   2 It is used to identify the bonds in the molecule.

   3 Each molecule has a unique pattern in the fingerprint region.

   A 1, 2 and 3   **B** Only 1 and 2

   C Only 2 and 3   **D** Only 3   (*1 mark*)

5 The molecular ion (M+) in the mass spectrum of a molecule tells you:

   A The abundance of the most common isotope

   B The relative molecular mass of the most abundant fragment ion

   C The relative mass of the most abundant isotope in the molecule

   D The relative molecular mass of the unfragmented molecule.   (*1 mark*)

6 **a** Ozone, O_3 is a gas present in both the stratosphere and the troposphere:

   i The depletion of ozone in the stratosphere has important implications for human health at the Earth's surface. Explain why.   (*2 marks*)

   ii Describe the processes that lead to CFC molecules causing ozone depletion in the stratosphere.   (*4 marks*)

b Ozone in the troposphere is regarded as a pollutant. One reason for this is that it is involved in a series of reactions that cause photochemical smog. One of these reactions is equation 5.1:

$$NO_2 \rightarrow NO + O$$

  i The reaction in equation 5.1 is caused by the action of sunlight on the NO_2 molecule. Suggest the type of bond breaking that occurs in this reaction. *(1 mark)*

  ii The NO species produced in the reaction is described as a radical. State the meaning of the term radical. *(1 mark)*

  iii The bond enthalpy of the N–O bond in NO_2 is $+305\,kJ\,mol^{-1}$. Calculate the minimum energy, in J, needed to break **one** N–O bond in the NO_2 molecule. *(2 marks)*

  iv Calculate the minimum frequency of light needed to break an N–O bond. *(2 marks)*

  v Suggest how the presence of O atoms enables ozone molecules to form in the troposphere. *(1 mark)*

7 A student oxidises propan-1-ol using sodium dichromate as an oxidising agent. She hopes to form propanal but suspects that the product of the reaction is propanoic acid.

a She obtains an infrared spectrum of the molecule, which shows peaks at the following wavenumbers

2510–$3490\,cm^{-1}$ (very broad)

$1715\,cm^{-1}$

$1415\,cm^{-1}$

$1240\,cm^{-1}$.

  i State what happens to a molecule when it absorbs infrared radiation. *(2 marks)*

  ii Describe two pieces of data that support the student's idea that the product is propanoic acid and not propanal. *(4 marks)*

b The student also obtains a mass spectrum of the product. Peaks are seen at the following *m/z* values: 75, 74, 57, and 45.

  i Explain why peaks are seen at both *m/z* = 74 and *m/z* = 75. *(2 marks)*

  ii Give the formula of the species responsible for the peaks at 57 and 45. *(2 marks)*

8 Which of these statements is true about high-resolution mass spectrometry?

A The *m/z* value of the molecular ion enables the structure of the molecule to be determined.

B The technique uses the fact that relative masses of isotopes are not exact integers.

C Different isomers will have different masses when measured to 4 decimal places.

D The technique can detect fragment species that do not have a charge.

(1 mark)

9 A molecule has the molecular formula C_4H_8O. A peak is observed in the mass spectrum with a *m/z* value of 29.

The species responsible for this peak could be:

A CH_3CH_2 **B** $C_4H_5^+$

C CHO^+ **D** C_2H_3O *(1 mark)*

10 A molecule has the structure shown below. Which row shows the correct number of environments in the ^{13}C and proton NMR spectra?

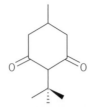

	Number of environments in the ^{13}C NMR spectrum	Number of environments in the proton NMR spectrum
A	7	7
B	10	10
C	9	6
D	7	5

11 In the proton NMR spectrum of propanal, CH_3CH_2CHO, the peak caused by the proton in the CHO environment will:

1 Have a chemical shift of $\delta = 3.5\,ppm$

2 Not be split

3 Have an area less than any of the other peaks

A 1, 2, and 3

B 1 and 2

C 2 and 3

D Only 3 *(1 mark)*

12 A molecule has the molecular formula $C_4H_8O_2$.

The proton NMR spectrum of this molecule shows the following peaks.

Chemical shift / ppm	Relative number of H atoms	Splitting
1.2	3	triplet
2.0	3	singlet
4.2	2	quartet

Suggest a structure for this molecule, explaining how the data from the spectrum links to the structure. *(4 marks)*

13 The structure of benzene and the dye molecule Disperse Red are shown below:

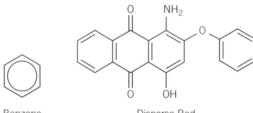

Benzene   Disperse Red

Benzene is a colourless liquid whilst Disperse Red is a solid with an intense red colour when in suspension.

Use ideas about the bonding in these substances to explain the differences in colour of these two substances. *(5 marks)*

7.1 Chemical equilibrium

Specification reference: ES(o)

Dynamic equilibrium

When the rate of the forward reaction is the same as the rate of the reverse reaction, a system is in dynamic equilibrium, represented by the symbol $\rightleftharpoons$. For example, if you are attempting to dry some wet socks on a radiator, the following process occurs:

$$H_2O(l) \rightarrow H_2O(g)$$

This is a physical change and results in the socks drying as the water evaporates. However, if the socks are in a sealed bag, the water vapour condenses on the inner surface of the bag. In other words, the reverse reaction occurs:

$$H_2O(g) \rightarrow H_2O(l)$$

At first, the rate of the forward reaction is greater than the rate of the reverse reaction, but gradually the rate of the reverse reaction increases. Eventually, both reactions occur at the same rate and a dynamic equilibrium is established. The socks will never fully dry.

$$H_2O(l) \rightleftharpoons H_2O(g)$$

Features of a dynamic equilibrium

- A dynamic equilibrium can only occur in a closed system. Gaseous reactions must take place in a sealed container. Reactions in solution act as a closed system, as long as no reactant can escape.

- Dynamic equilibrium can be established from either direction, starting with the reactants, the products or a mixture.

- Once a dynamic equilibrium has been established, the concentrations of reactants and products remain unchanged. However, the forward reaction and reverse reaction do not stop – their *rates* are equal.

- If most of the reactants become products before the reverse reaction increases sufficiently to establish equilibrium, we say that the position of equilibrium lies to the right. If little of the reactants have changed to products when the reverse reaction becomes equal to the rate of the forward reaction, we say that the position of equilibrium lies to the left.

A chemical equilibrium

When carbon dioxide dissolves in water, the following reversible chemical change occurs:

$$CO_2(aq) + H_2O(l) \rightleftharpoons HCO_3^-(aq) + H^+(aq)$$

This is another example of a dynamic equilibrium. Once equilibrium is established, all four species will be present. The concentrations of each depend on the position of equilibrium under those specific conditions.

> **Key term**
>
> **Dynamic equilibrium:** A system in which the rate of the forward and backward reactions is the same. The concentrations of the species in the equilibrium remains constant.

> **Revision tip**
>
> The concentrations of the substances are constant, but they are not necessarily equal to each other.

> **Key term**
>
> **Closed system:** A system in which no substances can escape, for example a sealed container. Reactions in solution are considered a closed system, provided the dissolved solutes cannot escape from the solution.

> **Revision tip**
>
> The convention is that the forward reaction for a given equation proceeds from left to right. The reverse reaction proceeds right to left.

Summary questions

1 Explain what is meant by the term 'dynamic equilibrium'. *(1 mark)*

2 Explain why a dynamic equilibrium can be established from either direction, referring to the rates of forward and reverse reaction. *(2 marks)*

3 Explain the effect on the equilibrium between liquid water and water vapour of increasing the temperature of the wet socks in a sealed bag. *(2 marks)*

7.2 Equilibrium constant K_c

Specification reference: ES(p)

Specification reference: ES(p)

Key term

Equilibrium constant: A mathematical expression incorporating the equilibrium concentrations of each substance present.

Revision tip

Remember – products divided by reactants!

Revision tip

a and b are the number of moles of reactants A and B; c and d are the number of moles of products C and D.

Revision tip

The square brackets indicate the concentration in mol dm^{-3}, of whatever is inside the brackets.

Synoptic link

You will find out how to calculate the value of K_c in Topic 7.3, Equilibrium constant and concentrations.

Synoptic link

There is more about factors that affect the value of K_c in Topic 7.4, Le Châtelier's principle.

Writing an equilibrium constant

Consider the general equilibrium reaction:

$$aA(aq) + bB(aq) \rightleftharpoons cC(aq) + dD(aq)$$

The equilibrium law states that:

$$K_c = \frac{[C]^c[D]^d}{[A]^a[B]^b}$$

This constant, K_c, is the equilibrium constant for the reaction at a specified temperature. The letter K is used to represent all equilibrium constants. When the expression is written in terms of concentrations, we write K_c.

For example, for the equilibrium reaction:

$$N_2(g) + 3H_2(g) \rightleftharpoons 2NH_3(g)$$
$$K_c = \frac{[NH_3]^2}{[N_2][H_2]^3}$$

K_c is a measure of how far a reaction proceeds. If an equilibrium mixture is composed largely of reactants, then the value of K_c is small; if the equilibrium mixture is composed largely of products, then the value of K_c is large.

What affects the value of K_c?

The *only* thing that affects the numerical value of K_c is a change in *temperature*.

	Exothermic reactions	Endothermic reactions
Temperature increases	value of K_c decreases	value of K_c increases
Temperature decreases	value of K_c increases	value of K_c decreases

Summary questions

1 Write expressions for K_c for the following reversible reactions. *(2 marks)*
   a $2SO_2(g) + O_2(g) \rightleftharpoons 2SO_3(g)$
   b $N_2O_4(g) \rightleftharpoons 2NO_2(g)$.

2 Ethanol is produced in industry by the hydration of ethene:
$$C_2H_4(g) + H_2O(g) \rightleftharpoons C_2H_5OH(g).$$
   The forward reaction is exothermic.
   a Write an expression for K_c for this reaction. *(1 mark)*
   b Explain how increasing the pressure would affect the position of equilibrium and the value of K_c. *(1 mark)*
   c Explain how increasing the temperature would affect the position of equilibrium and the value of K_c. *(1 mark)*
   d State the advantage of using a catalyst, in terms of the equilibrium. *(1 mark)*

7.3 Equilibrium constant and concentrations

Specification reference: ES(p)

Calculating an equilibrium constant

Recall the general expression for K_c:

$$K_c = \frac{[C]^c [D]^d}{[A]^a [B]^b}$$

Calculating the value of K_c is straightforward if the equilibrium concentrations are known.

> **Worked example: Calculating an equilibrium constant**
>
> Calculate the value of K_c at 763 K for the reaction $H_2(g) + I_2(g) \rightleftharpoons 2HI(g)$ given the following data:
>
> $[H_2(g)] = 1.92\ mol\ dm^{-3}$, $[I_2(g)] = 3.63\ mol\ dm^{-3}$ and $[HI(g)] = 17.8\ mol\ dm^{-3}$.
>
> **Step 1:** Write an expression for K_c.
>
> **Step 2:** Substitute the values for the equilibrium concentrations of each species:
>
> $$K_c = \frac{[HI]^2}{[H_2][I_2]} = \frac{17.8^2}{1.92\ 3.63} = 45.5.$$

> **Worked example: Composition of equilibrium mixtures**
>
> You can also calculate the composition of equilibrium mixtures:
>
> $$CH_3COOH(l) + C_2H_5OH(l) \rightleftharpoons CH_3COOC_2H_5(l) + H_2O(l).$$
>
> K_c for the above esterification reaction has a value of 4.10 at 25 °C.
>
> Calculate the equilibrium concentration of ethyl ethanoate, given that $[CH_3COOH(l)] = 0.255\ mol\ dm^{-3}$, $[C_2H_5OH(l)] = 0.245\ mol\ dm^{-3}$ and $[H_2O(l)] = 0.437\ mol\ dm^{-3}$.
>
> **Step 1:** Write an expression for K_c for the reaction.
>
> **Step 2:** Rearrange the K_c expression to make the unknown concentration the subject of the equation.
>
> **Step 3:** Substitute the values of the known concentrations and of K_c.
>
> $$K_c = \frac{[CH_3COOC_2H_5]\ [H_2O]}{[CH_3COOH]\ [C_2H_5OH]}$$
>
> Rearranging the expression:
>
> $$[CH_3COOC_2H_5] = \frac{K_c \times [CH_3COOH] \times [C_2H_5OH]}{[H_2O]}$$
>
> Substituting the values into the rearranged expression:
>
> $$[CH_3COOC_2H_5] = \frac{4.10 \times 0.255\ mol\ dm^{-3} \times 0.245\ mol\ dm^{-3}}{0.437\ mol\ dm^{-3}}$$
>
> $$= 0.586\ mol\ dm^{-3}.$$

Revision tip

Remember – if data is given to 3 significant figures, you should give your answer to 3 significant figures.

Summary questions

1   Calculate the value of K_c for the Haber process reaction:
    $3H_2(g) + N_2(g) \rightleftharpoons 2NH_3(g)$
    at 1 000 K, given
    $[H_2(g)] = 1.84\ mol\ dm^{-3}$,
    $[N_2(g)] = 1.36\ mol\ dm^{-3}$,
    and $[NH_3(g)] = 0.142\ mol\ dm^{-3}$. (*2 marks*)

2   Calculate the concentration of ethanol at equilibrium in the esterification reaction:
    $C_2H_5OH(l) + CH_3COOH(l) \rightleftharpoons CH_3COOC_2H_5(l) + H_2O(l)$
    given that $K_c = 4.10$,
    $[CH_3COOH] = 0.80\ mol\ dm^{-3}$,
    and $[CH_3COOC_2H_5] = [H_2O] = 3.0\ mol\ dm^{-3}$.
    (*2 marks*)

7.4 Le Châtelier's principle

Le Châtelier's principle

The position of the equilibrium can be altered by changing the concentration of solutions, the pressure of gases, or the temperature.

Le Châtelier's principle states that if a system is at equilibrium, and a change is made in any of the conditions, then the system responds to counteract the change as much as possible.

Concentration

Concentration change	Equilibrium shift
increasing reactant(s)	to the right (decreases reactants)
increasing product(s)	to the left (decreases products)
decreasing reactant(s)	to the left (increases reactants)
decreasing product(s)	to the right (increases products)

Pressure

Pressure change	Equilibrium shift
Increasing	to the side with fewer gas molecules – in the example below, to the right. $$CO(g) + 2H_2(g) \rightleftharpoons CH_3OH(g)$$ 3 molecules 1 molecule
Decreasing	to the side with more gas molecules – in the example below, to the right $$CH_4(g) + H_2O(g) \rightleftharpoons CO(g) + 3H_2(g)$$ 2 molecules 4 molecules

Temperature

Temperature change	Equilibrium shift
Increase	position of equilibrium shifts in the direction of the endothermic reaction
Decrease	position of equilibrium shifts in the direction of the exothermic reaction

Summary questions

1. $C_2H_4(g) + H_2O(g) \rightleftharpoons C_2H_5OH(g)$; $\Delta H = -46\,kJ\,mol^{-1}$.
   Predict, using le Châtelier's principle, the changes in the following that would move the position of equilibrium to the right:
   a concentration   *(1 mark)*
   b temperature   *(1 mark)*
   c pressure.   *(1 mark)*

2. $Fe^{3+}(aq) + SCN^-(aq) \rightleftharpoons [FeSCN]^{2+}(aq)$
   pale yellow   colourless   blood red
   For this reaction, what would you see if you added:
   a more $Fe^{3+}(aq)$ *(1 mark)*
   b more $[FeSCN]^{2+}(aq)$?   *(1 mark)*

3. Study the equilibrium reactions below. Predict the effect on the position of equilibrium of increasing the temperature and of increasing the pressure.
   a $N_2(g) + 3H_2(g) \rightleftharpoons 2NH_3(g)$; $\Delta H = -92\,kJ\,mol^{-1}$ *(2 marks)*
   b $N_2(g) + O_2(g) \rightleftharpoons 2NO(g)$; $\Delta H = +90\,kJ\,mol^{-1}$ *(2 marks)*

7.5 Equilibrium constant K_c, temperature, and pressure

Specification reference: CI(f), CI(h)

Equilibrium constant K_c

You have seen that K_c is a mathematical expression incorporating the equilibrium concentrations of each substance present, and that K_c is constant for a given reaction at a specified temperature:

$$K_c = \frac{[C]^c [D]^d}{[A]^a [B]^b}$$

Changing the pressure or concentration may affect the position of equilibrium and the equilibrium concentrations of the substances present, but the value of K_c remains constant.

Catalysts

Catalysts increase the *rate* of the reaction, so equilibrium is achieved more quickly, but adding a catalyst has no effect on the position of equilibrium or the numerical value of K_c.

Temperature

The *only* thing that affects the magnitude of K_c is a change in *temperature*:

- Increasing the temperature moves the position of equilibrium in the endothermic direction.

- Decreasing the temperature moves the position of equilibrium in the exothermic direction.

For example, in the Haber process, the following reaction takes place:

$$N_2(g) + 3H_2(g) \rightleftharpoons 2NH_3(g) \qquad \Delta H = -92\,kJ\,mol^{-1}$$

The forward reaction is exothermic, so increasing the temperature moves the position of equilibrium to the left. Therefore K_c decreases.

Decreasing the temperature moves the position of equilibrium to the right, in the exothermic direction, increasing the concentration of the product and increasing K_c.

▼ **Table 1** *The effects of temperature change*

	Exothermic reactions	Endothermic reactions
Temperature increases	value of K_c decreases	value of K_c increases
Temperature decreases	value of K_c increases	value of K_c decreases

How does K_c change?

▼ **Table 2** *Changes that affect K_c*

Change	Position of equilibrium	K_c	Rate
Concentrations	changed	unchanged	changed
Total pressure	may change	unchanged	may change
Temperature	changed	changed	changed
Catalyst used	unchanged	unchanged	changed

Key term

Catalysts: Catalysts speed up a reaction without getting used up. They do this by providing a route of lower activation enthalpy.

Common misconception: The effect of catalysts

Remember that catalysts do not affect the position of equilibrium.

Revision tip

Temperature is the only factor to alter K_c.

Synoptic link

You will come across other equilibrium constants, such as K_a in connection with weak acids in Topic 8.2, Strong and weak acids and pH.

Revision tip

Units for K_c need to be calculated in every case.

It may be that K_c has no units at all.

Key term

Stoichiometry: Stoichiometry refers to the mole ratio of reactants and products in a chemical equation.

Units of K_c

Units of K_c can be calculated by substituting the units of concentration (mol dm^{-3}) for each species into the K_c expression, raising to a power ($\text{mol}^2\,\text{dm}^{-6}$, $\text{mol}^3\,\text{dm}^{-9}$ etc.) where appropriate. It is then necessary to cancel from the top and bottom of the expression.

 Worked example: The units of K_c (1)

$$K_c = \frac{[NH_3]^2}{[N_2][H_2]^3}$$

Calculate K_c when $[NH_3] = 2.25 \times 10^{-5}\,\text{mol dm}^{-3}$, $[N_2] = 2.49 \times 10^{-3}\,\text{mol dm}^{-3}$, and $[H_2] = 9.91 \times 10^{-3}\,\text{mol dm}^{-3}$.

Step 1: $K_c = \dfrac{(2.25 \times 10^{-5}\,\text{mol dm}^{-3})^2}{(2.49 \times 10^{-3}\,\text{mol dm}^{-3})\,(9.91 \times 10^{-3}\,\text{mol dm}^{-3})^3}$

Step 2: $K_c = \dfrac{5.0625 \times 10^{-10}\,\text{mol}^2\,\text{dm}^{-6}}{2.49 \times 10^{-3}\,\text{mol dm}^{-3} \times 9.73 \times 10^{-7}\,\text{mol}^3\,\text{dm}^{-9}}$

Step 3: $K_c = 0.21\,\text{mol}^{-2}\,\text{dm}^6$

In this case, the units on the top are $\text{mol}^2\,\text{dm}^{-6}$ and the units on the bottom are $\text{mol}^4\,\text{dm}^{-12}$. Cancelling from the top and bottom leaves $\text{mol}^{-2}\,\text{dm}^6$.

 Worked example: Calculating K_c from initial concentrations

$$K_c = \frac{[CH_3CO_2C_2H_5]\,[H_2O]}{[CH_3COOH]\,[C_2H_5OH]}$$

The initial concentrations of CH_3COOH and C_2H_5OH were $0.500\,\text{mol dm}^{-3}$. No $CH_3CO_2C_2H_5$ and H_2O were present in the initial mixture. At equilibrium, the concentrations of $CH_3CO_2C_2H_5$ and H_2O were $0.333\,\text{mol dm}^{-3}$. Deduce the equilibrium concentrations of CH_3COOH and C_2H_5OH and calculate the value of K_c including units.

Step 1: As the concentrations of $CH_3CO_2C_2H_5$ and H_2O increased to $0.333\,\text{mol dm}^{-3}$, the concentrations of CH_3COOH and C_2H_5OH will have *decreased* by $0.333\,\text{mol dm}^{-3}$, due to the 1:1 mole ratio. Therefore the equilibrium concentrations are $0.500 - 0.333 = 0.167\,\text{mol dm}^{-3}$.

Step 2: Substitute the concentrations into the expression for K_c.

$$K_c = \frac{(0.333\,\text{mol dm}^{-3})(0.333\,\text{mol dm}^{-3})}{(0.167\,\text{mol dm}^{-3})(0.167\,\text{mol dm}^{-3})}$$

Step 3: $K_c = 3.98$ (no units)

In this case, K_c has no units because mol dm^{-3} cancels from the top and the bottom.

Techniques to determine equilibrium constants

To determine an equilibrium constant it is necessary to determine the concentrations of the substances involved. This can be done, for example, by titration if one of the substances is acidic. Then the concentrations of the other substances can be deduced from their mole ratio. Spectroscopic methods and calorimetric methods can also be used.

Summary questions

1 **a** Write K_c for the reaction $CO(g) + Cl_2(g) \rightleftharpoons COCl_2(g)$ ($\Delta H = -108\,\text{kJ mol}^{-1}$). *(1 mark)*

   b Describe the effect on K_c of: (i) increasing pressure, (ii) increasing temperature, (iii) adding a catalyst. *(3 marks)*

2 For the reaction $N_2O_4 \rightleftharpoons 2\,NO_2$ ($\Delta H = +54\,\text{kJ mol}^{-1}$), explain the effect on K_c of increasing the temperature. *(3 marks)*

3 Calculate K_c, including units, for the reaction $H_2 + I_2 \rightleftharpoons 2\,HI$, given that $[H_2] = 0.004\,\text{mol dm}^{-3}$, $[I_2] = 0.004\,\text{mol dm}^{-3}$ and $[HI] = 0.027\,\text{mol dm}^{-3}$. *(4 marks)*

7.6 Solubility equilibria
Specification reference: O(h)

Solubility product

K_{sp} is an equilibrium constant known as the solubility product.

A sparingly soluble ionic solid is in equilibrium with its constituent ions. For example:

$$AB(s) \rightleftharpoons A^+ (aq) + B^- (aq)$$

Because the concentration of a solid cannot be measured, the equilibrium constant for this reaction is the product of the concentrations of the two ions:

$$K_{sp} = [A^+] [B^-]$$

> **Worked example: Writing K_{sp}**
>
> **Step 1:** $AgCl (s) \rightleftharpoons Ag^+ (aq) + Cl^- (aq)$
>
> $\quad K_{sp} = [Ag^+] [Cl^-]$
>
> **Step 2:** Lead(II) iodide contains the ions Pb^{2+} and I^- in a 2 : 1 ratio, so the expression for K_{sp} is:
>
> $\quad K_{sp} = [Pb^{2+}] [I^-]^2$

Solubility product calculations

Given the value of K_{sp}, it is possible to predict whether a precipitate will form. If the product of the concentration of the ions (raised to a power as necessary) is *smaller than or equal to K_{sp}*, the ions will stay in solution and a precipitate will not form.

Techniques for determining solubility products

To determine a solubility product experimentally, it is necessary to measure the concentrations of the ions in solution. With $Ca(OH)_2$, for example, this can be done by titration to determine $[OH^-]$. It is necessary to filter the undissolved solid calcium hydroxide before titrating. The concentration of $[Ca^{2+}]$ can then be deduced by the mole ratio of the reaction since $[OH^-]$ will be twice as great as $[Ca^{2+}]$.

> **Worked example: Solubility product calculations**
>
> Will a precipitate form from a solution at 298 K in which $[Ag^+]$ and $[Cl^-] = 5.0 \times 10^{-6}\,mol\,dm^{-3}$?
>
> $K_{sp}(AgCl) = 2.0 \times 10^{-10}\,mol^2\,dm^{-6}$ at 298 K
>
> $[Ag^+]$ and $[Cl^-] =$
> $5.0 \times 10^{-6}\,mol\,dm^{-3} \times 5.0 \times 10^{-6}\,mol\,dm^{-3} = 2.5 \times 10^{-12}\,mol^2\,dm^{-6}$
>
> This answer is smaller than K_{sp}, so no precipitate will form.

Revision tip

A precipitate will not form if the product of the concentration of the ions is smaller than or equal to K_{sp}.

Summary questions

1 Write an expression for K_{sp} for:
   a AgBr
   b Ag_2S
   c Ag_2CrO_4 *(3 marks)*

2 Calculate K_{sp} for AgBr given that $[Ag^+]$ and $[Br^-] = 7.07 \times 10^{-7}\,mol\,dm^{-3}$.
                                                        (3 marks)

3 Will a precipitate form from a solution at 298 K in which $[Pb^{2+}]$ and $[SO_4^{2-}] = 1.45 \times 10^{-4}\,mol\,dm^{-3}$? $K_{sp}(PbSO_4) = 1.60 \times 10^{-8}\,mol^2\,dm^{-6}$ at 298 K.   *(2 marks)*

7.7 Gas–liquid chromatography

Specification reference: CD(n)

Stationary phase and mobile phase

Chromatography is a method of separating and identifying the components of a mixture.

In chromatography, equilibrium is established as components are distributed between the **stationary phase** and the **mobile phase**. Components with a greater affinity for the stationary phase consequently move more slowly than those with a lower affinity.

Gas–liquid chromatography

In gas–liquid chromatography (GLC), the stationary phase is a non-volatile liquid coated on the surface of finely divided solid particles. This material is packed inside a long thin **column**, which is coiled inside an oven. A **carrier gas**, which is unreactive, is the mobile phase and carries the mixture through the column.

A peak is recorded on the **chromatogram**, as each component emerges from the column. The *time* that a component takes to emerge is called the **retention time** and the *area* under each peak is proportional to the amount of that component in the mixture.

When the components emerge from the column they may pass into a mass spectrometer, which gives further information about the structure of the component, including its relative molecular mass.

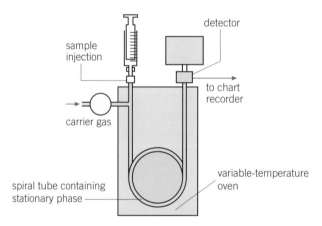

▲ **Figure 1** *GLC apparatus*

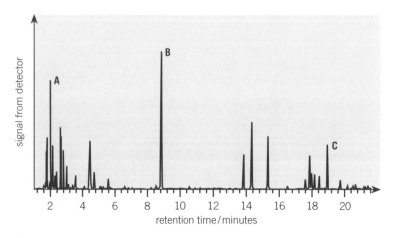

▲ **Figure 2** *A typical gas–liquid chromatogram*

Worked example: Interpreting a gas–liquid chromatogram

Referring to the chromatogram in Figure 2, identify:

a   the component which makes up the greatest proportion of the mixture,

b   the retention time of peaks A, B, and C,

c   which component, A, B, or C has least affinity for the mobile phase.

a   Component B is the tallest peak so it makes up the greatest proportion of the mixture. Because the peaks are very sharp, the height of the peak can be used instead of the area under the peak.

b   The retention time of A = 2.0 min; B = 8.8 min; C = 19.0 min

c   Substance C has the greatest retention time, so it spends more time in the column and consequently has least affinity for the mobile phase.

Summary questions

1   Often the outlet from a GLC instrument is connected to a mass spectrometer. Why is this beneficial? *(1 mark)*

2   What would be the effect on the retention time of increasing the flow rate of the carrier gas? *(1 mark)*

3   Sketch the output you would expect from a GLC instrument for a mixture containing compounds X, Y, and Z in the table below. *(1 mark)*

Compound	Retention time
X	3 min
Y	5.5 min
Z	6.4 min

Experimental techniques

Determining a solubility product, K_{sp}

The first stage in determining K_{sp} is to make a saturated solution of the salt in question, in distilled water.

1   Warm distilled water in a conical flask and add the salt. Keep adding the salt until no more dissolves. Allow the mixture to cool to room temperature, then filter the mixture, discarding the residue.

    Next, the concentration of one of the ions in solution must be determined. This allows K_{sp} to be determined. It is only necessary to determine the concentration of one of the ions as the concentration of the other ion will be proportional to it.

2   Take the temperature of the solution, as K_{sp} is temperature-dependent.

3   Decide the appropriate method for determining the concentration of the ion:

    a   for a basic solution such as calcium hydroxide, the concentration of hydroxide ions can be determined using a titration with hydrochloric acid.

    b   for a coloured solution such as one containing a transition metal ion, the concentration of the coloured ion can be determined using colorimetry.

Determining an equilibrium constant

1   Allow the mixture of reactants to reach equilibrium.

2   Record the temperature, as K_c is temperature-dependent.

3   Determine the concentration of one of the components. The method will depend on the substances involved, but could include titrations, colorimetry, or pH measurements.

4   The concentrations of the other components can be calculated from the concentration determined in step 3, using the chemical equation for the reaction. You will also need to know the initial concentrations of the substances.

5   Write an expression for K_c and substitute the values you have calculated. Determine the value of K_c including its units. Report the temperature.

Chapter 7 Practice questions

1 Which statement gives the best description of the term 'dynamic equilibrium'?

  A A reaction which goes in both directions.

  B A reversible reaction in which the concentrations of the reactants and the products are equal.

  C A reversible reaction in a closed system in which the rate of the forward reaction is equal to the rate of the reverse reaction.

  D A reaction which is static and has no change in the concentration of reactants and products. *(1 mark)*

2 Which expression shows the equilibrium concentration of PF_3 in the reaction:

$$PF_5 \rightleftharpoons PF_3 + F_2?$$

  A $[PF_3] = \dfrac{[PF_3][F_2]}{[PF_5]}$,   **B** $[PF_3] = \dfrac{K_c[F_2]}{[PF_5]}$,   **C** $[PF_3] = \dfrac{[F_2]}{K_c[PF_5]}$,   **D** $[PF_3] = \dfrac{K_c[PF_5]}{[F_2]}$

(1 mark)

3 Ammonia reacts with water as follows:

$$NH_3 + H_2O \rightleftharpoons NH_4^+ + OH^-$$

  a Suggest the likely pH of a solution of ammonia in water. *(1 mark)*

  b Use le Châtelier's principle to predict and explain how the position of equilibrium changes when an aqueous solution of an acid is added to the equilibrium mixture. *(2 marks)*

4 It is possible to produce liquid fuels from methane. This is only really feasible for countries that have large reserves of natural gas. The first stage is the production of methanol in a two-step process.

  Step 1 $CH_4(g) + H_2O(g) \rightleftharpoons CO(g) + 3H_2(g)$ $\Delta H = +205\,kJ\,mol^{-1}$

  Step 2 $CO(g) + 2H_2(g) \rightleftharpoons CH_3OH(g)$ $\Delta H = -125\,kJ\,mol^{-1}$

Step	Temperature (K)	Pressure (atm)	Catalyst used
1	1050	5	nickel
2	540	90	copper

  a **i** Describe and explain how temperature, pressure, and the presence of a catalyst affect the position of equilibrium for the reactions in **step 1** and **step 2**. *(6 marks)*

    ii Suggest and explain why the temperature used for the reaction shown in **step 2** is lower than for the reaction shown in **step 1**. *(3 marks)*

    iii Suggest and explain why the pressure used for the reaction shown in **step 2** is higher than for **step 1**. *(3 marks)*

  b Most of the methanol produced at the end of **step 2** is not used as a fuel, but is converted into a mixture of hydrocarbons in another two-step process.

    Step 3 $2CH_3OH(g) \rightleftharpoons CH_3OCH_3(g) + H_2O(g)$

    Step 4 $4CH_3OCH_3(g) \rightleftharpoons C_8H_{16}(g) + 4H_2O(g)$

    i Write an expression for the equilibrium constant, K_c, for the reaction shown in **step 3**. *(2 marks)*

    ii Use the data given below to calculate a value for K_c at 600 K for the reaction shown in **step 3**. *(2 marks)*

    $[CH_3OH(g)] = 0.050\,mol\,dm^{-3}$; $[CH_3OCH_3(g)] = 0.20\,mol\,dm^{-3}$;
    $[H_2O(g)] = 0.20\,mo\,dm^{-3}$

5 What is the only factor that affects the magnitude of K_c?

   A Concentration

   B Pressure

   C Surface area

   D Temperature   *(1 mark)*

6 Catalysts do **not**...

   A Speed up a reaction

   B Provide an alternative route of lower activation enthalpy

   C Alter the position of equilibrium

   D Remain unchanged at the end of the reaction   *(1 mark)*

7 What is the effect on K_c of increasing the temperature of an exothermic reaction?

   A K_c is unchanged

   B The position of equilibrium moves to the right

   C The value of K_c decreases

   D Collisions occur more frequently   *(1 mark)*

8 What is the correct expression for the solubility product of calcium carbonate?

   A $K_{sp} = \dfrac{[Ca^{2+}][CO_3^{2-}]}{[CaCO_3]}$

   B $K_{sp} = \dfrac{[CaCO_3]}{[Ca^{2+}][CO_3^{2-}]}$

   C $K_{sp} = [Ca^{2+}][CO_3^{2-}][CaCO_3]$

   D $K_{sp} = [Ca^{2+}][CO_3^{2-}]$ *(1 mark)*

9 **a** Write an expression for K_c for the reaction: $C_2H_5OH \rightleftharpoons C_2H_4 + H_2O$
      (1 mark)

   b Deduce the units of K_c for this reaction.   *(1 mark)*

10 The following equilibrium takes place during the production of sulfuric acid:

$$2SO_2 + O_2 \rightleftharpoons 2SO_3 \qquad \Delta H = -196\,kJ\,mol^{-1}$$

   a Write an expression for K_c for the reaction.   *(1 mark)*

   b Explain the effect on the position of equilibrium and the magnitude of K_c of:

     (i) increasing the pressure, (ii) adding a catalyst, (iii) increasing the temperature.   *(3 marks)*

   c Calculate the value of K_c, including units, when $[SO_2] = 7.45 \times 10^{-3}\,mol\,dm^{-3}$, $[O_2] = 3.62 \times 10^{-3}\,mol\,dm^{-3}$ and $[SO_3] = 1.80 \times 10^{-2}\,mol\,dm^{-3}$.   *(2 marks)*

11 **a** Write an expression for K_{sp} for lead(II) sulfide, PbS.   *(1 mark)*

   b If the concentration of Pb^{2+} ions is $1.14 \times 10^{-14}\,mol\,dm^{-3}$, what is the maximum concentration of S^{2-} ions that will not cause a precipitate to form? K_{sp} (PbS) $= 1.3 \times 10^{-28}\,mol^2\,dm^{-6}$.   *(2 marks)*

12 A 3:1 mixture of octane and decane was separated by gas chromatography. The retention time of octane was 5.5 minutes and the retention time of decane was 7.5 minutes.

   a Explain what is meant by 'retention time'.   *(1 mark)*

   b Describe how the areas under each peak will compare to each other.   *(1 mark)*

   c Explain how the identity of each component could be confirmed by mass spectrometry.   *(1 mark)*

8.1 Acids, bases, alkalis, and neutralisation

Specification reference: EL(t), EL(c (i))

Acids, bases, and alkalis

An acid is a compound that dissociates in water to produce hydrogen ions. For example, hydrochloric acid dissociates as follows:

$$HCl(aq) \rightarrow H^+(aq) + Cl^-(aq)$$

Bases are compounds that react with an acid to produce water and a salt. An alkali is a base that dissolves in water to produce hydroxide, OH^- ions. For example, sodium hydroxide is an alkali:

$$NaOH(aq) \rightarrow Na^+(aq) + OH^-(aq)$$

The properties of acids are a result of the transfer of H^+ ions to bases. Acids are sometimes referred to as proton donors and bases as proton acceptors.

Neutralisation reactions can be summarised using ionic equations. When hydrochloric acid reacts with sodium hydroxide the ions involved, and the products, are:

$$H^+(aq) + Cl^-(aq) + Na^+(aq) + OH^-(aq) \rightarrow H_2O(l) + Na^+(aq) + Cl^-(aq)$$

When the spectator ions have been removed, the ionic equation is:

$$H^+(aq) + OH^-(aq) \rightarrow H_2O(l)$$

Calculations of concentrations

Concentrations can be measured in grams per cubic decimetre ($g\,dm^{-3}$), but it is more usual to use moles per cubic decimetre ($mol\,dm^{-3}$).

Recall the formulae for concentration calculations:

$$C = \frac{n}{V}$$
$$n = c \times V$$
$$V = \frac{n}{c}$$

c is the concentration in $mol\,dm^{-3}$

V is the volume in dm^3

n is the amount in moles.

Any of the three quantities (concentration, amount, or volume) can be calculated by using the correct expression and the other two known values.

Common misconception: cm³ to dm³

Always remember to convert volumes to dm^3. Divide cm^3 by 1000 to get the volume in dm^3.

Key term

Acid: Produces H^+ ions in solution.

Key term

Base: A compound that reacts with an acid to produce water and a salt.

Revision tip

All acid/alkali neutralisation reactions have this ionic equation, whatever acid and alkali are involved.

Synoptic link

You can find more about calculating concentration in Topic 1.4, Concentrations of solutions.

Worked example: Concentration calculations

Concentration	Amount	Volume
$c = \dfrac{n}{V}$	$n = c \times V$ How many moles are in 20 cm³ of 0.1 mol dm⁻³ NaOH?	$V = \dfrac{n}{c}$
What is the concentration of a solution of 0.5 mole of NaOH in 100 cm³?	$n = 0.1 \times 0.02$	What volume of 0.2 mol dm⁻³ NaOH contains 0.1 mol?
$c = \dfrac{0.5}{0.1}$	$= 0.002\text{ mol}$	$V = \dfrac{0.1}{0.2}$
$= 5\text{ mol dm}^{-3}$		$= 0.5\text{ dm}^3$

Synoptic link

Details of how to carry out a titration calculation are given in Topic 1.4, Concentrations of solutions.

Using concentrations in calculations

The concentration of a solution can be determined using a titration. The *end-point* of the reaction is often detected by the use of an indicator that changes colour.

Making salts

Salts are produced by the reaction of bases or alkalis with acids. For example:

sodium carbonate + hydrochloric acid → sodium chloride + water + carbon dioxide

potassium hydroxide + nitric acid → potassium nitrate + water

The type of salt produced depends on the alkali and acid used. Hydrochloric acid produces chlorides, nitric acid produces nitrates, and sulfuric acid produces sulfates.

Salts produced in this way are generally soluble. The solid salt can be produced by evaporating the solution.

Some salts are insoluble and can be produced by precipitation reactions. For example, silver iodide can be produced by reacting silver nitrate and potassium iodide.

Summary questions

1 Convert the following volumes into dm³:
 a 50 cm³
 b 2500 cm³. (2 marks)
2 Calculate the concentration (in mol dm⁻³) of the following solutions:
 a 40 g of NaOH in 500 cm³ of solution (1 mark)
 b 10.6 g of Na_2CO_3 in 2000 cm³ of solution. (1 mark)
3 Calculate the mass of solute needed to make up the following solutions:
 a 100 cm³ of a 0.5 mol dm⁻³ solution of $MgCl_2$ (1 mark)
 b 2 dm³ of a 0.02 mol dm⁻³ solution of $KMnO_4$. (1 mark)
4 A volume of 20 cm³ sulfuric acid was used in a titration, delivered from a burette. The concentration of the solution was 0.100 mol dm⁻³.
 a Calculate the volume of acid used in dm³. (1 mark)
 b Calculate the number of moles of acid used. (1 mark)
 c Deduce how many moles of NaOH this acid would neutralise.
 $H_2SO_4(aq) + 2NaOH(aq) \rightarrow Na_2SO_4(aq) + 2H_2O(l)$ (1 mark)
 d If this number of moles of NaOH were in a 25.0 cm³ sample, calculate the concentration of the sodium hydroxide solution, in mol dm⁻³. (1 mark)

8.2 Strong and weak acids and pH

Specification reference: O(i), O(j), O(k), O(l), O(n)

Acids and bases

One way of characterising acids and bases is through their ability to transfer or accept H^+ ions. This is the Brønsted–Lowry theory.

Proton transfer

An acid is a proton (H^+) donor. In aqueous solution, an acid donates protons to water molecules to form oxonium ions (H_3O^+). These are often abbreviated to $H^+(aq)$.

Acid–base pairs

In many cases, the donation of a proton by an acid is reversible:

$$HA\ (aq) \rightleftharpoons H^+(aq) + A^-(aq)$$

HA donates protons and acts as an acid. In the reverse direction, A^- acts as a base as it gains a proton to reform HA. So HA and A^- are a **conjugate acid–base pair**.

Strength of acids and bases

Acids vary in their ability to donate protons. Acids which are powerful proton (H^+) donors are called **strong acids**. **Weak acids** are moderate or poor proton (H^+) donors. A strong acid has a weak conjugate base, and vice versa.

In strong acids, almost all of the acid molecules donate their protons – the acid undergoes **complete dissociation**. Examples of strong acids are hydrochloric acid (HCl), sulfuric acid (H_2SO_4), and nitric acid (HNO_3).

In weak acids, only a small proportion of the acid molecules donate their protons – the acid undergoes **incomplete dissociation**. Carboxylic acids such as ethanoic acid are weak acids:

$$CH_3COOH(aq) \rightleftharpoons CH_3COO^-(aq) + H^+(aq)$$

The acidity constant K_a

K_a is **the acidity constant** or **acid dissociation constant**. The greater the value of K_a, the stronger the acid:

$$HA(aq) \rightleftharpoons A^-(aq) + H^+(aq)$$

$$K_a = \frac{[A^-][H^+]}{[HA]}$$

When comparing weak acids, which can have *very* small K_a values, the K_a can be converted into a pK_a value for ease of use:

$$pK_a = -\log K_a$$

Calculating pH

$$pH = -\log[H^+(aq)]$$

Strong acids

Because a strong acid is fully dissociated, we can assume that the concentration of acid in the solution is the same as the concentration of the hydrogen ions in that solution if the acid is monoprotic.

Revision tip

Remember H^+ is a proton. An acid is a proton donor and a base is a proton acceptor.

Key term

The conjugate acid: This is so called because in the reverse reaction it is the species that donates protons.

Revision tip

Just as all acids have a conjugate base, so all bases have a conjugate acid.

Key terms

Strong acid: An acid which is fully ionised into H^+ and A^- ions.

Weak acid: An acid which has undergone incomplete ionisation into H^+ and A^- ions.

Revision tip

Sometimes the word 'dissociation' is used instead of ionisation when referring to acids.

Common misconception: Weak acids

Equations showing dissociation of weak acids must contain an equilibrium arrow $\rightleftharpoons$; strong acids have a normal arrow $\rightarrow$ as the dissociation is complete.

Revision tip

Remember that square brackets around a formula means the concentration of whatever is inside the brackets, in $mol\,dm^{-3}$.

Revision tip

pH has no units.

Worked example: The pH of a strong acid

Calculate the pH of a 0.001 mol dm^{-3} solution of hydrochloric acid.

Step 1: $HCl(aq) \rightarrow H^+(aq) + Cl^-(aq)$

Step 2: $[H^+] = 0.001$ mol dm^{-3}

Step 3: pH = $-\log 0.001 = 3.00$

 ### Worked example: The pH of a diprotic acid

Calculate the pH of a 0.005 mol dm^{-3} solution of sulfuric acid.

Step 1: $H_2SO_4(aq) \rightarrow 2H^+(aq) + SO_4^{2-}(aq)$

Step 2: $[H^+] = 0.005$ mol dm^{-3} $\times 2 = 0.01$ mol dm^{-3}

Step 3: pH = $-\log 0.01 = 2.00$

The concentration of H$^+$ is twice that of the H$_2$SO$_4$ because 1 mole of the acid dissociates to produce 2 moles of hydrogen ions.

Weak acids

Two assumptions are made:

- Assumption 1: the equilibrium concentration [HA] for a weak acid is the same as the initial concentration of the acid.
- Assumption 2: The equilibrium concentration [A$^-$] is equal to the equilibrium concentration [H$^+$]. A few protons will be provided by the water, but these are insignificant compared to those provided by the acid.

Worked example: The pH of a weak acid

Calculate the pH of 0.01 mol dm^{-3} CH$_3$COOH ($K_a = 1.7 \times 10^{-5}$ mol dm^{-3} at 298 K).

$$CH_3COOH(aq) \rightleftharpoons CH_3COO^-(aq) + H^+(aq)$$

Step 1: Write an expression for K_a: $K_a = \dfrac{[CH_3COO^-][H^+]}{[CH_3COOH]}$

Step 2: Assumption 1: We assume few molecules dissociate, so $[CH_3COOH(aq)] = 0.01$ mol dm^{-3}.

Step 3: Assumption 2: We can assume for every 1 mole of H$^+$ present there is 1 mole of CH$_3$COO$^-$, so $[H^+(aq)] = [CH_3COO^-(aq)]$.

Step 4: $K_a = \dfrac{[H^+]^2}{0.01} = 1.7 \times 10^{-5}$

Step 5: $[H^+] = \sqrt{1.7 \times 10^{-7}}$

Step 6: $[H^+] = 4.12 \times 10^{-5}$

Step 7: pH = $-\log_{10}(4.12 \times 10^{-5}) = 3.4$

Strong bases

For a strong base, we can assume that [OH$^-$(aq)] is equal to the concentration of the solution of the base, because it is fully dissociated in aqueous solution. To then calculate [H$^+$(aq)], the ionic product of water is used.

Water dissociates slightly and this gives rise to the equilibrium constant K_w:

$$H_2O(l) \rightleftharpoons H^+(aq) + OH^-(aq)$$

$$K_w = [H^+(aq)][OH^-(aq)] = 1 \times 10^{-14} \text{ mol}^2 \text{ dm}^{-6} \text{ at } 298 \text{ K}$$

Worked example: The pH of a strong base

Calculate the pH of a 0.01 mol dm⁻³ solution of NaOH.

Step 1: $NaOH(aq) \rightarrow Na^+(aq) + OH^-(aq)$

Step 2: The strong base is fully dissociated, so $[OH^-(aq)] = 0.01 \text{ mol dm}^{-3}$.

Step 3: $K_w = [H^+(aq)][OH^-(aq)]$

$$[H^+] = \frac{K_w}{[OH^-(aq)]} = \frac{1 \times 10^{-14} \text{ mol}^2 \text{ dm}^{-6}}{0.01 \text{ mol dm}^{-3}} = 1 \times 10^{-12} \text{ mol dm}^{-3}$$

Step 4: $pH = -\log[1 \times 10^{-12}] = 12.0$

The greenhouse effect

The earth is kept warm by the greenhouse effect, which is a natural phenomenon. However increased concentrations of CO_2, principally from burning fossil fuels, have led to an enhanced greenhouse effect. Solar energy reaches Earth mainly as visible and ultraviolet radiation. The Earth absorbs some of this energy and radiates some in the form of infrared radiation. Water vapour in the atmosphere absorbs certain wavelengths of infrared, but greenhouse gases such as carbon dioxide and methane in the troposphere absorb the re-radiated infrared in the 'IR window'. The 'IR window' refers to the wavelengths that are not absorbed by water molecules. The absorption of infrared by greenhouse gas molecules increases the vibrational energy of their bonds. This energy is transferred to other molecules by collisions, which increases their kinetic energy and raises the temperature. Greenhouse gas molecules also re-emit some of the absorbed infrared in all directions, which contributes to the heating of the Earth.

The oceans can absorb carbon dioxide and prevent it building up in the atmosphere. The carbon dioxide can react with the water, acting as a base as it accepts H⁺ ions to form hydrogencarbonate ions (see Reaction 1). Some hydrogencarbonate ions dissociate to form carbonate ions (see Reaction 2):

Reaction 1 $CO_2(aq) + H_2O(l) \rightleftharpoons HCO_3^-(aq) + H^+(aq)$

Reaction 2 $HCO_3^-(aq) \rightleftharpoons CO_3^{2-}(aq) + H^+(aq)$

However, these reactions are responsible for acidification of the oceans as H⁺ ions are formed.

Summary questions

1 In the reaction below, identify the two conjugate acid–base pairs:
$H_2SO_4 + OH^- \rightarrow HSO_4^- + H_2O$ *(2 marks)*

2 Calculate the pH of 0.10 mol dm⁻³ nitric acid. *(1 mark)*
3 Calculate the pH of 0.05 mol dm⁻³ potassium hydroxide solution. *(2 marks)*

4 a Give the expression for the K_a of methanoic acid (HCOOH). *(1 mark)*
  b Calculate the pH of 0.001 mol dm⁻³ methanoic acid $(K_a = 1.60 \times 10^{-4} \text{ mol dm}^{-3})$. *(2 marks)*

8.3 Buffer solutions

Specification reference: 0(m)

Key term

Buffer: A solution which minimises the change in pH on addition of small quantities of acid or alkali.

Synoptic link

You learned about conjugate acid–base pairs in Topic 8.2, Strong and weak acids and pH.

Synoptic link

You learned about equilibria in Topic 7.1, Chemical equilibrium.

What is a buffer solution?

Buffer solutions are solutions that have an almost constant pH, despite dilution or small additions of acid or alkali. Buffer solutions contain either:

- a weak acid and one of its salts, e.g. ethanoic acid and sodium ethanoate, or
- a weak base and one of its salts, e.g. ammonia and ammonium chloride.

All buffer solutions contain large amounts of a proton donor – a weak acid or conjugate acid – and large amounts of a proton acceptor, in other words a weak base or conjugate base. Any additions of acid or alkali react with these large amounts, and this keeps the pH constant within limits.

Take the buffer system comprising ethanoic acid and sodium ethanoate solution as an example. The weak acid (ethanoic) partially dissociates to produce its conjugate base and protons. In this case, the acid is a weak acid, so the position of equilibrium lies to the left:

$$CH_3COOH(aq) \rightleftharpoons CH_3COO^-(aq) + H^+(aq) \qquad \textbf{Equation 1}$$
$$\text{weak acid} \qquad\qquad \text{conjugate base}$$

If small amounts of alkali are added, the weak acid dissociates to produce more H^+ ions, which react with the added OH^- ions: the position of equilibrium in Equation 1 moves to the right, so maintaining the pH.

If small amounts of acid are added, the added H^+ ions react with the ethanoate ions present. However, it is not long before the ethanoate ions have all reacted. For the buffer to successfully resist a change in pH, larger quantities of ethanoate ions are required, which is why sodium ethanoate was added when the buffer is produced. Then the added H^+ ions can react with the large amounts of ethanoate ions from the salt – the position of the equilibrium in Equation 1 moves to the left, so maintaining the pH. In summary:

- H^+ ions are added.
- The position of equilibrium moves to the left.
- pH is maintained.
- There is a large concentration of the salt.

Calculations with buffers

Recall the K_a expression:

$$K_a = \frac{[A^-][H^+]}{[HA]}$$

We make two assumptions in buffer calculations:

- All the anions $[A^-]$ have come from the salt, so the contribution from the acid is negligible, i.e. $[A^-]$ = [salt].
- The concentration of the acid in solution [HA(aq)] is the same as the amount of acid put into the solution, in other words, ignore any dissociation, i.e. [HA] = [acid].

Using these assumptions, it can be shown that $K_a = [H^+]\dfrac{[\text{salt}]}{[\text{acid}]}$

Finding the pH of a buffer solution

If K_a of the weak acid is known, along with the concentrations of salt and weak acid, then the hydrogen ion concentration can be calculated, and hence the pH.

Revision tip

Remember pH $= -\log_{10}[H^+]$

 Worked example: Calculating the pH of a buffer

Calculate the pH of a buffer solution made by mixing equal volumes of 0.20 mol dm⁻³ ethanoic acid and 0.10 mol dm⁻³ sodium ethanoate solutions (for ethanoic acid, $K_a = 1.7 \times 10^{-5}$ mol dm⁻³ at 298 K).

Step 1: By mixing equal quantities, each original concentration will be halved, so:

$$[CH_3COOH(aq)] = 0.10 \text{ mol dm}^{-3} \text{ and } [CH_3COO^-(aq)] = 0.05 \text{ mol dm}^{-3}$$

Step 2: $K_a = [H^+]\dfrac{[\text{salt}]}{[\text{acid}]}$

Step 3: $[H^+] = K_a\dfrac{[\text{acid}]}{[\text{salt}]} = \dfrac{1.7 \times 10^{-5} \text{ mol dm}^{-3} \times 0.10 \text{ mol dm}^{-3}}{0.05 \text{ mol dm}^{-3}} = 3.4 \times 10^{-5} \text{ mol dm}^{-3}$

Step 4: pH $= -\log_{10}(3.4 \times 10^{-5}) = 4.5$

Buffers in action

Buffers are found in shampoos, and in food and drink where they are often referred to as 'acidity regulators'. Buffers in the blood protect us from changes in pH due to formation of CO_2 and H^+ in metabolic processes. Otherwise these pH changes could affect the action of enzymes, and have serious consequences for your health.

Summary questions

1   Calculate the pH of a solution containing equal amounts of benzoic acid and sodium benzoate, where the ratio $\dfrac{[\text{acid}]}{[\text{salt}]} = 1$. $K_a = 6.3 \times 10^{-5}$ mol dm⁻³   *(2 marks)*

2   Calculate the pH of a buffer in which $[CH_3COOH] = 0.001$ mol dm⁻³ and $[CH_3COONa] = 0.005$ mol dm⁻³. $K_a = 1.7 \times 10^{-5}$ mol dm⁻³   *(2 marks)*

3   Calculate the pH of a solution made by mixing equal volumes of 0.02 mol dm⁻³ methanoic acid and 0.012 mol dm⁻³ potassium methanoate solution. $K_a = 1.8 \times 10^{-4}$ mol dm⁻³   *(2 marks)*

Experimental techniques

Measuring pH

A pH electrode is used to measure pH accurately. It must be calibrated first using solutions of known pH.

1 Wash the electrode with distilled water, then transfer it to a buffer solution of pH 7.00. Check the bulb is completely immersed and, once the reading is stable, ensure it reads 7.00. Adjust if necessary.

2 If you wish to measure the pH of acidic solutions, further calibrate the electrode with an acidic buffer solution, such as a pH 4.00 buffer.

3 If you wish to measure the pH of alkaline solutions, further calibrate the electrode with an alkaline buffer solution, such as a pH 10.00 buffer.

4 If you wish to measure the pH of both acidic and alkaline solutions, calibrate the electrode with acidic and alkaline buffer solutions.

5 The pH electrode can then be used to measure the pH of the test solution, by immersing it in the solution to be measured.

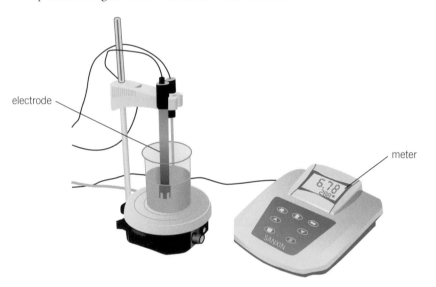

▲ **Figure 1** *Measurement of pH*

Chapter 8 Practice questions

1 Which of the following is **not** an acid?

   A HNO_3 **B** $Ca(OH)_2$

   C CH_3COOH **D** H_3PO_4 (*1 mark*)

2 Which is the general equation for neutralisation?

   A $H^+ + OH^- \rightarrow H_2O$

   B $HCl + NaOH \rightarrow H_2O + NaCl$

   C $H_2O \rightarrow H^+ + OH^-$

   D $HCl \rightarrow H^+ + Cl^-$ (*1 mark*)

3 What is the concentration of a solution containing 4 g of NaOH in $500 \, cm^3$ of water?

   A $2 \, g \, dm^{-3}$ **B** $4 \, g \, dm^{-3}$

   C $8 \, g \, dm^{-3}$ **D** $2000 \, g \, dm^{-3}$ (*1 mark*)

4 What is the concentration of a solution containing 23.8 g of KBr in $100 \, cm^3$ of water?

   A $0.002 \, mol \, dm^{-3}$ **B** $0.2 \, mol \, dm^{-3}$

   C $2 \, mol \, dm^{-3}$ **D** $20 \, mol \, dm^{-3}$ (*1 mark*)

5 $25.0 \, cm^3$ of $0.01 \, mol \, dm^{-3}$ HCl required $19.75 \, cm^3$ of a solution of NaOH for complete neutralisation.

   a Calculate the amount in moles of hydrochloric acid used.   (*1 mark*)

   b Deduce the amount in moles of sodium hydroxide used.   (*1 mark*)

   c Calculate the concentration of the sodium hydroxide in $mol \, dm^{-3}$.   (*1 mark*)

6 $25.0 \, cm^3$ of $0.2 \, mol \, dm^{-3}$ H_2SO_4 required $28.15 \, cm^3$ of a solution of KOH for complete neutralisation.

$$H_2SO_4 + 2KOH \rightarrow K_2SO_4 + 2H_2O$$

   a Calculate the amount in moles of sulfuric acid used.   (*1 mark*)

   b Deduce the amount in moles of potassium hydroxide used.   (*1 mark*)

   c Calculate the concentration of the potassium hydroxide in $mol \, dm^{-3}$.   (*1 mark*)

7 Household bleach solutions contain the chlorate(I) ion, ClO^-, as the active ingredient. The chlorate(I) ion content can be determined by titration. An analyst carries out a titration as follows:

A $5.00 \, cm^3$ sample of bleach is reacted with an excess of acidified potassium iodide. The solution goes deep brown.

$$ClO^- + 2I^- + 2H^+ \rightarrow Cl^- + I_2 + H_2O$$

This mixture is then titrated with $0.100 \, mol \, dm^{-3}$ sodium thiosulfate solution, $Na_2S_2O_3(aq)$. $25.40 \, cm^3$ of thiosulfate solution are needed to obtain the colourless end-point.

$$2S_2O_3^{2-} + I_2 \rightarrow S_4O_6^{2-} + 2I^-$$

   a **i** Calculate the number of moles of thiosulfate, $S_2O_3^{2-}$, used in the titration.   (*2 marks*)

       ii How many moles of iodine have reacted with the thiosulfate?   (*1 mark*)

       iii How many moles of chlorate(I) are in the sample of bleach?   (*1 mark*)

       iv What is the concentration of chlorate(I) in the bleach, in $mol \, dm^{-3}$?   (*3 marks*)

   b Calculate the concentration of the chlorate(I) ion, in $g \, dm^{-3}$, in the sample of bleach.   (*3 marks*)

8 A buffer solution...

   A Consists of a strong acid and its salt

   B Consists of a strong base and its salt

   C Is significantly affected by additions of acid or alkali

   D Contains approximately equal concentrations of acid (or base) and salt   *(1 mark)*

9 Which statement is correct?

   A The pH of a strong acid is equal to the concentration of the acid

   B $[H^+]$ for a strong acid is equal to the concentration of the acid

   C $[H^+]$ for a strong alkali is equal to $K_a \times \dfrac{[\text{acid}]}{[\text{salt}]}$

   D The pH of a weak acid is equal to $-\log_{10}[HA]$ *(1 mark)*

10 Look at the following statements about the reaction:

$$NH_3 + H_2O \rightleftharpoons NH_4^+ + OH^-$$

   1 NH_3 is acting as a base

   2 H_2O is acting as an acid

   3 OH^- is a conjugate acid

  Which statements are correct?

   A 1 only

   B 1 and 2 only

   C 2 and 3 only

   D 1, 2, and 3   *(1 mark)*

11 What is the pH of $0.05\,\text{mol dm}^{-3}\ H_3PO_4$?

   A −1.30

   B 0.05

   C 1.30

   D 0.82   *(1 mark)*

12 What is the pH of $0.05\,\text{mol dm}^{-3}$ propanoic acid ($K_a = 1.3 \times 10^{-5}\,\text{mol dm}^{-3}$)?

   A 8.06×10^{-4}

   B 0.05

   C 2.44

   D 3.09   *(1 mark)*

13 Explain why a solution of a weak acid on its own *does not* act as a buffer solution.   *(1 mark)*

14 A student prepares a sample of $0.01\,\text{mol dm}^{-3}$ benzoic acid, then adds an equal volume of $0.01\,\text{mol dm}^{-3}$ sodium benzoate. Calculate the pH of this buffer and explain how it minimises changes to pH.
K_a (benzoic acid) $= 6.3 \times 10^{-5}\,\text{mol dm}^{-3}$.   *(5 marks)*

9.1 Oxidation and reduction

Specification reference: ES(b), ES(d), ES(e), ES(f), ES(g)

Redox reactions

When an oxidation reaction and a reduction reaction occur simultaneously, this is called a **redox reaction**.

Oxidation and reduction can be defined in two different ways:

- **o**xidation **is** the **l**oss of electrons
- **r**eduction **is** the **g**ain of electrons

or

- an element is oxidised when its oxidation state is increased (becomes more positive)
- an element is reduced when its oxidation state is decreased (becomes more negative).

Electron transfer and half-equations

Sodium reacts with chlorine as follows:

$$2Na + Cl_2 \rightarrow 2NaCl$$

This can be written as two separate half-equations. From these you can decide which is the oxidation reaction and which is the reduction reaction:

$2Na \rightarrow 2Na^+ + 2e^-$ This is oxidation (electron loss)

$Cl_2 + 2e^- \rightarrow 2Cl^-$ This is reduction (electron gain)

Assigning oxidation states

Oxidation states help determine which species are oxidised and which are reduced, even when no ions are present.

You should learn the following rules:

- The atoms in elements are always in an oxidation state of zero.
- In compounds or ions, oxidation states are assigned to each atom or ion.

Since compounds have no overall charge, the oxidation states of all the constituents must add up to zero. In ions, the sum of the oxidation states is equal to the charge on the ion.

Some atoms rarely change their oxidation states in reactions. These can be used to help to assign oxidation states to other species. Examples are F is -1, O is -2 (except in O_2^{2-} and OF_2), H is $+1$, Cl is usually -1 (except when combined with O or F).

> **Key terms**
>
> **Redox reaction:** A reaction involving oxidation and reduction simultaneously.
>
> **Oxidation:** Oxidation is the loss of electrons.
>
> **Reduction:** Reduction is the gain of electrons.

> **Revision tip**
> Remember OIL RIG: Oxidation Is Loss; Reduction Is Gain.

 Worked example: Oxidation states

Assign the oxidation states for each element in the following compounds:

a CO_2

Step 1: O is -2; there are two Os so the total contribution of O to the oxidation state is $2 \times (-2) = -4$.

Step 2: To make the total of the oxidation states add up to zero, C must be $+4$.

b CH_4

Step 1: H is +1; there are four hydrogens, so the total contribution of H to the oxidation state is $4 \times (+1) = +4$.

Step 2: To make the total of the oxidation states add up to zero, C is −4.

c VO^{2+}

Step 1: O is −2.

Step 2: The charge on the ion is +2, so V must have an oxidation state of +4, because $+4 + (−2) = +2$.

Revision tip

You must always write down the sign of an oxidation state (+ or −), or you will lose marks. The sign is written *in front of* the oxidation number.

Key terms

Half-equation: An ionic equation showing the transfer of electrons from one type of atom. Two half-equations combine together to make the chemical equation for the reaction.

Oxidising agent: A species that causes another substance to become oxidised. Oxidising agents become reduced as they take electrons from the other substance.

Reducing agent: A species that causes another substance to become reduced. Reducing agents become oxidised as they provide electrons for the other substance.

Using oxidation states

The displacement reaction below can be used to extract iodine from iodide minerals present in seawater. A more reactive halogen, like Cl_2, is passed into a solution of iodide ions, which are less reactive:

$$Cl_2(aq) + 2I^-(aq) \rightarrow 2Cl^-(aq) + I_2(aq)$$

oxidation states: 0   −1   −1   0

- The oxidation state for chlorine decreases (from 0 to −1), so chlorine is reduced.
- The oxidation state for iodine increases (from −1 to 0), so iodine is oxidised.

This reaction can also be represented by two half-equations:

$$Cl_2(aq) + 2e^- \rightarrow 2Cl^-(aq) - \text{this is reduction (electron gain)}$$

$$2I^-(aq) \rightarrow I_2(aq) + 2e^- - \text{this is oxidation (electron loss)}$$

Common misconception: Oxidation

Remember the correct word is oxidation, not 'oxidisation'!

The chlorine molecules are acting as the oxidising agent, as they cause another species to be oxidised and in doing so are reduced themselves. The iodide ions are reducing agents as they cause another species to be reduced and in doing so are oxidised themselves.

Using oxidation states to balance redox equations

To balance redox equations, identify the change in oxidation state (if any) of each element present. Then balance the equation so the number of electrons lost is equal to the electrons gained.

 Worked example: Balancing a redox equation

Balance the following equation:

$$NH_3 + O_2 \rightarrow NO + H_2O$$

Step 1: Work out the change in oxidation state of each element.

Nitrogen is oxidised from −3 to +2 (a gain of 5 electrons).

Hydrogen remains as +1.

Oxygen is reduced from 0 to −2 (a loss of 2 electrons).

Step 2: Balance the equation so the number of electrons lost is equal to the electrons gained.

Two nitrogen atoms can gain a total of 10 electrons from 5 oxygen atoms. Each oxygen atom loses two electrons.

Two ammonia molecules are therefore required. This will produce two NO and three H_2O molecules.

The balanced equation is:

$$2NH_3 + 2\frac{1}{2}\,O_2 \rightarrow 2NO + 3H_2O$$

This can also be written as:

$$4NH_3 + 5\,O_2 \rightarrow 4NO + 6H_2O.$$

Oxidation states in names

Some compounds contain elements that can exist in more than one oxidation state. When this occurs, the systematic name for the compound includes the oxidation state of the element. For example, FeO is called iron(II) oxide and Fe_2O_3 is called iron(III) oxide.

Oxoanions are negative ions that contain oxygen and another element. Their names end with the letters '-ate', for example, chromate. The names of oxoanions should also include an oxidation state. For example:

Ion	Name	Oxidation state of N or S
NO_2^-	nitrate(III)	+3
NO_3^-	nitrate(V)	+5
SO_3^{2-}	sulfate(IV)	+4
SO_4^{2-}	sulfate(VI)	+6

Revision tip

The oxidation state for cations is written in Roman numerals, immediately *after* the element it refers to.

Summary questions

1 Write down the oxidation states of the elements in the following.

(6 marks)

    a KBr   **b** H_2O **c** CO
    d PO_4^{3-} **e** MnO_2 **f** $Cr_2O_7^{2-}$

2 Write down the formulae of the following.   *(4 marks)*

    a copper(II) chloride   **b** copper(I) oxide
    c lead(IV) chloride   **d** manganate(VII) ion

3 Write two half-equations for the following reaction, and identify which equation is the oxidation reaction and which is the reduction reaction.   *(4 marks)*

$$2Ca + O_2 \rightarrow 2CaO$$

4 Write a balanced equation for the following reaction by identifying the changes in oxidation states.   *(2 marks)*

$$Br^- + H^+ + H_2SO_4 \rightarrow Br_2 + SO_2 + H_2O.$$

9.2 Electrolysis as redox

Specification reference: ES(c)

Key terms

Cations: Positive ions are known as *cations*, as they are attracted to the *cathode*.

Anions: Negative ions are known as *anions*, as they are attracted to the *anode*.

Electrolysis of molten salts

Molten salts conduct electricity, because they contain ions which are free to move. The metal ions, which are positive, are attracted to the cathode where they gain electrons and turn into the metal:

$$Pb^{2+}(l) + 2e^- \rightarrow Pb(l)$$

The non-metal ions, which are negative, are attracted to the anode where they lose electrons and turn into the element.

$$2Br^-(l) \rightarrow Br_2(g) + 2e^-$$

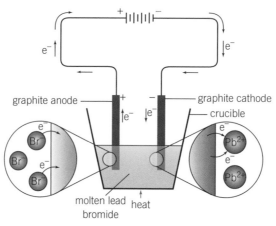

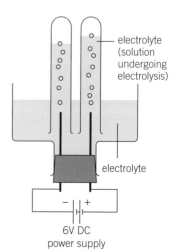

▲ **Figure 1** *Electrolysis of a molten salt*

▲ **Figure 2** *Electrolysis of a solution of a salt*

Summary questions

1 State the products of the electrolysis of these molten salts:
   a potassium chloride *(1 mark)*
   b calcium oxide *(1 mark)*
   c magnesium iodide. *(1 mark)*

2 State the products of the electrolysis of these aqueous solutions with graphite electrodes:
   a potassium chloride solution *(1 mark)*
   b sodium sulfate solution *(1 mark)*
   c copper nitrate solution. *(1 mark)*

3 Write half-equations for the electrolysis of:
   a molten aluminium chloride *(1 mark)*
   b aqueous lead nitrate solution *(1 mark)*
   c aqueous sodium chloride solution. *(1 mark)*

Reduction occurs at the cathode, because electrons are gained. Oxidation occurs at the anode.

Electrolyte	Cathode process	Anode process
Molten aluminium oxide	$Al^{3+}(l) + 3e^- \rightarrow Al(l)$	$2O^{2-}(l) \rightarrow O_2(g) + 4e^-$
Molten zinc iodide	$Zn^{2+}(l) + 2e^- \rightarrow Zn(l)$	$2I^-(l) \rightarrow I_2(g) + 2e^-$
Molten magnesium chloride	$Mg^{2+}(l) + 2e^- \rightarrow Mg(l)$	$2Cl^-(l) \rightarrow Cl_2(g) + 2e^-$

Electrolysis of solutions

When an electric current is passed through a solution of a salt, such as sodium chloride, positive ions are attracted to the cathode and negative ions are attracted to the anode. However, when predicting the products, you must take into account the presence of water, which can compete with the ions from the salt at the electrodes. If the solution contains a more-reactive metal such as sodium or potassium, hydrogen gas is produced. Less-reactive metals such as zinc or copper are deposited on the cathode. If the solution contains a halide ion, the corresponding halogen is produced at the anode. If the solution contains an anion such as sulfate or nitrate, oxygen gas is produced.

Using metal electrodes with metal ions in solution

If the anode is made of the same metal as the ions in solution, oxidation occurs and the anode gradually dissolves. For example:

$$Cu(s) \rightarrow Cu^{2+}(aq) + 2e^-$$

The ions produced are then deposited at the cathode.

9.3 Redox reactions, cells, and electrode potentials

Specification reference: DM(c), DM(d)

Redox reactions

You have seen that oxidation is the loss of electrons and reduction is gain of electrons. A redox reaction is a reaction in which oxidation and reduction occur at the same time. The overall equation for the reaction can be split into two parts, called half-equations, which show the species that gain or lose electrons. Half-equations involve ions and electrons.

Combining half-equations

Adding together the two half-equations gives the overall equation for the redox reaction.

> **Synoptic link**
>
> Oxidation and reduction are introduced in Topic 9.1, Oxidation and reduction.

> **Key term**
>
> **Redox reaction:** A reaction involving simultaneous oxidation and reduction.

 Worked example: Writing redox equations

Write the combined equation for the displacement reaction when chlorine reacts with bromide ions.

Step 1: Write the half-equations for the oxidation and reduction reactions:

$$2Br^- \rightleftharpoons Br_2 + 2e^- \quad \textbf{oxidation half-equation}$$

$$Cl_2 + 2e^- \rightleftharpoons 2Cl^- \quad \textbf{reduction half-equation}$$

Step 2: Make sure the number of electrons is the same in each half-equation. Here there are two electrons in each half-equation.

Step 3: Add the two half-equations together, cancelling electrons from the left- and right-hand sides.

$$2Br^- + Cl_2 \rightarrow Br_2 + 2Cl^-$$

Model answer: Writing redox equations

Write the combined equation for the reaction of chromate(VI) ions, $Cr_2O_7^{2-}$, with sulfate(IV) ions, SO_3^{2-}.

Step 1: Write the half-equations for the oxidation and reduction reactions:

$$SO_3^{2-}(aq) + H_2O(l) \rightleftharpoons SO_4^{2-}(aq) + 2H^+(aq) + 2e^- \quad \textbf{oxidation half-reaction}$$

$$Cr_2O_7^{2-}(aq) + 14H^+(aq) + 6e^- \rightleftharpoons 2Cr^{3+}(aq) + 7H_2O(l) \quad \textbf{reduction half-reaction}$$

Step 2: Make sure the number of electrons is the same in each half-equation. In this case, there needs to be 6 electrons in each half-equation:

$$3SO_3^{2-}(aq) + 3H_2O(l) \rightleftharpoons 3SO_4^{2-}(aq) + \textbf{6}H^+(aq) + \textbf{6}e^-$$

$$Cr_2O_7^{2-}(aq) + 14H^+(aq) + \textbf{6}e^- \rightleftharpoons 2Cr^{3+}(aq) + 7H_2O(l)$$

Step 3: Add the two half-equations together, cancelling electrons and hydrogen ions from the left- and right-hand sides:

$$Cr_2O_7^{2-}(aq) + 8H^+(aq) + 3SO_3^{2-}(aq) \rightarrow 2Cr^{3+}(aq) + 4H_2O(l) + 3SO_4^{2-}(aq)$$

> The oxidation half-equation must be multiplied by a factor of 3 to produce 6 electrons, which are accepted by the reduction half-equation.

> **Revision tip**
>
> Electrode potentials allow you to decide which is the oxidation half-reaction and which is the reduction half-reaction.

Electrode potentials

If you place a strip of metal in an aqueous solution of its ions, an electrode potential, or potential difference, is created. An equilibrium is established between the metal and the ions. For example:

$$Cu^{2+}(aq) + 2e^- \rightleftharpoons Cu(s)$$

$$Zn^{2+}(aq) + 2e^- \rightleftharpoons Zn(s)$$

Like all equilibria, these are affected by temperature and the concentration of the ions, and the conditions should always be stated.

Metals differ in their tendency to release electrons. Metals which release electrons more readily have a more negative electrode potential. Altering the temperature or the concentration of ions in solution alters the value of the electrode potential.

Electrochemical cells

An electrochemical cell consists of two half-cells joined together, with a high-resistance voltmeter measuring the maximum potential difference between two half-cells. To complete the circuit a salt bridge provides an ionic connection between two half-cells. It is often made from a strip of filter paper soaked in a saturated solution of potassium nitrate.

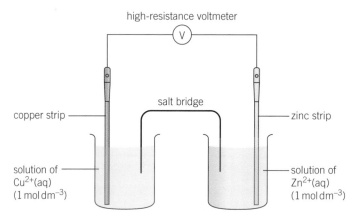

▲ **Figure 1** A copper–zinc cell

It is not possible to measure the electrode potential of a single half-cell. To measure the electrode potential of a metal/ion system, the two half-cells must be joined together to form an electrochemical cell.

The potential difference between the two half-cells is called the cell potential, E_{cell}. Electrons flow through the wires from the negative terminal to the positive terminal:

● The half-cell with the more negative electrode potential forms the negative terminal of the cell.

● The half-cell with the more positive electrode potential forms the positive terminal of the cell.

Standard electrode potentials

It is necessary to have a standard set-up to compare the electrode potentials of all metal/ion half-cells. To do this a standard hydrogen half-cell is used. Its electrode potential under standard conditions is defined as 0.00 V.

The half-reaction occurring in the standard hydrogen half-cell is:

$$H^+(aq) + e^- \rightleftharpoons \frac{1}{2}H_2(g)$$

The **standard electrode potential**, $E^{\ominus}$, of a half-cell is the potential difference between the half-cell and a standard hydrogen half-cell.

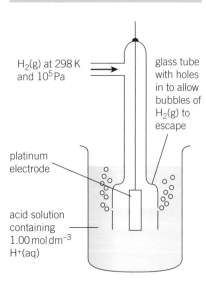

▲ **Figure 2** The standard hydrogen half-cell

To measure a standard electrode potential, the half-cell being investigated is connected to a standard hydrogen half-cell.

For half-cells such as $I_2/2I^-$ or Fe^{3+}/Fe^{2+}, which do not feature a metal in elemental form, an inert electrode such as platinum is dipped into a solution containing all the ions and molecules involved in the half-reaction.

 Worked example: Calculating $E^{\varnothing}_{cell}$

What is $E^{\varnothing}_{cell}$ when the Fe^{2+}/Fe and Cu^{2+}/Cu half-cells are connected?

Step 1: Look up the standard electrode potentials for the two half-reactions:

$$Fe^{2+}(aq) + 2e^- \rightleftharpoons Fe(s) \qquad E^{\varnothing} = -0.44\,V$$

$$Cu^{2+}(aq) + 2e^- \rightleftharpoons Cu(s) \qquad E^{\varnothing} = +0.34\,V$$

Step 2: Calculate $E^{\varnothing}_{cell} = E^{\varnothing}$ [most positive electrode] $- E^{\varnothing}$ [most negative electrode]

$$E^{\varnothing}_{cell} = +0.34\,V - (-0.44\,V) = 0.78\,V$$

Predicting the direction of a reaction

Electrons are released from the half-cell with the more negative electrode potential, where oxidation occurs. Electrons are accepted by the half-cell with the more positive electrode potential, where reduction occurs.

It is possible to use this idea to predict the feasibility of a reaction, by calculating $E^{\varnothing}_{cell}$ for the proposed reaction.

Model answer: Predicting the feasibility of a reaction

Explain whether the reaction between aqueous chlorine and aqueous iodide ions is feasible.

Step 1: Look up the half-reactions and their standard electrode potentials:

$$I_2(aq) + 2e^- \rightleftharpoons 2I^-(aq) \qquad E^{\varnothing} = +0.54\,V \qquad \text{half-reaction 1}$$

$$Cl_2(g) + 2e^- \rightleftharpoons 2Cl^-(aq) \qquad E^{\varnothing} = +1.36\,V \qquad \text{half-reaction 2}$$

Step 2: Construct an electrode potential chart.

> Although both half-cells have a positive $E^{\varnothing}$, the I_2/I^- is the more negative half-cell.

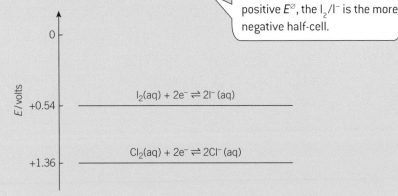

Step 3: Use the electrode potential chart to predict whether a reaction could occur.

Electrons flow to the positive terminal, which is the $Cl_2(aq)/Cl^-(aq)$ half-cell, which has an electrode potential of +1.36 V.

A reduction reaction will occur in the positive half-cell:

$$Cl_2(g) + 2e^- \rightleftharpoons 2Cl^-(aq)$$

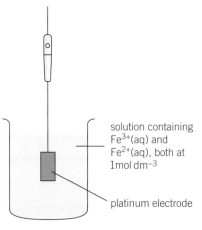

▲ **Figure 3** *A standard half-cell for the $Fe^{3+}(aq)/Fe^{2+}(aq)$ half-reaction*

Revision tip

The half-equations for $I_2/2I^-$ and Fe^{3+}/Fe^{2+} are:

$$I_2 + 2e^- \rightleftharpoons 2I^-$$

$$Fe^{3+} + e^- \rightleftharpoons Fe^{2+}$$

Revision tip

Remember that all concentrations in half-cells are $1.00\,mol\,dm^{-3}$.

Revision tip

Both half-cells might have negative $E^{\varnothing}$ values. In this case, the one with the least negative value has the most positive $E^{\varnothing}$.

Revision tip

$E^{\varnothing}$ cell values do not have a sign as they represent the *difference* in potential.

Revision tip

Oxidation is the loss of electrons, which occurs at the negative electrode. Reduction is the gain of electrons, which occurs at the positive electrode.

Step 4: The other half-cell must supply electrons and is the negative terminal of the cell. Oxidation occurs.

$$2I^-(aq) \rightleftharpoons I_2(aq) + 2e^-$$

Step 5: Use the half-equations to give an overall equation.

$$2I^-(aq) + Cl_2(g) \rightarrow I_2(aq) + 2Cl^-(aq)$$

Therefore the reaction that occurs is that of iodide ions with aqueous chlorine. The reverse reaction does not occur, so chloride ions *do not* react with aqueous iodine.

Revision tip

Changing the conditions such as concentration and/or temperature will change the $E^\varnothing$ cell values, which may also cause the reaction to happen.

Remember that this method predicts the feasibility of a reaction occurring under standard conditions. If the activation enthalpy is high, the reaction may not actually happen. However a catalyst might enable the reaction to occur.

Summary questions

1 Describe the experimental set-up and conditions required to determine $E^\varnothing_{cell}$ for a copper–silver cell. *(5 marks)*

2 Write a balanced equation for the reaction of manganate(VII) ions with iron(II) ions. The half-equations are:
   $MnO_4^-(aq) + 8H^+(aq) + 5e \rightleftharpoons Mn^{2+}(aq) + 4H_2O(aq)$
   and $Fe^{3+}(aq) + e^- \rightleftharpoons Fe^{2+}(aq)$ *(2 marks)*

3 Calculate $E^\varnothing_{cell}$ for the reaction in question 2, and explain why Mn^{2+} ions do not reduce Fe^{3+} ions. $E^\varnothing(MnO_4^-/Mn^{2+})$ $= +1.51\,V$; $E^\varnothing(Fe^{3+}/Fe^{2+}) = +0.77\,V$ *(2 marks)*

9.4 Rusting and its prevention

Specification reference: DM(f)

Rusting

Corrosion of iron, commonly known as rusting, is a widespread problem, and it can have serious economic consequences. It occurs due to electrochemical processes, and certain methods of rust prevention also rely on electrochemistry.

Half-equations in rusting

Rusting occurs in a series of steps. The half-equations involved are:

$$Fe^{2+}(aq) + 2e^- \rightleftharpoons Fe(s) \qquad E^\varnothing = -0.44\,V$$

$$O_2(g) + 2H_2O(l) + 4e^- \rightleftharpoons 4OH^-(aq) \qquad E^\varnothing = +0.40\,V$$

$$E^\varnothing_{cell} = +0.84\,V$$

The Fe/Fe^{2+} half-cell is more negative, so it releases electrons. This means that Fe atoms are oxidised into Fe^{2+} ions. The electrons flow to the half-cell comprising oxygen dissolved in a water droplet, which accepts the electrons and produces hydroxide ions.

Two further reactions then occur. First the Fe^{2+} and OH$^-$ ions react to form solid iron(II) hydroxide:

$$Fe^{2+}(aq) + 2OH^-(aq) \rightarrow Fe(OH)_2(s)$$

Secondly the iron(II) hydroxide reacts with further oxygen to form hydrated iron(III) oxide, Fe$_2$O$_3 \cdot x$H$_2$O(s), which is rust.

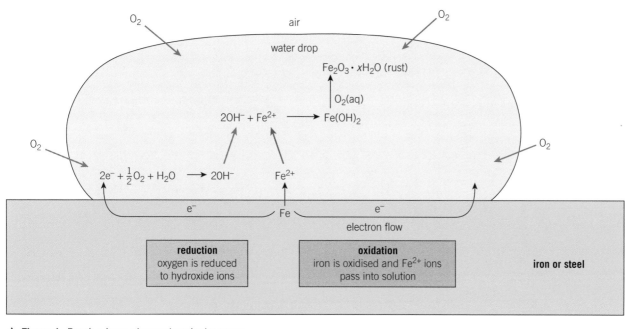

▲ **Figure 1** *Rusting is an electrochemical process*

Preventing rusting

A common means of protecting iron objects from rusting is to paint, grease, or oil them. This provides a barrier between the metal and atmospheric oxygen.

The iron can be covered with a layer of another metal, such as zinc. This is known as **galvanising**. Although the zinc reacts with oxygen, it produces a hard layer of zinc oxide which ultimately protects the iron underneath.

If the zinc coating is damaged, it will still protect against rusting. This is because $E^{\ominus}$ for the zinc half-cell is more negative than $E^{\ominus}$ for the iron half-cell:

$$Zn^{2+}/Zn \quad E^{\ominus} = -0.76\,V$$

$$Fe^{2+}/Fe \quad E^{\ominus} = -0.44\,V$$

As a result, the zinc corrodes in preference to the iron. This is known as **sacrificial protection**. Large iron objects are sometimes protected by having blocks of metals such as zinc or magnesium, which have more negative electrode potentials, applied to them.

Key term

Sacrificial protection: The use of a metal with a more negative electrode potential to prevent rusting. The other metal corrodes first.

Summary questions

1 Explain why rusting only occurs in the presence of oxygen and water. (*1 mark*)

2 Identify which species are oxidised and reduced in the half-equations for rusting, by stating the change in oxidation number. (*2 marks*)

3 Calculate $E^{\ominus}_{cell}$ for the reaction of zinc with the O_2/OH^- half-cell, and suggest why zinc corrodes before iron. (*2 marks*)

Experimental techniques

Specification reference: ES (c)

Electrolysis of molten compounds

Solid ionic compounds do not conduct electricity since ions in a giant ionic lattice are not free to move. On melting, the ions are free to move and the liquid can carry a current. For electrolysis to occur, an electric current is passed through the electrolyte, using a power supply and electrodes (anode and cathode) which are commonly made of graphite.

Electrolysis of aqueous solutions

For electrolysis to occur in an aqueous solution an electrical circuit needs to be set up using a dc power supply, as in the electrolysis of a molten salt. The dc power supply may be a powerpack or batteries. Graphite is commonly used for the electrodes as it is cheap and inert. Platinum may be used instead, but it is much more expensive.

If the products of electrolysis are gases, an alternative arrangement can be used to collect the gases (see Figure 1). The test tubes are filled with water and the gaseous products displace the water and fill the test tubes.

> ### Key term
>
> **Electrolyte:** An ionic liquid or solution that conducts electricity.

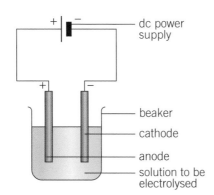

▲ **Figure 1** *Electrolysis of aqueous solutions*

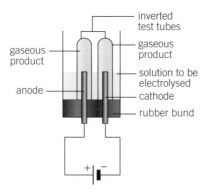

▲ **Figure 2** *Apparatus to collect gaseous products in electrolysis*

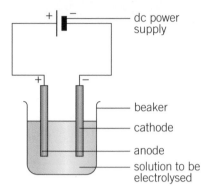

▲ **Figure 1** *Apparatus for electrolysis*

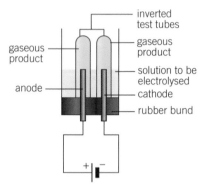

▲ **Figure 2** *Apparatus to collect gaseous products in electrolysis*

Revision tip

Standard temperature and concentration for electrochemical cells is 298 K and 1.0 mol dm^{-3}.

Electrolysis of aqueous solutions

An electric current is passed through an electrolyte. Different set-ups are required, depending whether the products of electrolysis are solids or gaseous.

The test tubes are filled with water at the start of the reaction, and the gaseous products displace the water.

Electrolysis can be used to purify a metal, in which case the anode is made of the impure metal. The electrolyte must contain ions of the metal and the cathode must be made of the pure metal.

Electrochemical cells

The potential of an electrochemical cell can be determined by connecting two half-cells. Standard electrode potentials are measured by connecting the half-cell in question to a standard hydrogen half-cell, or a calibrated reference half-cell.

1   Start by constructing the half-cell of interest:

- For a metal/metal ion half-cell such as Cu^{2+}/Cu, the electrode is a strip of metal dipping into a 1.0 mol dm^{-3} solution of the metal ion.

- For a half-cell containing two ions of the same element such as Fe^{3+}/Fe^{2+}, the electrode is a platinum (or graphite) rod dipping into a mixture of the two ions. The concentration must be 1.0 mol dm^{-3}.

- The temperature must be 298 K.

2   Connect the half-cell to the reference cell using a salt bridge and a high-resistance voltmeter.

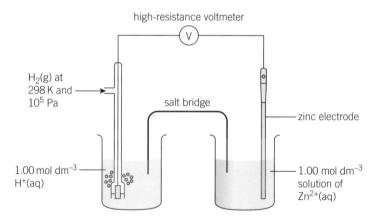

▲ **Figure 3** *An example of a standard electrochemical cell*

3   Check that the reading on the voltmeter is positive. This shows that the half-cell connected to the positive terminal of the voltmeter is the positive electrode.

4   If the reading is negative, change the connections round on the voltmeter to give a positive reading.

5   Record the voltmeter reading.

Chapter 9 Practice questions

1 Which statement is correct about the reaction below?

$Mg + Cl_2 \rightarrow MgCl_2$

  A Magnesium is oxidised

  B Magnesium gains electrons

  C Chlorine is oxidised

  D Chlorine's oxidation state does not change.   (*1 mark*)

2 Which species contains phosphorus in oxidation state +5?

  A P_5 **B** PF_6^-

  C H_3PO_3 **D** PH_3 (*1 mark*)

3 What is the oxidiation state of sulfur in H_2SO_4?

  A +2   **B** +4

  C +6   **D** +8   (*1 mark*)

4 What is the name of $NaNO_2$?

  A sodium nitrate(I)   **B** sodium nitrate(II)

  C sodium nitrate(III)   **D** sodium nitrate(V)   (*1 mark*)

5 What are the products of electrolysis of aqueous potassium nitrate?

  A potassium at the cathode, nitrogen at the anode

  B nitrogen at the anode, potassium at the cathode

  C oxygen at the cathode, nitrogen at the anode

  D hydrogen at the cathode, oxygen at the anode   (*1 mark*)

6 Which substances produce copper at the cathode and oxygen at the anode during electrolysis?

  1 Aqueous copper sulfate

  2 Aqueous copper nitrate

  3 Molten copper chloride

  A 1 only   **B** 1 and 2 only

  C 2 and 3 only   **D** 1, 2, and 3   (*1 mark*)

7 Balance the equation below using oxidation numbers:

$$I^- + H^+ + MnO_2 \rightarrow I_2 + H_2O + Mn^{2+}$$ (*2 marks*)

8 Identify which species are oxidised and which are reduced in the equation below:

$$2\,I_2 + N_2H_4 \rightarrow 4\,HI + N_2$$ (*2 marks*)

9 The water in the Dead Sea is particularly rich in bromide ions, and an important chemical industry has grown up in Israel to exploit this natural resource.

  The process involves oxidising bromide ions into bromine molecules.

  a Describe the colour change you would expect to see when Br^- is converted to Br_2.   (*2 marks*)

  b Explain why the process in which bromide ions, Br^- are changed into bromine, Br_2 is referred to as *oxidation*.   (*1 mark*)

  c Chlorine is used to oxidise the bromide ions. Chlorine is manufactured on site by the electrolysis of brine. Write a half-equation for the production of chlorine from brine.   (*1 mark*)

  d Write an equation for the reaction of chlorine with bromide ions.   (*1 mark*)

  e Explain why chlorine is described as an oxidising agent, referring to changes in the oxidation state of chlorine.   (*2 marks*)

10 In the equation below, what is the change in oxidation state of S?

$$S_2O_3^{2-} + I_2 \rightarrow S_4O_6^{2-} + 2I^-$$

A +4 to +6   **C** +2 to +4

B +6 to +12   **D** 2– to 2– (no change in oxidation state) (*1 mark*)

11 What is the correct full equation for the two half-equations given below?

$$V^{3+} + e^- \rightarrow V^{2+} \quad\quad E^\varnothing = -0.26\,V$$
$$Zn^{2+} + 2e^- \rightarrow Zn \quad\quad E^\varnothing = -0.76\,V$$

A $V^{2+} + Zn^{2+} \rightarrow Zn + V^{3+}$ **C** $V^{3+} + Zn^{2+} \rightarrow Zn + V^{2+}$

B $2V^{2+} + Zn \rightarrow Zn^{2+} + 2V^{3+}$ **D** $2V^{3+} + Zn \rightarrow Zn^{2+} + 2V^{2+}$ (*1 mark*)

12 What is E_{cell} for the vanadium/zinc cell given in question 2?

A 0.24 V   **C** 0.50 V

B 0.26 V   **D** 1.02 V   (*1 mark*)

13 What occurs at the more negative electrode of an electrochemical cell? (*1 mark*)

A Oxidation, because electrons are gained

B Oxidation, because electrons are lost

C Reduction, because electrons are gained

D Reduction, because electrons are lost

14 Which of the following are correct explanations of different methods of preventing rusting?   (*1 mark*)

1 Oiling the item reduces contact with water

2 Painting the item reduces contact with atmospheric oxygen

3 Sacrificial protection involves more reactive metals corroding in preference to iron

     A 1 only   **C** 3 only

     B 1 and 2 only   **D** 1, 2, and 3

15 A student investigated the redox activity of vanadium compounds by converting VO_2^+ to VO^{2+} using zinc.

a Give the oxidation states of VO_2^+ and VO^{2+}.   (*2 marks*)

b Explain whether the vanadium was being oxidised or reduced. (*1 mark*)

c Balance the half-equation for the conversion:

$VO_2^+ + ...H^+ + ...e^- \rightleftharpoons VO^{2+} + ...H_2O$ (*2 marks*)

d $E^\varnothing$ for the half-reaction in part **c** is +1.00 V. $E^\varnothing$ for Zn/Zn^{2+} is –0.76 V. Calculate E_{cell} for the reaction that the student investigated.   (*1 mark*)

e Write the balanced ionic equation for the redox reaction that the student carried out.   (*1 mark*)

f Vanadium can be further reduced to V^{3+} and V^{2+}. The half-cell values are $E^\varnothing(VO^{2+}/V^{3+})$ = +0.34 V and $E^\varnothing(V^{3+}/V^{2+})$ = –0.26 V.

Explain why zinc is able to reduce the vanadium species as far as V^{2+}.   (*1 mark*)

g $E^\varnothing$ for Pb/Pb^{2+} is –0.13 V. What will be the final oxidation state of vanadium when VO_2^+ reacts with Pb? Give a reason.   (*2 marks*)

16 Give full, ionic, or half-equations for the following processes that occur during rusting:

a the oxidation of iron metal to Fe^{2+} ions   (*1 mark*)

b the reduction of oxygen in the presence of water   (*1 mark*)

c the formation of iron(II) hydroxide   (*1 mark*)

d the sacrificial oxidation of zinc metal   (*1 mark*)

10.1 Factors affecting reaction rates

Specification reference: OZ(f)

Rates of reaction

Rates of reaction can be affected by a number of factors:

- concentration
- pressure
- a catalyst
- temperature
- surface area
- particle size
- intensity of radiation.

Collision theory

Reactions occur when particles of reactants collide with a certain *minimum* kinetic energy:

- At higher concentrations and higher pressures, the particles are in closer proximity to each other, encouraging more frequent collisions.
- At higher temperatures, a much higher proportion of colliding particles have sufficient energy to react and more collisions have a greater energy than the activation enthalpy.
- When a solid is more finely divided, there is a larger surface area on which the reactions can take place, so the greater the frequency of successful collisions.
- Heterogeneous catalysts provide a surface where reacting particles may break and make bonds.

All of the above serve to increase the rate of a chemical reaction.

Measuring rate of reaction

To measure the reaction rate you need to measure how quickly a reactant is used up, or how quickly a product is made. This can be done by:

- Measuring volume of gas produced. This can be done using a gas syringe, or by displacement of water. The more gas produced per unit time, the faster the reaction.
- Measuring mass changes. Reactions which give off a gas involve mass changes. The bigger the change in mass per unit time, the faster the reaction.
- Colorimetry. Measuring the change of intensity as a coloured chemical is used up or produced. For example, a reaction involving the production of bromine will involve a gradually darkening brown colouration. The rate of change of intensity of the brown colouration can be measured using a colorimeter.
- pH changes. If an acid or alkali is used up or produced, a pH meter can monitor the rate of change of pH. The faster the change, the faster the reaction.

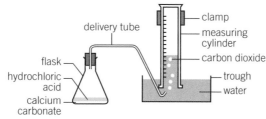

▲ **Figure 1** *Measuring volume changes*

Summary questions

1. Suggest three ways to increase the rate of the following reaction:
   $Mg(s) + H_2SO_4(aq) \rightarrow MgSO_4(aq) + H_2(g)$. *(3 marks)*

2. Suggest three ways of measuring the rate of the reaction in question 1. *(3 marks)*

3. Use the collision theory to explain why the rate of a chemical reaction is fastest at the beginning. *(2 marks)*

10.2 The effect of temperature on rate

Specification reference: OZ[e], OZ [f]

Activation enthalpy

Rates of reaction do not just depend on how frequently particles collide, but also on how much energy they have when a collision takes place.

Collision theory states that reactions occur when molecules collide with a certain *minimum* kinetic energy. This minimum kinetic energy is called the **activation enthalpy**. The energy needed to overcome the energy barrier is called the activation energy barrier.

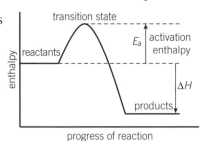

▲ **Figure 1** *Enthalpy profile for an exothermic reaction*

Molecular energies

As the temperature increases, the rate of a chemical reaction also increases. This is because of the distribution of energies among the reacting particles – this distribution is called the **Maxwell–Boltzmann distribution**.

There needs to be enough molecules with sufficient energy for a reaction to take place. The molecules need to have a combined kinetic energy higher than the activation enthalpy.

Reactions go faster at higher temperatures, because a larger proportion of the colliding particles have the minimum activation enthalpy needed to react.

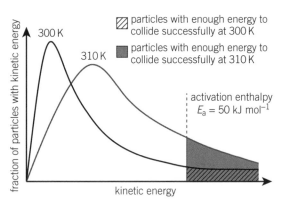

▲ **Figure 2** *Maxwell-Boltzmann distribution for a reaction at two different temperatures*

Referring to the diagram above:

- the peak of the number of collisions at 300 K is at a lower kinetic energy value than the peak at 310 K (i.e. the most probable kinetic energy for a particle is lower at lower temperatures)
- at the kinetic energy value of 50 kJ mol⁻¹, the number of collisions at 310 K is almost twice as many as at 300 K
- for reactions with an activation enthalpy around 50 kJ mol⁻¹, when the temperature rises by 10 °C (10 K), the rate of reaction approximately doubles.

Summary questions

1 Explain what E_a represents. *(1 mark)*

2 Sketch a graph showing a typical distribution of energy among the molecules of a reaction mixture. Shade the area of the graph representing the number of collisions with sufficient energy to lead to a reaction. *(2 marks)*

3 Suggest a reason for the following observations:
   a N_2 and O_2 do not react at room temperature, but in a car engine the temperature is high enough for them to form NO. *(1 mark)*
   b The reaction between NO and O_2 to make NO_2 occurs easily at room temperature. *(1 mark)*

10.3 Catalysts

Specification reference: OZ(f), DF(h)

Catalysts and activation energy

Catalysts speed up the rate of a chemical reaction without getting used up.

In heterogeneous catalysis, the reactants and catalysts are in different physical states. The effectiveness of heterogeneous catalysts can be reduced by catalyst poisons. In homogeneous catalysis, the reactants and catalysts are in the same physical state. Homogeneous catalysts work by forming an intermediate compound with the reactants. In the first step, an *intermediate* is formed in a reaction with lower activation enthalpy. In the second step, this intermediate breaks down to give a product and reform the catalyst.

For a chemical reaction to proceed, a pair of reacting molecules must collide, with a combined energy greater than the activation enthalpy for the reaction, in order to make a successful collision. With a catalyst, successful collisions can take place at a lower energy.

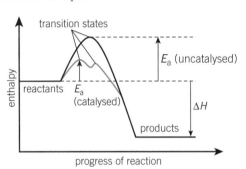

▲ **Figure 1** *The effect of a catalyst on the enthalpy profile of a reaction*

The reaction profiles for a catalysed reaction and an uncatalysed reaction are shown in Figure 1.

Catalysts work by providing an alternative reaction pathway for the breaking and making of bonds. This alternative path has a lower activation enthalpy than the uncatalysed pathway.

Maxwell-Boltzmann distribution

You can see in the Maxwell-Boltzmann distribution shown in Figure 2 that E_a for the catalysed reaction is lower than for the uncatalysed reaction. Therefore more particles have energy greater than the activation enthalpy and consequently the rate of reaction is greater.

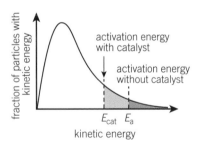

▲ **Figure 2** *Maxwell-Boltzmann distribution showing activation enthalpies for a catalysed and an uncatalysed reaction*

Summary questions

1 Hydrogen peroxide slowly decomposes into water and oxygen.
   a Explain why the activation enthalpy is lower for the decomposition of hydrogen peroxide solution in the presence of manganese(IV) oxide. *(1 mark)*
   b A solution of the enzyme catalase can also be used to decompose hydrogen peroxide solution. Explain what *type* of catalyst is
     i   manganese(IV) oxide *(1 mark)*
     ii catalase. *(1 mark)*

2 Look again at the reaction profile in Figure 1. Explain why there is a trough in the pathway for the catalysed reaction. *(2 marks)*

3 Suggest whether the value of the enthalpy change, ΔH, is affected by the use of a catalyst. *(2 marks)*

10.4 How do catalysts work?

Revision tip

Heterogeneous catalysts must have a large surface area in order to adsorb reactants.

They are used in a finely divided form, sometimes supported on a porous material.

Key term

Catalyst poison: A substance which binds irreversibly to the catalyst surface and stops the catalyst functioning properly.

Revision tip

An example of a catalyst poison is lead, which poisons the metals platinum and rhodium found in catalytic converters in cars.

Heterogeneous catalysts

Heterogeneous catalysts are in a different physical state to the reactants and products. For example, the hydrogenation of ethene to produce ethane uses a nickel catalyst. Nickel is a solid, whereas hydrogen, ethene, and ethane are gases.

Heterogeneous catalysts provide a surface on which a reaction may take place, thus lowering the energy needed for a successful collision – this lowers the activation energy barrier.

This happens in four stages:

1   The reactant molecules are adsorbed onto the surface of the catalyst. This weakens the bonds in the reactant.
2   Bonds in the reactant break.
3   New bonds form. This creates the products.
4   The products are released from the catalyst surface and diffuse away.

Catalyst poisons

Catalysts may be poisoned if a molecule becomes adsorbed more strongly than the reactant molecules. This means that fewer reactant molecules can be adsorbed, and the catalyst becomes less efficient.

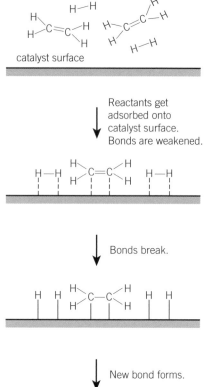

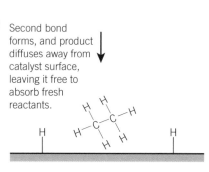

▶ **Figure 1** *A model of heterogeneous catalysis*

Summary questions

1   Explain which of the following are examples of heterogeneous catalysis.   (*4 marks*)
    a   the use of iron oxides in the reaction of nitrogen and hydrogen to make ammonia.
    b   breakdown of proteins into amino acids by the protease enzymes.
    c   the use of vanadium oxides in the oxidation of sulfur dioxide to form sulfur trioxide in the production of sulfuric acid.
    d   the use of sulfuric acid in the formation of esters from carboxylic acids and alcohols.

2   Describe the stages of heterogeneous catalysis.   (*4 marks*)

3   Suggest why leaded petrol must not be used in a car fitted with a catalytic converter.   (*1 mark*)

10.5 Rates of reactions

Specification reference: CI(a), CI(c)

Reaction kinetics

The study of reaction kinetics is an important area of chemistry. Understanding the rates of reactions allows chemists to identify factors which affect the rate, and gives insights into the mechanism of reactions.

During a reaction, reactants are used up as products are made. Rate of reaction is calculated by measuring the decrease in concentration of a reactant, or the increase in concentration of a product over a certain time period.

$$\text{rate of reaction} = \frac{\text{change in concentration of reactant or product}}{\text{time taken}}$$

The units of rate of reaction are usually **$mol\,dm^{-3}\,s^{-1}$**.

To measure the rate of reaction it is necessary to identify a characteristic that can be easily monitored. These include:

- volume of gas, using a gas syringe
- mass changes, using a balance
- colour changes, using a colorimeter
- pH changes, using a pH meter.

These can be measured directly.

It is also possible to quench a reaction by adding a large quantity of water, rapidly cooling the reaction mixture, or by neutralising any acid reactant or catalyst. Then the concentration of one of the reactants or products can be determined by titration.

Summary questions

1 How is rate of reaction calculated? *(1 mark)*

2 Suggest how to measure the rate of reaction of magnesium reacting with hydrochloric acid. *(3 marks)*

3 Explain why quenching is necessary when continuous methods of measuring rates are not appropriate. *(1 mark)*

Synoptic link

You will learn more about using kinetics to propose the mechanism of a reaction in Topic 10.7, Finding the order of reaction with experiments.

Revision tip

The unit $mol\,dm^{-3}\,s^{-1}$ measures the change in concentration $(mol\,dm^{-3})$ per second (s^{-1}).

Synoptic link

You first learned about methods of measuring rates of reaction in Topic 10.1, Factors affecting reaction rates.

Synoptic link

You will learn about using concentrations to determine rate of reaction, including orders of reaction, in Topic 10.6, Rate equation of a reaction.

Key term

Quench: To effectively stop a reaction occurring by slowing down its rate suddenly.

10.6 Rate equation of a reaction

Specification reference: CI(a), CI(d)

▼ **Table 1** Orders of reaction

Order with respect to reactant	Effect of doubling the concentration of a reactant on the rate of reaction
zero	No effect (rate $\propto [A]^0$)
first	Rate is doubled when concentration is doubled (rate $\propto [A]$)
second	Rate is quadrupled when concentration is doubled (rate $\propto [A]^2$)

What is a rate equation?

A rate equation is a mathematical expression which enables the rate of reaction (in $mol\,dm^{-3}\,s^{-1}$) to be calculated.

For a general reaction in which A and B are the reactants:

$$A + B \rightarrow products$$

the rate equation is **rate = $k[A]^m [B]^n$**

[A] and [B] are the concentrations of reactants A and B, and m and n are the powers to which [A] and [B] are raised. m and n are the orders of reaction with respect to reactant A and B respectively. The overall order of reaction is $(m + n)$.

Orders of reactions usually have values of 0, 1, or 2. They indicate the effect of changing the concentration of the reactant on the rate of reaction.

It is very important to note that you cannot predict the rate equation for a reaction from its balanced equation. The only way of determining a rate equation is from experimental data.

Examples of rate equations

Iodine reacts with propanone as shown below. An acid catalyst (H^+) is used:

$$CH_3COCH_3(aq) + I_2(aq) \xrightarrow{H^+} CH_3COCH_2I(aq) + H^+(aq) + I^-(aq)$$

Experimental data show that the rate equation for this reaction is **rate = $k[CH_3COCH_3][H^+]$**

You can see the following with regard to the rate equation:

- It is first order with respect to CH_3COCH_3 because $[CH_3COCH_3]$ is raised to the power 1.
- It is first order with respect to H^+ – a catalyst can appear in the rate equation.
- It is zero order with respect to I_2 – changing the concentration of I_2 does not affect the rate of reaction, so $[I_2]$ does not appear in the rate equation.
- It is second order overall because the orders of reaction of individual components total 2.

A second example is the reaction of 2-bromo-2-methylpropane with hydroxide ions:

$$(CH_3)_3CBr(aq) + OH^-(aq) \rightarrow (CH_3)_3COH(aq) + Br^-(aq)$$

Experimental data show that the rate equation is **rate = $k[(CH_3)_3CBr]$**

Note the following:

- The reaction is first order with respect to $(CH_3)_3CBr$.
- It is zero order with respect to OH^-, as $[OH^-]$ does not appear in the rate equation.
- The overall order of reaction is first order.

Calculations involving rate equations

It is possible to calculate k by rearranging a rate equation. The units of k must also be calculated as they differ from equation to equation.

Model answer: Calculating the value of k

The rate equation for the reaction of chloroethane with hydroxide ions is determined to be:

$$\text{rate} = k[C_2H_5Cl][OH^-]$$

Calculate k when $[C_2H_5Cl] = 0.002 \, \text{mol dm}^{-3}$, $[OH^-] = 0.010 \, \text{mol dm}^{-3}$, and rate $= 8.75 \times 10^{-2} \, \text{mol dm}^{-3} \text{s}^{-1}$. Give the units of k.

Rearrange the rate equation with k as the subject.

$$k = \frac{\text{rate}}{[C_2H_5Cl][OH^-]}$$

Substitute the values, then cancel units from the top and bottom.

$$k = \frac{8.75 \times 10^{-2} \, \text{mol dm}^{-3} \text{s}^{-1}}{0.002 \, \text{mol dm}^{-3} \times 0.010 \, \text{mol dm}^{-3}}$$

$$k = 4375 \frac{\cancel{\text{mol dm}^{-3}} \, \text{s}^{-1}}{\cancel{\text{mol dm}^{-3}} \times \text{mol dm}^{-3}}$$

$$k = 4375 \, \text{mol}^{-1} \text{dm}^3 \text{s}^{-1}$$

Rate constant k and temperature changes

$$\text{rate} = k[A]^m[B]^n$$

Increasing the temperature increases the rate of a chemical reaction. Since increasing the temperature does not affect [A] or [B], a rise in temperature must increase the value of the rate constant, k.

The Arrhenius equation

The Arrhenius equation is $k = Ae^{-\frac{E_a}{RT}}$, where

- k is the rate constant
- T is the temperature in kelvin
- R is the gas constant $8.314 \, \text{J K}^{-1} \text{mol}^{-1}$
- E_a is the activation energy in joules per mole
- e is a mathematical constant.
- A is the frequency factor.

A, e, and R are constants, but E_a and T can vary. It is possible to use the Arrhenius equation to calculate E_a and A by taking the natural logarithm of both sides of the equation:

$$\ln k = \ln A - \frac{E_a}{RT}$$

This can be rearranged as $\ln k = -\frac{E_a}{R} \times \frac{1}{T} + \ln A$, which is in the form of the equation of a straight line, $y = mx + c$.

Therefore, plotting a graph of $\ln k$ on the y-axis against $\frac{1}{T}$ on the x-axis gives a straight line. The gradient of the line is $-\frac{E_a}{R}$ and the intercept is $\frac{1}{T}$.

Summary questions

1   Identify the orders of reaction with respect to each reactant, and the overall order of reaction, in the rate equation rate $= k[A]^2[B]$. (*3 marks*)

2   Explain how the order of reaction can be determined from a rate–concentration graph. (*3 marks*)

3   For a particular reaction, rate $= k[X]^2$. Calculate the value of k, including units, when $[X] = 0.01 \, \text{mol dm}^{-3}$ and rate $= 4.2 \times 10^{-3} \, \text{mol dm}^{-3} \text{s}^{-1}$. (*3 marks*)

Revision tip

Catalysts are not classed as reactants in chemical equations but they **may** appear in a rate equation.

Revision tip

Remember: $\frac{1}{mol \, dm^{-3}} = \text{mol}^{-1} \, \text{dm}^3$

Revision tip

Increasing the temperature increases the magnitude of k.

Revision tip

The frequency factor A includes factors such as the frequency of collisions and their orientation.

Synoptic link

You met the gas constant R when you learned about the ideal gas equation $pV = nRT$ in Topic 1.5, Calculations involving gases.

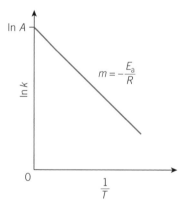

▲ **Figure 1** *Graph of* ln k *against* $\frac{1}{T}$.

Revision tip

In order to plot a graph of $\ln k$ against $\frac{1}{T}$ it is necessary to determine the rate constant k at several different temperatures.

10.7 Finding the order of reaction with experiments

Specification reference: CI(b), CI(c), CI(e)

Experimental determination of reaction rate

It is not possible to deduce the rate equation from the balanced chemical equation for a reaction. Instead, it must be determined from a series of experiments.

The concentrations of the reactants [A], [B], etc., are varied one at a time, and the rate of reaction for those particular conditions is determined. The rate can be measured by measuring the change in a particular property, such as colour or pH.

The **initial rate** can be measured by plotting a graph of the change of this property over time, then drawing a tangent at time $t = 0$. The gradient of the tangent gives you the rate, as shown in the example below.

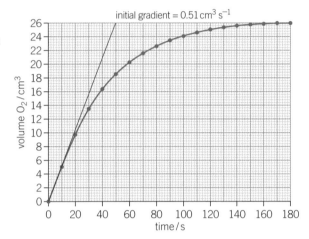

▲ **Figure 1** *Drawing a tangent at time t = 0 to determine initial rate*

The progress curve method can be used instead. A graph showing the change in concentration of a reactant is plotted over the course of the reaction. A series of tangents are taken at different concentrations. The gradient gives the rate of reaction at different times, as shown below.

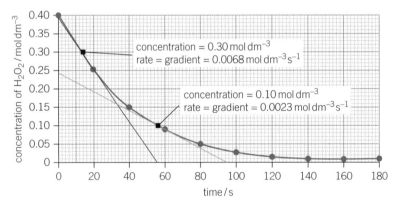

▲ **Figure 2** *The progress curve method*

Deducing the order of reaction for each reactant

Once a series of experiments are carried out, you will know how changing the concentration of a given reactant affects the rate of reaction. It is then possible to deduce the order of reaction:

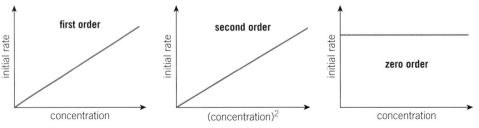

▲ **Figure 3** *The effect of concentration on rate for first, second, and zero order reactants (note concentration² as the variable on the x axis in the second order graph)*

Remember:

- If the rate is doubled when concentration is doubled (rate $\propto$ [A]), it is first order.
- If the rate is quadrupled when concentration is doubled (rate $\propto$ $[A]^2$), it is second order.
- If changing the concentration does not affect the rate, it is zero order.

Model answer: Deducing orders of reaction

Table 1 shows some data for the reaction of nitrogen(II) oxide and oxygen at 40 °C.

▼ **Table 1** *Results from Experiments 1–3*

Experiment	Initial NO concentration (mol dm^{-3})	Initial O$_2$ concentration (mol dm^{-3})	Initial rate (mol dm^{-3} s^{-1})
1	0.20	0.20	1.2×10^{-4}
2	0.20	0.40	2.4×10^{-4}
3	0.40	0.20	4.8×10^{-4}

Deduce the orders of reaction with respect to each reactant and calculate the value of k, including its units.

> Compare Experiments 1 and 2 – the rate doubles.

Doubling the concentration of O$_2$ doubles the rate of reaction. Therefore it is first order with respect to oxygen.

Doubling the concentration of NO quadruples the rate of reaction. Therefore it is second order with respect to NO.

> Compare Experiments 1 and 3 – the rate quadruples when the concentration is doubled.

The rate equation is rate = k [NO]^2[O$_2$].

The value of $k = \dfrac{\text{rate}}{[\text{NO}]^2[\text{O}_2]} = \dfrac{1.2 \times 10^{-4}\,\text{mol dm}^{-3}\text{s}^{-1}}{(0.20\,\text{mol dm}^{-3})^2 \times (0.20\,\text{mol dm}^{-3})}$

> Substitute the values from any of the experiments in the table.

$k = \dfrac{1.2 \times 10^{-4}\,\text{mol dm}^{-3}\text{s}^{-1}}{(0.008\,\text{mol}^3\,\text{dm}^{-9})} = 0.015\,\text{mol}^{-2}\,\text{dm}^6\,\text{s}^{-1}$

> Cancel units from the top and the bottom.

The half-lives method

A rate–concentration graph shows how the concentration of a reactant changes over time. It is possible to use such a graph to determine the successive half-lives of the reactant, and this can be used to determine the order of reaction with respect to that reactant.

During the successive half-lives, the concentration decreases to half its original value, then one-quarter and then one-eighth. If the half-lives are a constant time, as in the graph below, the reaction is first order with respect to that reactant. Otherwise the reaction is second or zero order.

Key term

Half-life ($t_{1/2}$): The time taken for the concentration of a reactant to halve.

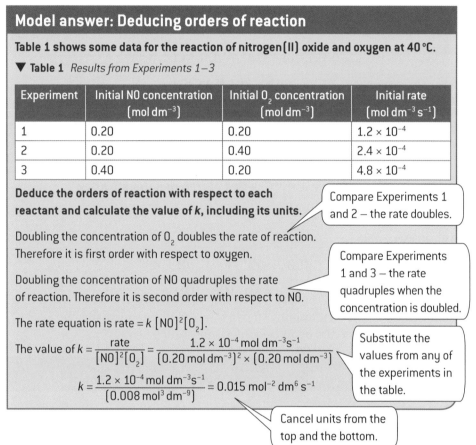

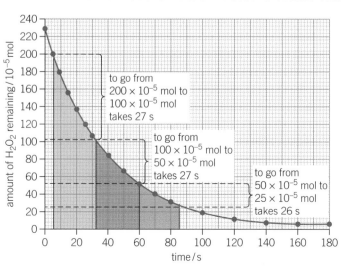

▲ **Figure 4** *Calculating half-life from a graph*

Reaction mechanisms

By studying rate equations and orders, chemists can deduce a mechanism for a reaction. This describes the steps involved in the chemical reaction. The slowest step is known as the **rate-determining step**.

Rate-determining step

The rate equation tells us which particles are involved in the rate-determining step. For example:

$$(CH_3)_3CBr + OH^- \rightarrow (CH_3)_3COH + Br^-$$

$$\textbf{rate} = \textit{k}\textbf{[(CH}_3\textbf{)}_3\textbf{CBr]}$$

The reaction is first order with respect to 2-bromo-2-methylpropane and zero order with respect to hydroxide ions. Therefore the rate-determining step involves only 2-bromo-2-methylpropane. It involves the heterolytic fission of the C–Br bond to form an intermediate:

$$(CH_3)_3CBr \rightarrow (CH_3)_3C^+ + Br^- \qquad \text{rate-determining step}$$

$$(CH_3)_3C^+ + OH^- \rightarrow (CH_3)_3COH \qquad \text{fast}$$

Another example is:

$$C_2H_5Br + OH^- \rightarrow C_2H_5OH + Br^-$$

$$\textbf{rate} = \textit{k}\textbf{[C}_2\textbf{H}_5\textbf{Br][OH}^-\textbf{]}$$

In this case the rate-determining step involves both bromoethane and hydroxide ions. It occurs by nucleophilic substitution.

Key term

Intermediate: A chemical formed and then destroyed during the course of a reaction.

Key term

Rate-determining step: The slowest step of a reaction mechanism.

Summary questions

1 Sketch a graph showing successive half-lives for a first order reaction. *(3 marks)*

2 Use the data below to determine the rate equation for the acid-catalysed reaction A + B $\rightarrow$ products. *(3 marks)*

[A] (mol dm^{-3})	[B] (mol dm^{-3})	[H$^+$] (mol dm^{-3})	Rate (mol dm^{-3} s^{-1})
0.01	0.01	0.01	4
0.02	0.01	0.01	8
0.02	0.02	0.02	16
0.02	0.02	0.04	32

3 For the reaction in question 2, a student proposes a two-step mechanism:
Slow step: A + H$^+$ $\rightarrow$ AH$^+$
Fast step: AH$^+$ + B $\rightarrow$ products
Explain whether or not this mechanism is consistent with the rate equation. *(1 mark)*

10.8 Enzymes

Specification reference: PL(f), PL(g)

What are enzymes?

Enzymes are proteins that act as catalysts in the body. An enzyme reacts with a substrate, and each enzyme has a high **specificity** for a particular substrate.

The active site of an enzyme is a region of the tertiary structure which is an exact fit for the structure of the substrate. All enzymes have an **active site**, where the tertiary structure of the enzyme exactly matches the structure of its substrate. The substrate binds to the active site, forming an enzyme–substrate complex. A reaction can then occur, after which the products leave the active site.

Key term

Substrate: A molecule that fits into the active site of an enzyme, and reacts with it.

Key term

Active site: A cleft in the enzyme surface where substrate molecules can bind and react.

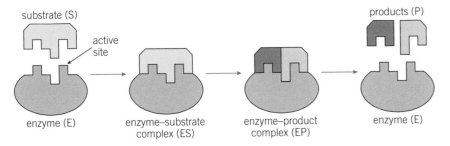

▲ **Figure 1** *A model of enzyme catalysis*

Sometimes a molecule that is a similar shape to the substrate binds strongly to the active site. This means that the usual substrate cannot enter, and the usual reaction does not occur. This is known as **competitive inhibition**.

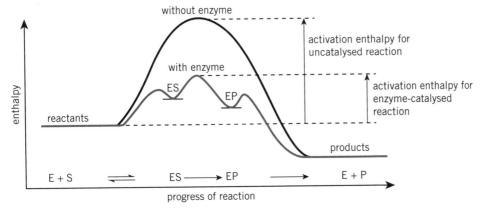

▲ **Figure 2** *How an enzyme provides a route of lower activation energy*

Kinetics of enzyme reactions

When the concentration of substrate is low, the rate equation is **rate = k[E][S]**. Under these conditions there are plenty of active sites for the substrate, so doubling the substrate concentration doubles the rate of reaction.

However when the substrate concentration is higher, at any given time all the active sites are occupied. The reaction is then zero order with respect to the substrate, as increasing the concentration of the substrate any further has no effect on the rate. Under these conditions, the rate equation is **rate = k[E]**.

Synoptic link

Rate equations are covered in Topic 10.6, Rate equation of a reaction.

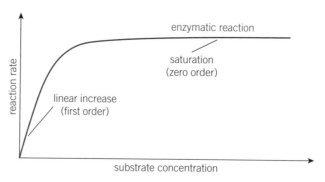

▲ **Figure 3** *How rate of reaction of an enzyme reaction varies with substrate concentration*

Key term

Denatured: When the active site of the enzyme changes shape due to changes in pH, or increased temperature, the enzyme is denatured, and the substrate no longer fits.

Changes to the shape of the active site

The shape of the active site depends on the tertiary structure of the protein that makes up the enzyme. The tertiary shape is determined by interactions between the amino acid side groups, such as hydrogen bonding and ionic bonds. Increasing the temperature can break hydrogen bonds and pH changes can disrupt ionic interactions; this leads to a change in the shape of the active site.

Consequently the enzyme is **denatured**. The substrate can no longer fit into the active site and the enzyme loses its activity.

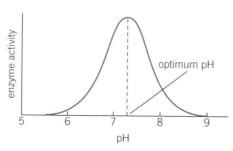

▲ **Figure 4** *Graph showing how the rate of a typical enzyme reaction varies with pH*

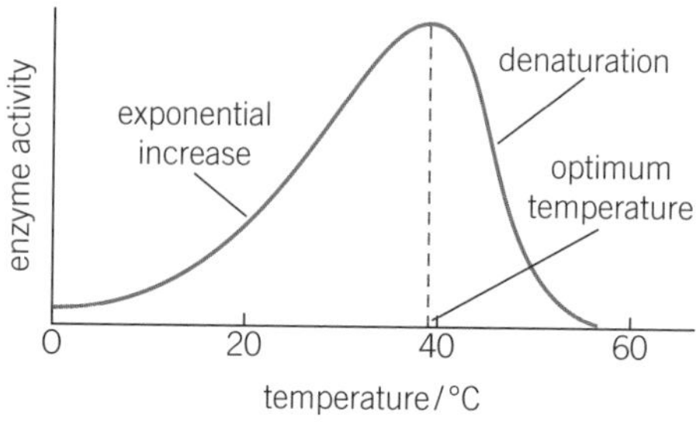

▲ **Figure 5** *Graph showing how the rate of an enzyme reaction varies with temperature*

Summary questions

1 What is: **a** a substrate **b** an inhibitor? *(2 marks)*

2 Why are enzymes denatured at high/low pH and at high temperatures? *(2 marks)*

3 Why does the rate equation for enzyme kinetics depend on the substrate concentration? *(1 mark)*

Experimental techniques

Specification reference: OZ (f)

Measuring rate of reaction

'Rate of reaction' is a measure of how quickly a product is made, or a reactant is used up. To determine the rate of a reaction you need to measure the concentration of the reactant or product in question, or, a property that changes during the reaction and that is proportional to the concentration of the substance.

$$rate\ of\ reaction = \frac{Change\ in\ property}{time\ taken}$$

The unit of rate of reaction is $mol\,dm^{-3}\,s^{-1}$.

Measuring volumes of gases evolved

If a reaction produces a gas that can be collected, the volume produced can be used to follow the reaction rate. This can be done using a gas syringe, or by collecting the gas over water. Collecting gas over water is only suitable for gases that are insoluble in water, such as H_2 or O_2 (but not CO_2).

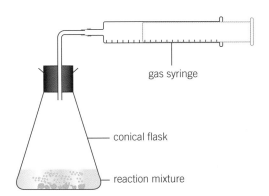

▲ **Figure 1** *Measuring rate of reaction using a gas syringe*

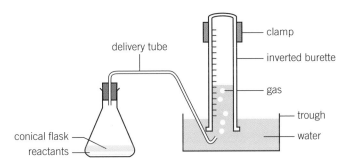

▲ **Figure 2** *Collecting gas over water*

Measuring mass changes

Another way of measuring the rate of a reaction that produces a gas is by recording the mass lost (in the form of gas) from the reaction.

pH measurement

Some reactions involve H⁺(aq) or OH⁻(aq) ions as a reactant, or produce them as a product. During these reactions, the concentration of the ions will change and the pH will consequently change. Measuring the rate of change of pH of the reaction mixture is a way of following the rate of reaction.

Colorimetry

A colorimeter measures the change in colour of a reaction. Many reactions result in a change of colour and the changing intensity of colour can be followed using a colorimeter.

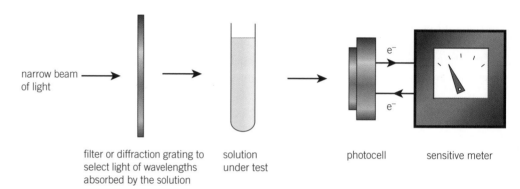

narrow beam of light

filter or diffraction grating to select light of wavelengths absorbed by the solution

solution under test

photocell

sensitive meter

▲ **Figure 3** *A simple colorimeter*

Chemical analysis

Chemical analysis involves taking samples of the reaction mixture at regular intervals, and 'quenching' it before analysis. Quenching stops the reaction in the sample, or significantly slows it down.

For example, reactions involving an acid catalyst can be quenched by the addition of sodium hydrogen carbonate which neutralises the acid catalyst, effectively stopping the reaction.

Other quenching techniques include suddenly decreasing the temperature using an ice bath, or adding a large quantity of distilled water to dilute the substances. Both of these techniques decrease the rate of reaction significantly allowing time for the sample to be analysed by titration or other means.

Chapter 10 Practice questions

1 What is the best explanation of why increasing the concentration increases the rate of reaction?

    A The particles are closer together, so they collide less frequently.

    B The particles are further apart, so there are fewer collisions.

    C The particles are closer together, so they collide more frequently.

    D The particles are further apart, so there are more collisions.   (*1 mark*)

2 Which of the following methods of measuring rate of reaction would be suitable for the reaction: $CuSO_4 + Zn \rightarrow ZnSO_4 + Cu$?

    A colorimetry   **B** measuring pH changes

    C measuring mass changes   **D** measuring volume changes   (*1 mark*)

3 What is the best definition of activation enthalpy?

    A The minimum kinetic energy required for a reaction to occur.

    B The energy released by a reaction.

    C The energy absorbed by the particles during a reaction.

    D The proportion of molecules with sufficient energy to react.   (*1 mark*)

4 How do catalysts increase the rate of reaction?

    A They decrease the activation energy of the reaction.

    B They increase the activation energy of the reaction.

    C They increase the energy of the colliding particles.

    D They provide an alternative pathway of lower activation energy. (*1 mark*)

5 The following stages explain the action of heterogeneous catalysis:

    1 bonds in the reactant break

    2 new bonds form

    3 reactant molecules are adsorbed onto the surface of the catalyst

    4 the products are released from the catalyst surface and diffuse away
       What is the correct order of these stages?

      A 1 – 3 – 4 – 2,   **B** 3 – 2 – 1 – 4

      C 4 – 2 – 1 – 3,   **D** 3 – 1 – 2 – 4   (*1 mark*)

6 In a petrochemical refinery, the reforming process converts straight-chain alkanes into ring compounds. Hot alkane vapours are passed over a catalyst. The catalyst used in the process is platinum metal, which is finely dispersed on a solid support of aluminium oxide.

    a Name the type of catalyst used in the reforming process.   (*1 mark*)

    b Explain why high temperatures are used.   (*2 marks*)

    c Explain why the catalyst is finely dispersed.   (*2 marks*)

    d Explain why the catalyst speeds up the rate of reaction.   (*2 marks*)

    e Describe how the catalyst works.   (*4 marks*)

7 Nitrogen monoxide, NO, is one of a number of radicals that catalyse the breakdown of ozone in the stratosphere.

    a Explain whether NO is a heterogeneous or homogeneous catalyst.

                                       (*1 mark*)

    b Explain how NO molecules can act as catalysts for the breakdown of ozone, referring to intermediates, activation energy, and reaction pathways.   (*3 marks*)

    c Explain why the breakdown of ozone occurs faster in the hot, top layer of the stratosphere.   (*3 marks*)

8 What are the units of rate of reaction?

    A there are no units   **C** s^{-1}

    B it depends on the rate equation   **D** $mol\,dm^{-3}\,s^{-1}$ *(1 mark)*

9 What is the overall order of reaction if rate = $k\,[X]^2[Y]$?

    A −1   **C** 2

    B 1   **D** 3   *(1 mark)*

10 What conclusions can be drawn from the rate equation rate = $k\,[CH_3Br][OH^-]$?

    1 The reaction is first order with respect to both reactants.

    2 The reaction is first order overall.

    3 The rate determining step is likely to involve CH_3Br and OH^-.

        A 1 and 2 only   **C** 2 and 3 only

        B 1 and 3 only   **D** 3 only   *(1 mark)*

11 The statements below describe competitive inhibition of an enzyme by a molecule. Which is the most logical order of the statements?   *(1 mark)*

    1 The molecule has a similar shape to the substrate.

    2 The molecule binds to the active site of the enzyme.

    3 The usual reaction cannot occur.

    4 The usual substrate cannot bind to the active site.

        A 4 − 2 − 3 − 1   **C** 1 − 2 − 3 − 4

        B 1 − 2 − 4 − 3   **D** 3 − 2 − 4 − 1

12 What is the rate equation for an enzyme reaction when the substrate concentration is high?

    A Rate = k **C** Rate = $k\,[S]$

    B Rate = $k\,[E]$ **D** Rate = $k\,[E]\,[S]$ *(1 mark)*

13 A particular reaction has the rate equation rate = $k\,[M]\,[N]^2$, where M and N are the reactants.

    a What will be the effect on the rate of doubling [M]?   *(1 mark)*

    b What will be the effect on the rate of doubling both [M] and [N]? *(1 mark)*

    c Calculate the value of k, including its units, when rate = $0.025\,mol\,dm^{-3}\,s^{-1}$ and [M] and [N] are both equal to $0.1\,mol\,dm^{-3}$.

                                                            (3 marks)

    d What is the effect on k of increasing the temperature?   *(1 mark)*

    e How will the half-lives of [M] and [N] differ in this reaction? (Assume that, in each case, all other reactants are in excess.)   *(2 marks)*

14 Look at the table of data for the reaction: P + Q → S + T

[P] ($mol\,dm^{-3}$)	[Q] ($mol\,dm^{-3}$)	Rate ($mol\,dm^{-3}\,s^{-1}$)
0.01	0.01	0.75×10^{-3}
0.02	0.01	1.50×10^{-3}
0.03	0.01	2.25×10^{-3}
0.03	0.02	4.50×10^{-3}

    a What is the order with respect to P and with respect to Q?   *(2 marks)*

    b Explain how the initial rate could be determined from a graph of concentration [P] against time.   *(1 mark)*

    c Comment on whether the following reaction mechanism, involving an intermediate R, is supported by the rate equation.   *(1 mark)*

              Slow step: P → R + S

              Fast step: R + Q → T

11.1 Periodicity

Specification reference: EL(f), EL(m), EL(n), [EL(q)]

Periodicity

Periodicity is exhibited when there is a *regular pattern* in a property as you go across a period and the regular pattern is *repeated* in other periods.

Melting point – an example of periodicity

The pattern is for the melting point to increase and then decrease across a period. This pattern is repeated in more than one period. There is periodicity in melting points.

Melting point increases and then decreases across Period 2 (atomic numbers 3–10). There is the same increase and decrease in Period 3 (atomic numbers 11–18).

Metals are on the left-hand side of a period and have metallic bonding, with high melting points. Some Group 4 elements have covalent network structures and so tend to have high melting points. Other non-metals found on the right-hand side of a period have simple molecular structures and so tend to have lower melting points.

How does ionisation enthalpy vary across a period?

First ionisation enthalpy

The first ionisation enthalpy is the energy needed to remove one electron from each of *one mole* of *isolated gaseous* atoms of an element. *One mole* of *gaseous ions* with one positive charge is formed:

$$X(g) \rightarrow X^+(g) + e^-$$

Figure 1 shows the periodic pattern of first ionisation enthalpy for elements 1–56.

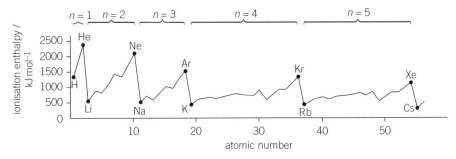

▲ **Figure 1** *First ionisation enthalpy of elements 1–56*

You should notice the following patterns:

- The general trend is one of *increasing* first ionisation enthalpy across each period.
- The elements at the peaks are all in Group 0. It is difficult to remove an electron from these atoms with full outer shells and the elements are all very unreactive.
- The elements at the troughs are all in Group 1. These atoms have only one outer shell electron and are relatively easy to ionise. They are all very reactive elements.

Ionisation enthalpies and reactivity

s-block elements are more reactive than p-block elements, because the formation of an M^+ or M^{2+} ion only requires input of energy equivalent to the first ionisation enthalpy (for M^+) or first and second ionisation enthalpy (for M^{2+}). Loss of electrons from p-block elements is more difficult.

Electronic configurations of s- and p-block elements

Group 1 and 2 elements all have one or two electrons, respectively, in their outermost sub-shell, which is an s-orbital. These are known as the s-block elements. For example, magnesium is $1s^2 2s^2\ 2p^6 3s^2$.

Group 3, 4, 5, 6, 7, and 0 elements all have three, four, five, six, seven, or eight electrons, respectively, in their outermost sub-shell, which are p-orbitals. They are known as the p-block elements. For example, selenium is $1s^2\ 2s^2\ 2p^6\ 3s^2\ 3p^6\ 4s^2\ 3d^{10}\ \mathbf{4p^4}$.

Elements in the d-block and f-block have electronic configurations where the outermost sub-shell is a d-orbital or an f-orbital. For example, nickel is in the d-block and is $1s^2\ 2s^2\ 2p^6\ 3s^2\ 3p^6\ 4s^2\ \mathbf{3d^8}$.

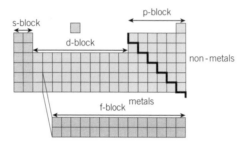

▲ **Figure 2** s-, p-, d-, and f-blocks of the periodic table

Summary questions

1 Predict the melting point of fluorine given the data below about other halogens. *(1 mark)*

Element	Melting point (°C)
Fluorine	?
Chlorine	−102
bromine	7
Iodine	114

2 Name the elements which have these electronic configurations. *(3 marks)*

 a $1s^2\ 2s^2\ 2p^5$

 b $1s^2\ 2s^2\ 2p^6\ 3s^2\ 3p^6\ 4s^2$

 c $1s^2\ 2s^2\ 2p^6\ 3s^2\ 3p^6\ 4s^2\ 3d^{10}$

3 Classify the elements in Question 2 as s-, p-, or d-block elements. *(3 marks)*

4 Given below are the values of the first four ionisation enthalpies of calcium and aluminium.

 1st I.E. (Mg) = 736 kJ mol⁻¹   1st I.E. (Al) = 577 kJ mol⁻¹

 2nd I.E. (Mg) = 1450 kJ mol⁻¹   2nd I.E. (Al) = 1820 kJ mol⁻¹

 3rd I.E. (Mg) = 7740 kJ mol⁻¹   3rd I.E. (Al) = 2740 kJ mol⁻¹

 4th I.E. (Mg) = 10 500 kJ mol⁻¹   4th I.E. (Al) = 11 600 kJ mol⁻¹

 a Use this information to explain why magnesium forms Mg^{2+} ions, whilst aluminium forms Al^{3+} ions. *(2 marks)*

 b Calculate the total energy required to form Mg^{2+} and Al^{3+}, and explain why magnesium is a more reactive element than aluminium. *(3 marks)*

11.2 The s-block: Groups 1 and 2

Specification reference: EL(p), EL(r), EL(u)

Reactions of the elements in Group 2

The Group 2 elements are Be, Mg, Ca, Sr, Ba, and Ra.

The metals react with water to give the metal hydroxide and hydrogen:

$$\text{metal} + \text{water} \rightarrow \text{metal hydroxide} + \text{hydrogen}$$
$$\mathbf{M}(s) + 2H_2O(l) \rightarrow \mathbf{M}(OH)_2(s) + H_2(g)$$

As is typical for metals, the reactions become more vigorous as you go down the group.

The oxides

The oxides react with water to produce an alkaline solution of the hydroxide:

$$\text{metal oxide} + \text{water} \rightarrow \text{metal hydroxide}$$
$$\mathbf{M}(s) + H_2O(l) \rightarrow \mathbf{M}(OH)_2(s)$$

The oxides react with acids and therefore act as bases, accepting a proton:

$$\text{metal oxide} + \text{acid} \rightarrow \text{salt} + \text{water}$$
$$\mathbf{M}O(s) + H_2SO_4(aq) \rightarrow \mathbf{M}SO_4(aq) + H_2O(l)$$
$$\mathbf{M}O(s) + 2HNO_3(aq) \rightarrow \mathbf{M}(NO_3)_2(aq) + H_2O(l)$$

The hydroxides

The hydroxides become more soluble as you go down the group. The solutions produced are alkaline, since the solutions contain $OH^-(aq)$ (pH > 7).

The hydroxides react with acids to produce a salt and water:

$$\text{metal hydroxide} + \text{acid} \rightarrow \text{salt} + \text{water}$$
$$\mathbf{M}(OH)_2(s) + 2HCl(aq) \rightarrow \mathbf{M}Cl_2(aq) + 2H_2O(l)$$

The carbonates

The carbonates become less soluble as you go down the group.

The carbonates undergo thermal decomposition on heating to give the metal oxide and carbon dioxide:

$$\text{metal carbonate} \rightarrow \text{metal oxide} + \text{carbon dioxide}$$
$$\mathbf{M}CO_3(s) \rightarrow \mathbf{M}O(s) + CO_2(g)$$

Thermal stability increases down the group. This means that barium carbonate breaks down at a higher temperature than magnesium carbonate.

Common misconception: Thermal decomposition

Avoid incorrectly stating that barium carbonate will be more likely to break down on heating as barium is more reactive.

The M^{2+} ions get larger as you go down Group 2, so their charge density is lower. Ions such as Ba^{2+}, with a low charge density, polarise the carbonate ion less than ions such as Mg^{2+}, which has a high charge density. The more polarised the carbonate ion, the more likely it is to break up during thermal decomposition into an oxide ion and CO_2.

Revision tip
In these equations, M is any Group 2 metal.

Revision tip
In these equations, the state of M(OH)$_2$ could be (s) or (aq), depending on the solubility of the metal hydroxide.

Revision tip
Remember nitric acid is monobasic, so two moles of HNO$_3$ react with each mole of the oxide.

Revision tips
Group 2 elements have similar chemical reactions, as they all have two electrons in their outer shell.

Group 2 metals are less reactive than Group 1 metals. For example, Na reacts with cold water, whereas Mg reacts with steam.

Revision tips
Oxides and hydroxides of Group 2 metals are basic.

The reaction of oxides and hydroxides with acids are very similar.

Key terms

Polarised ions: Large ions such as CO_3^{2-} can have their electron distribution altered by small, highly-charged ions. This is known as polarisation.

Charge density: Small, highly charged ions, such as Mg^{2+}, are said to have a high charge density. Larger ions with a lower charge, such as I^-, are said to have a low charge density.

Revision tip

If an ion is more polarised, it is more likely to break up on thermal decomposition. A similar pattern is seen with Group 2 nitrates.

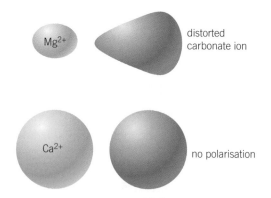

distorted carbonate ion

no polarisation

Summary questions

1 Write a balanced equation with state symbols for the reaction of strontium with water. *(1 mark)*

2 Write a balanced equation, with state symbols, for the thermal decomposition of calcium carbonate. *(1 mark)*

3 Which one in each pair is the most soluble?
  a $Mg(OH)_2$ or $Ba(OH)_2$ *(1 mark)*
  b $MgCO_3$ or $BaCO_3$ *(1 mark)*

4 Write a balanced equation, with state symbols, for the reaction of:
  a magnesium oxide with hydrochloric acid. *(1 mark)*
  b magnesium hydroxide and sulfuric acid. *(1 mark)*

11.3 The p-block: Group 7
Specification reference: ES(h), ES(i), ES(j), ES(k)

Properties of the halogens
Physical properties

You need to be able to recall the following physical properties of the halogens:

	Fluorine (F_2)	Chlorine (Cl_2)	Bromine (Br_2)	Iodine (I_2)
Appearance and state at room temperature	pale yellow gas	green gas	dark red liquid	shiny grey/black solid
Volatility	gas	gas	liquid quickly forms brown gas on warming	sublimes on warming to give a purple vapour
Solubility in water	reacts with water	slightly soluble to give pale green solution	slightly soluble to give orange-yellow solution	barely soluble, gives a brown solution
Solubility in organic solvents	soluble	soluble to give a pale green solution	soluble to give a orange/brown/red solution	soluble to give a violet solution

Common misconception: Formulae of the halogens

Remember that the elements are diatomic molecules, so should be written as F_2, etc.

All the halogens exist as elements as *diatomic* non-polar molecules (F_2, Cl_2, Br_2 and I_2). The *intramolecular* bonds are covalent and the *intermolecular* bonds are instantaneous dipole–induced dipole bonds. Fluorine is the most volatile halogen, as it has the smallest molecules with the fewest electrons. As the size of the molecule and number of electrons increases, so does the strength of the intermolecular bonds. This explains why the physical state of the halogens changes from gas to liquid to solid as you go down the group.

The elements become darker in colour as you go down the group.

Reactivity and redox reactions of the halogens

Halogens are all reactive. They tend to remove electrons from other elements – they are oxidising agents:

$$X_2 + 2e^- \rightarrow 2X^-$$

Halogen oxidation state 0 is reduced to oxidation state −1.

Halogens and halide ions undergo displacement reactions. For example, when a solution containing chlorine is added to a solution containing iodide ions, a brown colour appears as iodine is produced:

$$Cl_2(aq) + 2I^-(aq) \rightarrow I_2(aq) + 2Cl^-(aq)$$

This occurs because chlorine is a stronger oxidising agent than iodine. Chlorine causes the iodine to be oxidised (the oxidation state of iodine changes from −1 to 0 in the half-equation $2I^- \rightarrow I_2 + 2e^-$). Chlorine has been reduced (its oxidation state changes from 0 to −1 in the half-equation $Cl_2 + 2e^- \rightarrow 2Cl^-$).

- Fluorine is the strongest oxidising agent in Group 7. Fluorine displaces chlorine, bromine, and iodine.

- Chlorine displaces bromine and iodine.

- Bromine displaces iodine.
- Iodine does not displace bromine, chlorine, or fluorine.

An organic solvent such as hexane can be added to detect the halogen present at the end of the reaction. For example, in the reaction of chlorine with iodide ions shown above, the hexane layer would be violet due to the presence of iodine.

Precipitation reactions with silver ions

The general reaction is:

$$Ag^+(aq) + X^-(aq) \rightarrow AgX(s)$$

Chloride ions give a white precipitate of silver chloride, AgCl, which is soluble in dilute ammonia. Bromide ions give a cream precipitate of silver bromide, AgBr, which is soluble in concentrated ammonia. Iodide ions give a yellow precipitate of silver iodide, AgI, which is insoluble in ammonia.

These precipitation reactions can be used as tests to identify halides in solution.

Preparation of hydrogen halides

Hydrogen chloride

Hydrogen chloride is prepared by adding concentrated sulfuric acid to a solid chloride (for example sodium chloride)

$$H_2SO_4(l) + NaCl(s) \rightarrow HCl(g) + NaHSO_4(g)$$

Steamy fumes of hydrogen chloride are observed.

Hydrogen bromide and hydrogen iodide

It is not possible to prepare pure samples of hydrogen bromide and hydrogen iodide by adding concentrated sulfuric acid to solid halides.

Concentrated sulfuric acid is able to oxidise bromide and iodide ions to form halogens. The sulfuric acid is reduced to sulfur dioxide (SO_2) or hydrogen sulfide (H_2S).

Substances mixed	Observations	Gaseous products formed
Concentrated sulfuric acid and sodium bromide	Orange vapour, steamy fumes	HBr, Br_2, SO_2
Concentrated sulfuric acid and sodium iodide	Purple vapour, steamy fumes	HI, I_2, H_2S

Because of these reactions, hydrogen bromide and hydrogen iodide are prepared by adding concentrated phosphoric acid (H_3PO_4) to sodium bromide or sodium iodide.

Properties of hydrogen halides

Thermal stability

Hydrogen halides can undergo thermal decomposition to form hydrogen and a halogen.

$$2HI(g) \rightarrow H_2(g) + I_2(g)$$

The thermal stability of the hydrogen halides decreases down Group 7. Hydrogen iodide decomposes most easily, while hydrogen chloride does not decompose.

Revision tip

Look back at the colours of halogens in organic solvents in the table above.

Revision tip

Remember to include state symbols in precipitation reactions.

Key term

Precipitate: An insoluble solid formed when two ionic solutions react together.

Revision tip

You may need to be able to balance the equation for the redox reactions involving sulfuric acid and bromide or iodide ions by using changes in oxidation state.

Synoptic link

The strategy for balancing redox reactions using changes in oxidation states was described in Topic 9.1, Oxidation and reduction.

Revision tip

You should be able to link the trend in thermal stability to the bond enthalpy of the hydrogen –halogen bond. The H-I bond is the weakest bond and so hydrogen iodide decomposes most easily.

Reaction with ammonia

All hydrogen halide gases react with ammonia vapour to form a white cloud of ammonium halide.

$$HCl(g) + NH_3(g) \rightarrow NH_4Cl(s)$$

Reaction with concentrated sulfuric acid

Hydrogen halides react in a similar way to halide ions. With hydrogen bromide, bromine and sulfur dioxide are formed. With hydrogen iodide, iodine and hydrogen sulfide are formed.

Summary questions

1 Describe how the appearance of iodine differs depending on the conditions present. (*4 marks*)

2 Write an ionic equation for the reaction of potassium iodide solution with silver nitrate. State what you would see. (*2 marks*)

3 Chlorine gas is bubbled through potassium bromide solution.
   a Write an ionic equation for the reaction that occurs. (*1 mark*)
   b Describe and explain what you would see. (*2 marks*)
   c Describe and explain what you would see if cyclohexane (an organic solvent) was added after the reaction. (*2 marks*)
   d Explain why no reaction would occur if bromine were added to potassium chloride solution. (*1 mark*)

11.4 The d-block

Oxidation states of transition metals

Transition metals often have variable oxidation states. They display a range of different oxidation states, which often are different in colour.

▼ **Table 1** *The common oxidation states of iron and of copper*

Oxidation state	Ion	Electron configuration	Colour of aqueous ion
Iron(II)	Fe^{2+}	$1s^2\,2s^2\,2p^6\,3s^2\,3p^6\,3d^6$	green
Iron(III)	Fe^{3+}	$1s^2\,2s^2\,2p^6\,3s^2\,3p^6\,3d^5$	orange/brown
Copper(I)	Cu^+	$1s^2\,2s^2\,2p^6\,3s^2\,3p^6\,3d^{10}$	N/A
Copper(II)	Cu^{2+}	$1s^2\,2s^2\,2p^6\,3s^2\,3p^6\,3d^9$	blue

Chromium and copper

In the ground state of an atom, electrons are arranged to give the lowest total energy. The energies of the 3d and 4s orbitals are very close together in period 4, and this leads to the electron configurations in Cr and Cu being different to what you might expect. In chromium, putting one electron in each 3d and 4s orbital gives a lower energy than having the usual two in 4s. In copper, putting two electrons in the 4s orbital would give a higher energy than filling up the 3d orbital.

- The electron configuration of Cr is **$1s^2\,2s^2\,2p^6\,3s^2\,3p^6\,3d^5\,4s^1$** (not $3d^4\,4s^2$).
- The electron configuration of Cu is **$1s^2\,2s^2\,2p^6\,3s^2\,3p^6\,3d^{10}\,4s^1$** (not $3d^9\,4s^2$).

Identifying transition metal ions

Fe^{2+}, Fe^{3+}, and Cu^{2+} ions can be readily identified through test tube reactions as they form precipitates when aqueous sodium hydroxide is added:

Fe^{2+} (aq) + 2OH⁻ (aq) → $Fe(OH)_2$ (s)   Iron(II) hydroxide is a green precipitate.

Fe^{3+} (aq) + 3OH⁻ (aq) → $Fe(OH)_3$ (s)   Iron(III) hydroxide is a brown precipitate.

Cu^{2+} (aq) + 2OH⁻ (aq) → $Cu(OH)_2$ (s)   Copper(II) hydroxide is a blue precipitate.

If ammonia solution is added, the same reactions occur, as ammonia is a source of OH⁻ ions. However, on addition of excess ammonia, the pale blue precipitate of $Cu(OH)_2$ then dissolves to give a deep blue solution of a copper/ammonia complex ion, $[Cu(NH_3)_4(H_2O)_2]^{2+}$. Iron(II) hydroxide and iron(III) hydroxide do not form complexes with ammonia.

Complex ions

A complex ion contains a transition metal ion surrounded by a number of ligands. Complex ions are written using square brackets, for example: $[Fe(H_2O)_6]^{2+}$, $[NiCl_4]^{2-}$, $[Ni(CN)_4]^{2-}$, or $[Ag(NH_3)_2]^+$.

Monodentate, bidentate, and polydentate ligands

Revision tip

The electron configuration of elemental iron is $1s^2\,2s^2\,2p^6\,3s^2\,3p^6\,3d^6\,4s^2$. The electron configuration of elemental copper is $1s^2\,2s^2\,2p^6\,3s^2\,3p^6\,3d^{10}\,4s^1$.

Revision tip

Remember when d-block metals lose electrons, they are removed from the 4s sub-shell before they are removed from the 3d sub-shell.

Synoptic link

You learned about electron configuration in Topic 2.3, Shells, sub-shells, and orbitals.

Revision tip

Remember to include state symbols for precipitation reactions.

Key term

Complex ion: A transition metal ion surrounded by a number of ligands.

Key term

Ligand: A negatively charged ion, or a neutral molecule with a lone pair of electrons, surrounding a transition metal ion.

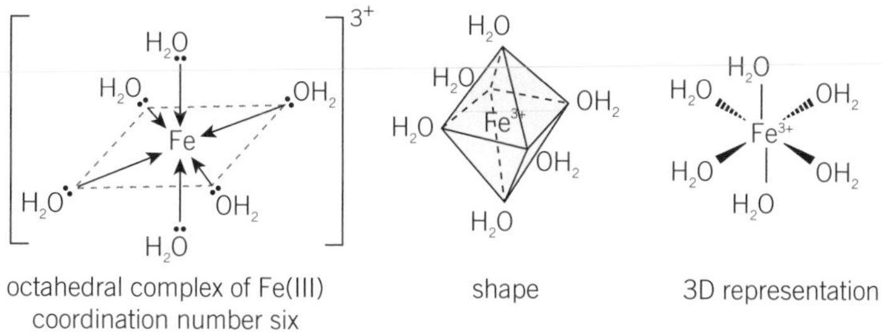

octahedral complex of Fe(III)
coordination number six

shape

3D representation

▲ **Figure 1** *Some examples of complex ions*

> **Revision tip**
>
> Complexes can have a positive charge, negative change, or they can have no charge.

> **Key term**
>
> **Monodentate ligand:** A ligand that attaches to a transition metal through one atom only.

Model answer: The charge of a complex ion

Work out the charge of the complex ion formed from a a copper(II) ion and six water molecules b a copper(II) ion and four chloride ions.

a The copper ion has a charge of 2+.
   The water molecules have no charge.

   The total charge is $(2+) + (6 \times 0) = 2+$

Therefore the charge of the complex ion is 2+.

b The copper ion has a charge of 2+.
   The chloride ions have a charge of 1−.

   The total charge is $(2+) + (4 \times 1-) = 2-$

Therefore the charge of the complex ion is 2−.

> **Key term**
>
> **Bidentate ligand:** A ligand with two atoms with lone pairs or negative charges, which forms two bonds to a metal ion.

Ligands attach to transition metal ions through dative covalent bonding with the lone pairs on the ligands. Some ligands, such as H_2O and NH_3 can only bond through a single atom, and are called **monodentate ligands**. Some ligands, such as the ethanedioate ion and the 1,2-diaminoethane molecule, can form two dative covalent bonds to the metal ion, and are known as **bidentate ligands**. **Polydentate ligands** can form several dative covalent bonds as they contain several atoms with lone pairs or negative charges. $EDTA^{4-}$ is an example of a polydentate ligand.

> **Key term**
>
> **Polydentate ligand:** A ligand which forms several bonds to a metal ion.

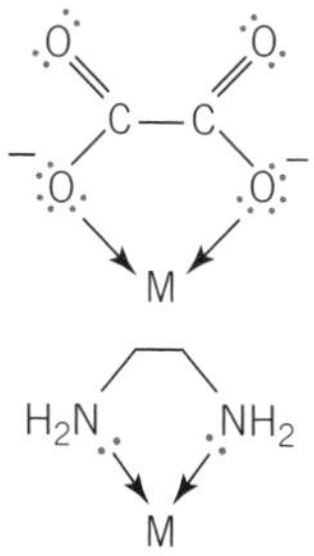

▲ **Figure 2** *Some bidentate ligands*

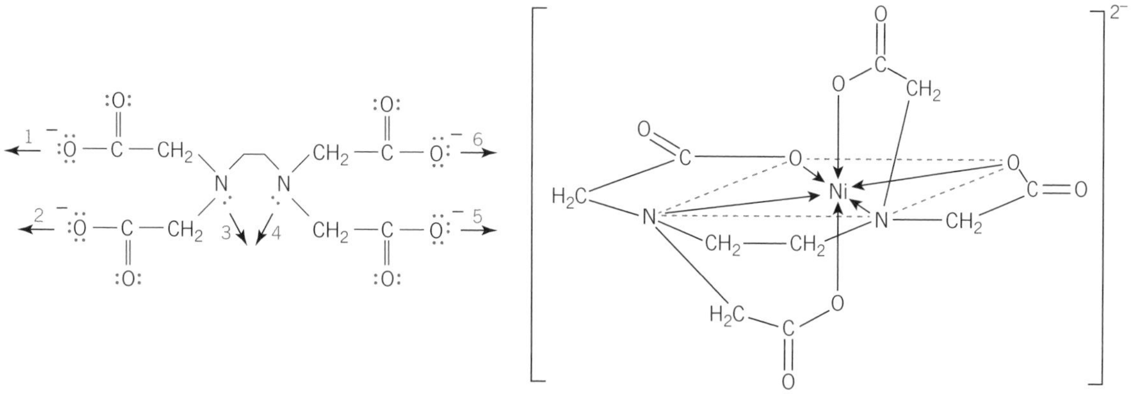

▲ **Figure 3** *A polydentate ligand (*$EDTA^{4-}$*)*

Coordination number

The number of bonds made between a metal ion and ligands is known as the **coordination number**. The most common coordination numbers are six and four. When a bidentate or polydentate ligand forms a complex, each metal–ligand bond is counted, so, for example, $[Ni(OOC-COO)_3]^{4-}$ has a coordination number of six.

> **Revision tip**
>
> $^-OOC-COO^-$ is an ethanedioate ion, a bidentate ligand. This is the only structure of a polydentate ligand you need to learn. Others will be given if needed.

▼ **Table 2** *The shapes of some complex ions*

Coordination number	Shape of complex	Example
6	octahedral	$[Fe(CN)_6]^{3-}$
4	tetrahedral	$[NiCl_4]^{2-}$
4	square planar	$[Ni(CN)_4]^{2-}$
2	linear	$[Ag(NH_3)_2]^+$

Revision tip

h is Planck's constant, 6.63×10^{-34} J s.

v is the frequency in s^{-1}.

Revision tip

Remember to mention the *gap* between energy levels (ΔE).

Synoptic link

You first learned about the relationship between frequency and energy in Topic 6.1, Light and electrons.

Revision tip

Transition metals *transmit* (or reflect) the complementary colour. They do not *emit* the complementary colour – they are not light sources.

Revision tip

Remember that the electrons *do not* fall back down, releasing light.

Colours of d-block ions

The presence of ligands causes the d sub-shell to split into higher and lower energy levels. When visible light is absorbed, electrons can be excited to a higher energy level. The frequency of light absorbed is proportional to the **gap** between the energy levels and is given by $\Delta E = hv$.

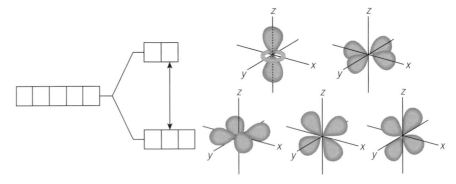

▲ **Figure 4** *Splitting of d-orbitals*

The complementary colour is transmitted, and is the colour we observe. For example, a Ti^{3+} ion absorbs yellow-green light, so it transmits the complementary colour, which is blue-violet. Therefore it appears a blue-violet colour.

Colorimetry

A colorimeter is used to find the concentration of a coloured solution. The absorbance of a solution is proportional to its concentration, so a more concentrated solution will absorb more strongly (see Experimental techniques).

Summary questions

1 Work out the charge of a complex ion formed from a cobalt(II) ion and four carbon monoxide molecules. *(1 mark)*

2 Explain how aqueous ammonia can be used to identify copper(II) ions. Give formulae for the species produced. *(4 marks)*

3 Explain how the presence of H_2O causes aqueous iron(III) ions to appear orange. *(4 marks)*

11.5 Nitrogen chemistry

Specification reference: CI(j)

Nitrogen compounds

Nitrogen forms a range of compounds and ions. Many of these are important in soil chemistry: oxides such as N_2O, NO, and NO_2, oxoanions such as NO_3^-, NO_2^- and ammonia, NH_3 and the ammonium ion NH_4^+.

Oxides of nitrogen

▼ **Table 1** *Formula and appearance of oxides of nitrogen*

Systematic name	Oxidation state of nitrogen	Molecular formula	Appearance
dinitrogen oxide	+1	N_2O	Colourless gas
nitrogen oxide	+2	NO	Colourless gas
nitrogen dioxide	+4	NO_2	Brown gas

Ammonia and ammonium ions

▼ **Table 2** *Bonding in ammonia and ammonium ions*

Name	Oxidation state of N	Formula	Dot-and-cross diagram	Shape of molecule or ion	Bond angle
Ammonia	−3	NH_3	H ⦁× N ×⦁ H or more simply H—N—H	*Trigonal pyramid*	107°
Ammonium	−3	NH_4^+	H ⦁× N ×⦁ H or more simply H—N⁺—H a dative covalent bond	*Tetrahedral*	$109\frac{1}{2}°$

The lone pair of electrons on the N atom of the ammonia molecule allows it to act as a base. So it accepts H^+ ions from acids to become the ammonium ion.

$$NH_3 + H^+ \rightleftharpoons NH_4^+$$

Oxoanions of nitrogen (nitrate ions)

▼ **Table 3** *Name and formula of nitrate ions*

Name	Oxidation state of N	Formula
Nitrate(III)	+3	NO_2^-
Nitrate(V) [often just called nitrate]	+5	NO_3^-

Tests for ions containing nitrogen

All compounds containing nitrate(V) or ammonium ions are soluble. So it is not possible to use a test involving a precipitation reaction to positively identify these ions. Instead, tests involving the formation of ammonia gas are used.

Ammonium ions

Warm the substance being tested (solid or solution) with sodium hydroxide solution. Ammonia gas is formed. Ammonia gas causes moist red litmus paper, placed in the top of the test tube, to turn blue.

+ Go further: Nitrogen oxide

Nitrogen oxide, NO_2 is in equilibrium with a dimer, N_2O_4.

$2NO_2 \rightleftharpoons N_2O_4$ $\Delta H = -58.0\,kJ\,mol^{-1}$

N_2O_4 is a pale yellow gas, whereas NO_2 is dark brown. Use ideas about changes in equilibrium position to predict what happens to the colour of a sample of NO_2 gas in a sealed container if it is cooled from 298 K to 273 K.

Revision tip

Although you are only expected to **know** the dot-and-cross diagrams for ammonium and nitrogen (N_2), you should be able to *deduce* possible dot-and-cross diagrams for oxides and oxoanions of nitrogen, given suitable information about the species, such as the presence of multiple or dative bonds.

Synoptic link

Acid–base equilibria are covered in Topic 8.2, Strong and weak acids and pH.

Nitrate(V) ions

Sodium hydroxide is added to a solution containing a nitrate, followed by a spatula of powdered Devarda's alloy (a mixture of copper, aluminium, and zinc). Ammonia is formed (test using indicator paper as above). The aluminium metal reduces the nitrate(V) to ammonia.

Interconversion of nitrogen compounds

As shown in Tables 1–3, nitrogen can have oxidation states ranging from −3 to +5. Nitrogen compounds can take part in a variety of redox reactions, allowing ions and compounds containing nitrogen atoms to be interconverted. You can use ideas about oxidation states of nitrogen to write a half-equation for any of these interconversions. Half-equations may involve water molecules and H+ ions as well as electrons.

🖩 Worked example: Writing half-equations for the interconversion of nitrogen compounds

Balance this half-equation for the reduction of nitrate(V) ions to nitrogen(II) oxide:

$$NO_3^- + \_H^+ + \_e^- \rightarrow NO + \_H_2O$$

Step 1: Deduce the oxidation states of nitrogen in the reactant and product species: +5 in NO_3^-, +2 in NO.

Step 2: Calculate the number of electrons needed to cause this change in oxidation states: +5 to +2 requires the gain of $3e^-$.

So the equation becomes $NO_3^- + \_H^+ + 3e^- \rightarrow NO + \_H_2O$

Step 3: Balance the number of O atoms by adding in the appropriate number of H_2O molecules. There are 3O atoms in NO_3 and 1 in NO so there must be $2H_2O$:

$$NO_3^- + \_H^+ + 3e^- \rightarrow NO + 2H_2O$$

Step 4: Balance the number of H atoms in these water molecules by adding in the appropriate number of H^+ ions: There are 4H atoms in $2H_2O$, so $4H^+$ ions are needed.

$$NO_3^- + 4H^+ + 3e^- \rightarrow NO + 2H_2O$$

Step 5: Check that the charges balance on the RHS and LHS of the equation: right-hand side and left-hand side are both 0, so the equation is likely to be correct.

Many of these interconversions occur naturally in the soil and are important in maintaining the balance of nutrient ions such as nitrate(V) in the soil.

Summary questions

1 Give the systematic names for the following species:
   a N_2O **b** NO_2^- **c** NO_2 *(3 marks)*

2 A student suggests that a white solid might be ammonium nitrate. Outline how they could carry out chemical tests to confirm this. *(4 marks)*

3 The formation of nitrate(V) ions is an important reaction that occurs in agricultural soils. Deduce a half-equation for the conversion of ammonium ions to nitrate(V) ions. *(3 marks)*

Experimental techniques

Colorimetry

Colorimetry is used to determine the concentration of a coloured solution. Coloured solutions absorb specific wavelengths of light. The concentration is proportional to the amount of this light that is absorbed or transmitted.

- Make up a series of standard solutions of the test solution of known concentration.
- Select a filter with the complementary colour to the test solution.
- Zero the colorimeter with a tube of pure solvent.
- Measure the absorbance of the standard solutions and plot a calibration curve, showing concentration against absorbance.
- Measure the absorbance of the test solution and read off the concentration from the calibration curve.

A visible spectrophotometer can be used instead of a colorimeter. It works in a similar way except that it can give data for absorption or transmission for any given value in the visible spectrum. A colorimeter only gives information about a specific number of wavelengths, determined by the filter chosen.

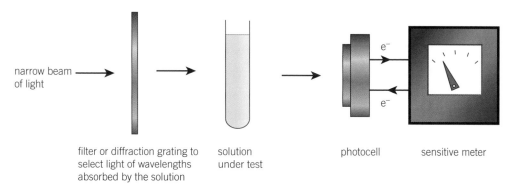

narrow beam of light

filter or diffraction grating to select light of wavelengths absorbed by the solution

solution under test

photocell

sensitive meter

▲ **Figure 1** *Colorimetry*

Chapter 11 Practice questions

1   How are elements arranged in the modern periodic table?

 A increasing relative atomic mass

 B increasing atomic radius

 C increasing reactivity

 D increasing atomic number   (*1 mark*)

2   Which element has the electronic configuration $1s^2\ 2s^2\ 2p^6\ 3s^2\ 3p^6\ 4s^2\ 3d^2$?

 A calcium   **B** scandium

 C gallium   **D** titanium   (*1 mark*)

3   Which of the following equations represents the third ionisation enthalpy of silicon?

 A $Si(g) \rightarrow Si^{3+}(g) + e^-$ **B** $Si(s) \rightarrow Si^{3+}(s) + 3e^-$

 C $Si^{2+}(s) \rightarrow Si^{3+}(s) + 3e^-$ **D** $Si^{2+}(g) \rightarrow Si^{3+}(g) + e^-$ (*1 mark*)

4   Which elements appear at the peaks of a plot of first ionisation energy for successive elements in the periodic table?

 A Group 0 elements   **B** Group 1 elements

 C Group 7 elements   **D** Transition elements   (*1 mark*)

5   Which statement(s) are true for Group 2 elements?

 1 The hydroxides become more soluble down the group.

 2 Group 2 elements are more reactive than Group 1 elements.

 3 The carbonates become more soluble down the group.

 A 1 only   **B** 1 and 2 only

 C 2 and 3 only   **D** 1, 2, and 3   (*1 mark*)

6   Which statement is true about thermal decomposition of Group 2 carbonates?

 A $MgCO_3$ is the most thermally stable Group 2 carbonate.

 B $BaCO_3$ will decompose more easily than $SrCO_3$ on heating.

 C The carbonate ion in $SrCO_3$ is more polarised than in $CaCO_3$.

 D Ca^{2+} has a higher charge density than Sr^{2+}.   (*1 mark*)

7   Which halogens will displace bromine?

 A fluorine, chlorine, and iodine

 B iodine only

 C fluorine and chlorine only

 D potassium bromide.   (*1 mark*)

8   In the following reaction: $Br_2(aq) + 2I^-(aq) \rightarrow I_2(aq) + 2Br^-(aq)$

 A bromine is oxidised

 B bromine is a weaker oxidising agent than iodine

 C iodine is an oxidising agent

 D iodine is oxidised.   (*1 mark*)

9   Which statement is correct about the volatility of halogens?

 A Fluorine is the most volatile as it has the weakest intermolecular forces.

 B Fluorine is the most volatile as it has the strongest intermolecular forces.

 C Fluorine is the least volatile as it has the weakest intermolecular forces.

 D Fluorine is the least volatile as it has the strongest intermolecular forces.   (*1 mark*)

10 Which substance is a yellow solid, insoluble in concentrated ammonia?

   A silver bromide

   B silver chloride

   C silver iodide

   D silver nitrate   (*1 mark*)

11 **a** Special precautions must be taken when transporting bromine. Emergency breathing apparatus and protective suits must be carried by the lorry driver. Describe two properties of bromine that make this necessary.   (*2 marks*)

   b Explain, in terms of intermolecular bonds, why bromine is a liquid at room temperature and pressure but chlorine is a gas. (*4 marks*)

12 Magnesium oxide has a very high melting point and is used for making furnace bricks. Explain this in terms of the charge densities of the ions involved.   (*2 marks*)

13 The Romans discovered that thermal decomposition of limestone, $CaCO_3$ can be used to make a product that neutralises acid soil.

   a Write an equation for the thermal decomposition of limestone. (*1 mark*)

   b Identify a Group 2 carbonate that decomposes more readily than $CaCO_3$. Explain why it decomposes more readily.   (*2 marks*)

   c Name the product of the thermal decomposition of limestone. (*1 mark*)

   d Show how the product in (c) reacts with an acid. Use the general symbol HA to represent any acid.   (*2 marks*)

12.1 Alkanes

Specification reference: DF(m), DF(l), DF(r), DF(c)

Straight-chain alkanes

- have the general formula C_nH_{2n+2}
- name ends in -ane
- are **saturated** – all the bonds between carbon atoms are *single bonds*
- are **aliphatic** – they do not contain benzene rings.

Naming alkanes

- Choose the longest carbon chain and name it.
- Use prefixes in alphabetical order for any alkyl side chains.
- Use *di, tri, tetra* before the prefix if the side chains are identical.
- Show the position of any side chains by using numbers which are as low as possible.

Examples of alkane names

Put a comma between numbers and a hyphen between a number and a letter:

$CH_3CH(CH_3)CH_3$ is methylpropane

$CH_3CH(CH_3)CH_2CH_2CH_3$ is 2-methylpentane

$CH_3C(CH_3)_2CH_3$ is 2,2-dimethylpropane

$CH_3CH(CH_3)CH_2CH(CH_3)CH_2CH(CH_3)CH_3$ is 2,4,6-trimethylheptane.

Different types of formulae

Full structural formulae show all the bonds and all the atoms in the molecule. Shortened structural formulae abbreviate groups such as CH_3 or CH_2, but still clearly show the arrangement of groups of atoms in the molecule.

▼ **Table 1** *Different types of formulae*

Name	Molecular formula	Full structural formula	Shortened structural formula	Further shortened to
methane	CH_4	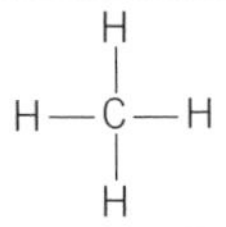 H—C—H (with H above and below)	CH_4	
ethane	C_2H_6	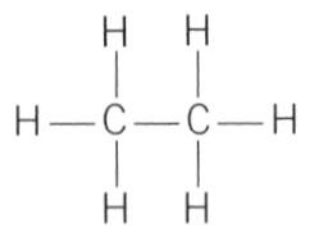 H—C—C—H (with H atoms)	CH_3-CH_3	CH_3CH_3
propane	C_3H_8	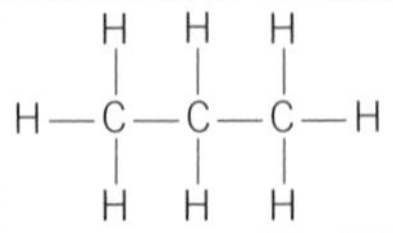 H—C—C—C—H (with H atoms)	$CH_3-CH_2-CH_3$	$CH_3CH_2CH_3$

Skeletal formulae use lines to represent carbon–carbon bonds. Hydrogen atoms attached to C atoms are not drawn in skeletal formulae. The skeletal formulae of butane and methylbutane are shown in Figures 1 and 2.

Key terms

Alkane: An alkane is a saturated hydrocarbon.

Structural formula: A representation of the atoms, bonds, and groups in a molecule.

Skeletal formula: A representation of a molecule using lines to represent C—C bonds.

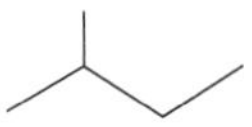

▲ **Figure 2** *Skeletal formula of methylbutane*

▲ **Figure 1** *Skeletal formula of butane*

Structural and skeletal formulae do not accurately represent the 3-dimensional shapes of molecules. To get round this problem, chemists use wedge and dotted bonds:

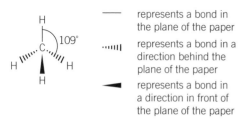

—   represents a bond in the plane of the paper

▫▫▫▫   represents a bond in a direction behind the plane of the paper

◀━   represents a bond in a direction in front of the plane of the paper

▲ **Figure 3** *Three-dimensional structure of methane*

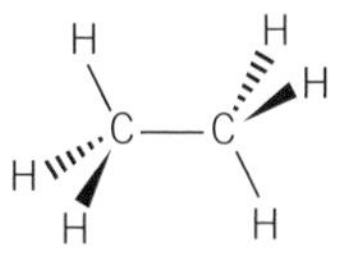

▲ **Figure 4** *Three-dimensional structure of ethane*

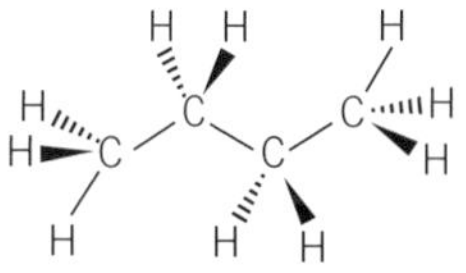

▲ **Figure 5** *Three-dimensional structure of butane*

The carbon chain found in alkanes can be substituted by different functional groups such as −OH, −Cl, −COOH. This gives different homologous series – a group of molecules with the same functional group but different carbon chain lengths. Members of homologous series have chain lengths differing by a CH_2 unit.

Cycloalkanes

- have the general formula C_nH_{2n}
- name ends in *-ane*
- are **saturated** – all the bonds between carbon atoms are *single bonds*
- are *not* **aromatic** – they don't have a benzene ring
- are aliphatic.

Summary questions

1   Draw the skeletal formulae for 3-methylhexane and 2,2,4-trimethylheptane.   *(2 marks)*

2   Give the molecular formula of (i) an alkane and (ii) a cycloalkane with 10 carbon atoms.   *(2 marks)*

3   Describe what the molecular formulae of hexene, 2-methylpentene, and cyclohexane have in common.   *(1 mark)*

Revision tip

Remember that wedge bonds project towards you, out of the plane of the paper. Dotted bonds project away from you.

Revision tip

All the angles in alkanes are the same, 109.5°. The bonds are arranged tetrahedrally.

Key term

Cycloalkane: A saturated hydrocarbon, where the carbon atoms are joined in a ring.

12.2 Alkenes

Unsaturated hydrocarbons

Alkenes are a homologous series of hydrocarbons and have the following characteristics:

- the general formula C_nH_{2n}
- name ends in *-ene*
- are **unsaturated** – there are one (or more) *double bonds* between carbon atoms in a molecule
- are **aliphatic** – they don't have delocalised benzene ring structures.

Naming alkenes

The names of alkenes end in *-ene*. The number preceding the -ene indicates the position of the double bond. But-1-ene is $CH_3-CH_2-CH=CH_2$ and but-2-ene is $CH_3-CH=CH-CH_3$.

Shape of alkenes

All bond angles around the C=C double bond are 120°, since there are three groups of electrons around each carbon atom (two single bonds and one double bond). These groups of electrons repel each other as far as possible.

A C=C bond contains a sigma (σ) bond and a pi (π) bond. A σ-bond is an area of increased electron density between the carbon atoms. A π-bond consists of two areas of negative charge. One of these is above the line of the atoms, and the other is below.

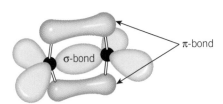

▲ **Figure 1** *Sigma- and pi-orbitals*

Reactions of alkenes

Alkenes undergo **electrophilic addition** reactions. In the following examples we will use ethene as a typical alkene. There are four electrons in the double bond of ethene – these give the region between the two carbon atoms a high negative charge density. Electrophiles are attracted to this negatively charged region in an alkene and accept a pair of electrons from the double bond at the start of the reaction. When electrophiles react, they accept a pair of electrons. A general scheme for the reactions is shown in Figure 2, using bromine as an electrophile.

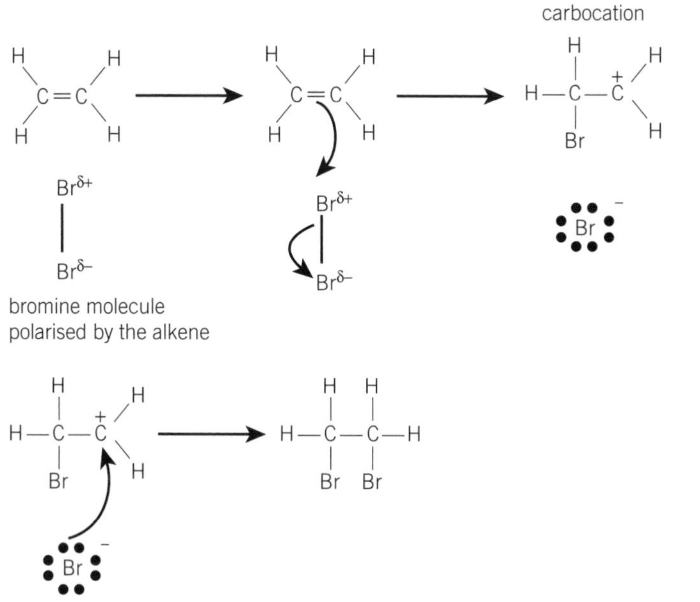

▲ **Figure 2** *Electrophilic addition*

- When a bromine molecule approaches an alkene, it becomes **polarised.**
- The electrons in the bromine molecule are repelled back along the molecule.
- The electron density is *unequally distributed.*
- The bromine atom nearest to the alkene becomes slightly positively charged.
- It now acts as an **electrophile.** A pair of electrons from the alkene moves towards the slightly charged bromine atom and a C–Br bond is formed.
- The carbon species is now positively charged; it is a **carbocation.**
- The other bromine, now negatively charged, moves in rapidly to make another bond.

The overall process is addition by an electrophile across a double bond – it is **electrophilic addition.** This model for the mechanism for electrophilic addition (*via* a carbocation) can be supported by experimental evidence. If chloride ions, Cl-, are present when ethene reacts with bromine, the molecule $BrCH_2CH_2Cl$ forms, as well as the expected $BrCH_2CH_2Br$. This is because both chloride and bromide ions can attack the intermediate carbocation.

Alkenes can react with a number of electrophiles.

Electrophile	Product	Conditions
Br_2	CH_2BrCH_2Br 1,2-dibromoethane	room temperature and pressure
HBr(aq)	CH_3CH_2Br bromoethane	aqueous solution, room temperature and pressure
H_2O (H–OH)	CH_3CH_2OH ethanol	phosphoric acid / silica at 300 °C / 60 atm or with conc. H_2SO_4, then H_2O at 1 atm
H_2	CH_3CH_3 ethane	Pt catalyst, room temperature and pressure, or Ni catalyst at 150 °C / 5 atm

Addition polymerisation

Alkenes are the basic hydrocarbon units of many polymers:

- –A–A–A–A–A–A– polymers are made from one type of alkene monomer
- –A–B–A–B–A–B– polymers can be made from more than one type of alkene monomer.

Key terms

Alkene: Alkenes are unsaturated hydrocarbons containing one or more C=C bonds.

Electrophiles: Positive ions or molecules with a partial positive charge on one of the atoms.

Electrophilic addition: A reaction in which an electrophile joins onto an alkene. No atoms are removed from the alkene.

Polarised: A covalent bond is said to be polarised if the electrons are unevenly distributed between the atoms.

Common misconception: Carbocations

The correct pronunciation of carbocation is 'carbo – cat – ion'.

Revision tip

When bromine approaches an alkene it becomes polarised, because the electrons in the Br–Br bond are repelled from the electron-dense C=C bond.

Key term

Carbocation: A carbon-containing positive ion, formed as an intermediate in electrophilic addition.

Revision tip

Make sure you know the reaction conditions for all the reactions in the table.

Revision tip

Bromine is used as a test for unsaturation. The brown bromine becomes decolourised.

Key term

Monomer: A molecule which is the starting material for a polymer.

Revision tip

In addition polymerisation, the monomers are generally alkenes.

Key term

Polymer: A long-chain molecule formed when many monomers join together.

Under the right conditions, alkenes can undergo addition polymerisation. The small *unsaturated* starting molecules are called **monomers** and they join together to form a long chain *saturated* **polymer**. No other product is formed.

$CH_2{=}CHCH_3 + CH_2{=}CHCH_3 + CH_2{=}CHCH_3 \rightarrow$
$-CH_2-CH(CH_3)-CH_2-CH(CH_3)-CH_2-CH(CH_3)-$

propene (monomers) *poly(propene)* (polymer)

This may be written as:

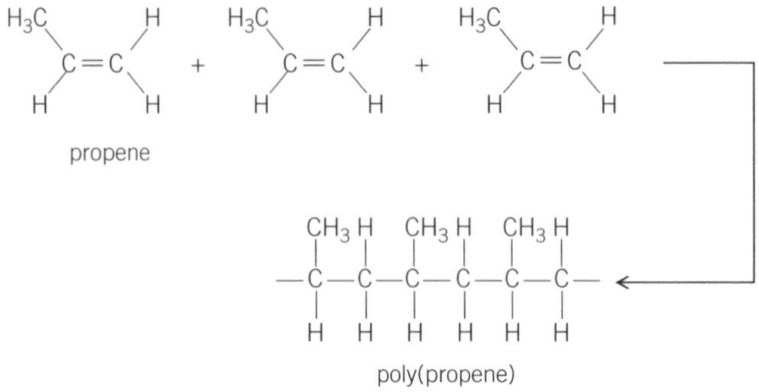

▲ **Figure 3** *Addition polymerisation*

The repeat unit is written as:

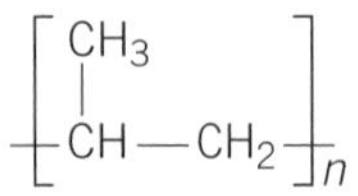

▲ **Figure 4** *Repeat unit of poly(propene)*

The polymer is named by putting the name of the monomer in brackets and prefixing with 'poly', for example, choroethene monomer gives poly(chloroethene) as the polymer. Note, however, that the polymer is *not* an alkene.

Summary questions

1. Draw the full structural, shortened structural, and skeletal formulae of pent-1-ene. *(3 marks)*
2. Draw a skeletal formula of the product when $H_2(g)$ reacts with pent-2-ene. *(1 mark)*
3. Draw the mechanism of the electrophilic addition of HBr to but-2-ene. *(4 marks)*
4. Draw the structure of the polymer formed when the monomer CHCl=CHCl undergoes addition polymerisation, showing three repeat units. *(2 marks)*

12.3 Structural isomerism and E/Z isomerism

Specification reference: DF(s), DF(t)

Structural isomers

Chain isomerism, in which the chain lengths are different because of branching, is often seen in alkanes. Butane has a four-carbon chain, whereas methylpropane has three carbon atoms in its longest chain, but both have the molecular formula C_4H_{10}.

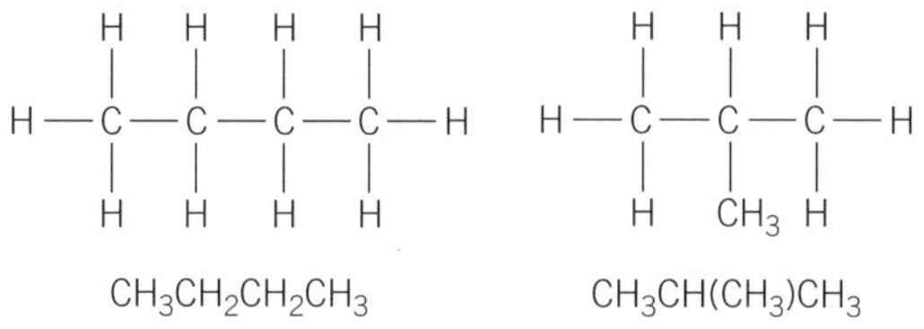

▲ **Figure 1** *Isomers of C_4H_{10}*

Position isomerism, in which the same functional group appears in different positions, is often seen in alcohols. For example, propan-1-ol, $CH_3CH_2CH_2OH$, and propan-2-ol, $CH_3CH(OH)CH_3$, both have the molecular formula C_3H_8O.

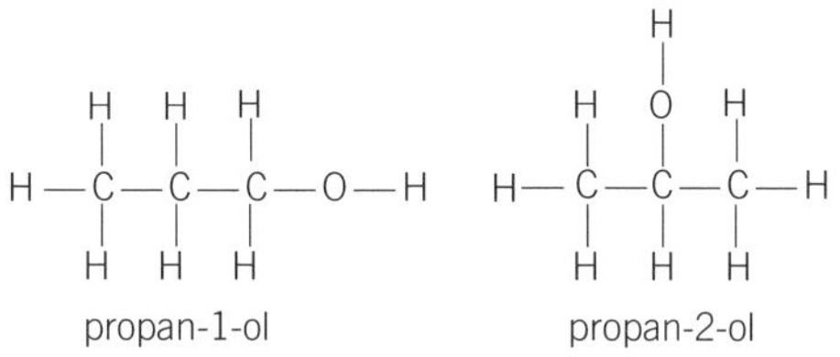

▲ **Figure 2** *Isomers of C_3H_8O*

Functional group isomerism, in which the molecular formulae are the same but the functional groups are different, can be seen in alcohols and ethers, for example ethanol, C_2H_5OH, and methyoxymethane, CH_3OCH_3.

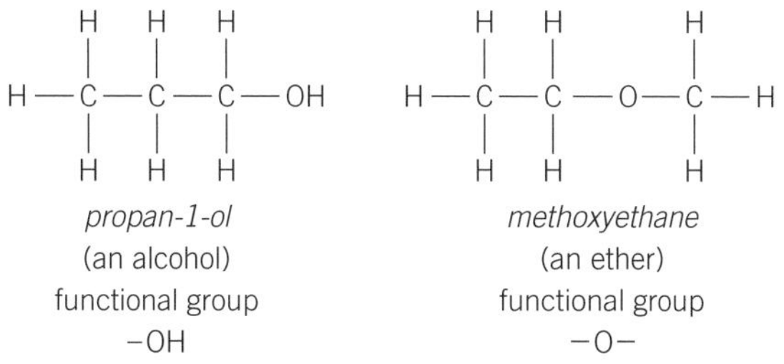

▲ **Figure 3** *Functional group isomerism*

Stereoisomerism

E/Z isomerism is one type of stereoisomerism. In stereoisomerism, the atoms are bonded in the same order, but are arranged differently in space in each isomer.

The isomer of but-2-ene that has the two methyl groups on the same side of the double bond is called the *Z* isomer, that is Z-but-2-ene. It is also called *cis*-but-2-ene. The isomer that has the two methyl groups on opposite sides of the double bond is called the *E* isomer, that is E-but-2-ene. It is also called *trans*-but-2-ene.

The reason why these two isomers exist is that to turn one form into the other you need to break one of the bonds in the carbon-carbon double bond. There is not enough energy at room temperature to enable this to occur. As a result, interconversion of the two isomers does not occur.

Key term

Structural isomers: Structural isomers have the same molecular formula, but different structural formulae.

Key term

Molecular formula: This shows the numbers of each type of atom in the molecule.

Common misconception: Molecular formulae

Molecular formulae do not indicate functional groups. For example, the molecular formula of ethanol is C_2H_6O, not C_2H_5OH.

Revision tip

'*E*' (from the German *entgegen*) means *opposite*; '*Z*' (from the German *zusammen*) means *together*.

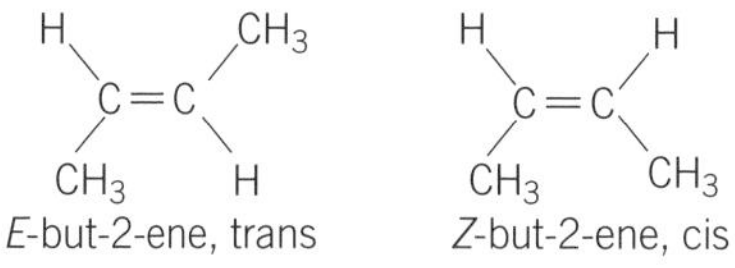

▲ **Figure 4** *E/Z isomers of but-2-ene*

Revision tip

You do not get isomers if there are two identical groups at one end of a double bond. E/Z isomers only occur if each carbon in the double bond has two different groups attached to it.

Summary questions

1 Draw the isomers of $C_2H_4Br_2$. *(2 marks)*

2 Draw the skeletal formulae of 2-methylpropan-2-ol and 2-methylpropan-1-ol and explain whether or not they are isomers. *(3 marks)*

3 Explain why 2-methylpropene does not exhibit E/Z isomerism. *(1 mark)*

a   b

▲ **Figure 1** *The two representations of benzene* **a** *the delocalised structure* **b** *the Kekulé structure, showing three double bonds*

> **Revision tip**
> Benzene rings in questions may be shown using either the delocalised or the Kekulé representation. Unless you are told otherwise, you should use the delocalised representation when you are drawing benzene rings.

Benzene

The simplest arene is benzene, C_6H_6.

Representing benzene

The structure and bonding in benzene can be represented in two ways:

- The first is based on the current model of benzene and uses a circle to show the delocalisation of electrons in the benzene ring.
- The second is based on the model of structure and bonding first proposed by Kekulé in the 19th century; this shows three separate double bonds in the molecule.

Delocalisation

The benzene ring contains delocalised electrons, represented by the circle in Figure 1 a.

The delocalisation can be described as follows:

- Six electrons are involved.
- These electrons come from the p orbitals of the C atoms in the benzene ring; one electron comes from each carbon.
- The six p orbitals overlap to form rings of electron density above and below the carbon atoms of the benzene ring.

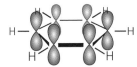

▲ **Figure 2** *The overlapping of p orbitals to form delocalised rings of electron density in a benzene ring*

The shape of the benzene molecule

- Benzene is a planar molecule.
- The carbon atoms are arranged in the form of a regular hexagon; all the C—C bond lengths are identical.
- All the bond angles in the molecule are 120°.

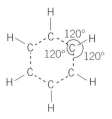

▲ **Figure 3** *The bond angles in benzene*

Relating shape to ideas about bonding

Both the Kekulé and the delocalised model of bonding would predict a planar hexagonal structure for benzene.

However, the Kekulé structure, with its alternating double and single bonds, would be an irregular hexagon with two different C–C bond lengths; with the C=C bond being shorter than the C–C bond.

The delocalised model, with electrons evenly spread over all the C atoms, correctly predicts the equal C–C bond lengths in the regular hexagonal structure.

Further evidence for the delocalised model

The Kekulé model suggests that the six electrons in p orbitals form three double bonds. Scientists used this idea, along with data for hydrogenation reactions of alkenes, to predict a value for the hydrogenation of benzene.

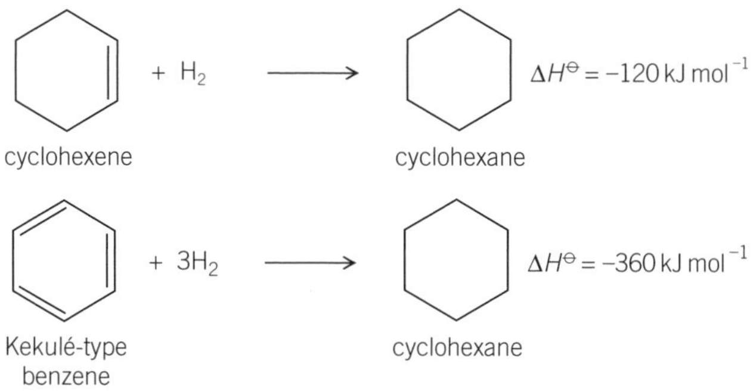

cyclohexene + H$_2$ $\longrightarrow$ cyclohexane $\Delta H^\ominus = -120\,\text{kJ mol}^{-1}$

Kekulé-type benzene + 3H$_2$ $\longrightarrow$ cyclohexane $\Delta H^\ominus = -360\,\text{kJ mol}^{-1}$

▲ **Figure 4** *The enthalpy changes for the hydrogenation of cyclohexene and the prediction for benzene using the Kekulé model*

The experimental value for benzene is $-208\,\text{kJ mol}^{-1}$, which is significantly less negative than the prediction using the Kekulé model.

This can be explained using the delocalised model. Delocalisation increases the strength of bonding in a molecule (and hence makes it more stable). More energy is needed to break the delocalised bonds and so the overall enthalpy change is less negative.

Evidence from the pattern of reactions for benzene

Benzene takes part in a range of substitution reactions and undergoes addition only under extreme conditions. This can be explained using the delocalised model; substitution reactions result in a product that also has a delocalised system and so the stability of this system is preserved.

The Kekulé model would be expected to undergo mainly addition reactions is it contains separate double bonds.

Aromatic compounds

Many derivatives of benzene can be synthesised, and the word **aromatic** is used to describe these compounds.

> ### ➕ Go further
>
> 'Aromatic' is strictly used to describe molecules that contain a cyclic structure with electrons that are delocalised in the same way as benzene.
>
> This delocalisation will occur if there are alternating double and single bonds and if there are $4n + 2$ electrons ($n = 1,2$ etc.).
>
> Decide whether the structures in Figure 5 are aromatic or not: 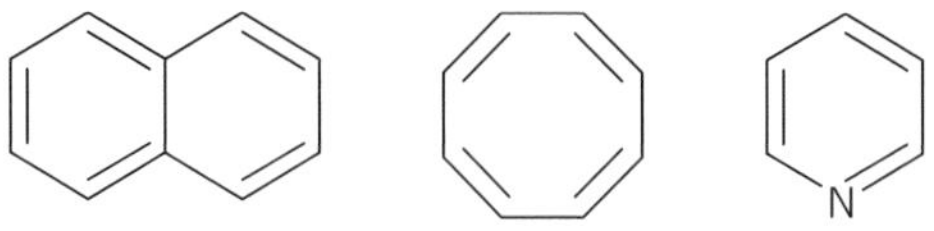
>
>
>
>   a   b   c
>
> ▲ **Figure 5**

Naming arenes and aromatic compounds

Simple arenes and aromatic compounds

Substituted benzene rings can be named using the same ideas as cycloalkanes. The numbering of the carbon atoms starts at the group which comes first in alphabetical order.

> ### Key terms
>
> **Arene:** A hydrocarbon molecule that contains a benzene ring.
>
> **Aromatic compound:** A molecule that contains a benzene ring.
>
> **Delocalised electrons:** Electrons that are shared between more than two atoms.

> ### Synoptic link
>
> You also encountered the idea of delocalised electrons when studying the structure of metals in Topic 5.2, Bonding, structure, and properties.

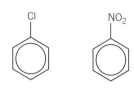

chlorobenzene   nitrobenzene

▲ **Figure 7** *Names of some of simple aromatic compounds*

  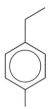

methylbenzene   1,3-dimethylbenzene   1-ethyl-4-methylbenzene

▲ **Figure 6** *Names of some of simple arenes*

Benzene rings containing halogens or nitro ($-NO_2$) groups occur commonly in reaction sequences.

More difficult names

Some names of aromatic compounds do not follow the simple naming pattern and are named in a less predictable way:

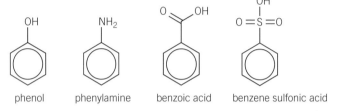

phenol   phenylamine   benzoic acid   benzene sulfonic acid

▲ **Figure 8** *The names of some other important aromatic compounds*

The phenyl group

Sometimes it makes more sense to think of the benzene ring as a side group in a structure. In these cases, the word 'phenyl' is used to describe the benzene ring.

Revision tip

You should learn the names of these common aromatic compounds, and be prepared to use other unpredictable names when they are introduced in a question.

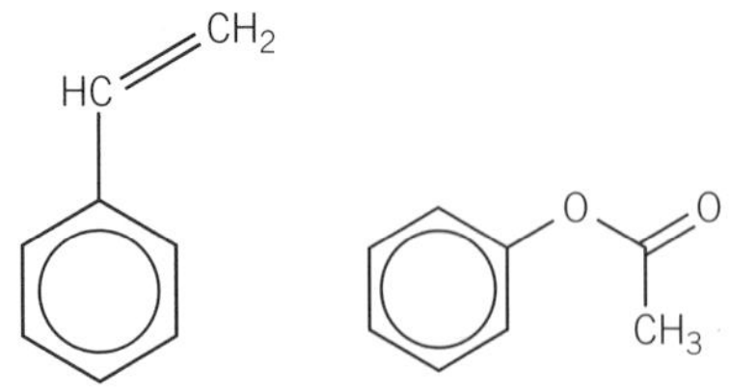

phenylethene   phenyl ethanoate

▲ **Figure 9** *Phenyl groups used in naming an alkene and an ester*

Synoptic link

If phenol is used to form esters, then the resulting ester will contain a phenyl group. The formation of esters is covered in Topic 13.4, Carboxylic acids and phenols.

Summary questions

1 Describe the current model of the bonding between the carbon atoms in benzene. *(5 marks)*

2 a Name the molecules shown below: *(2 marks)*

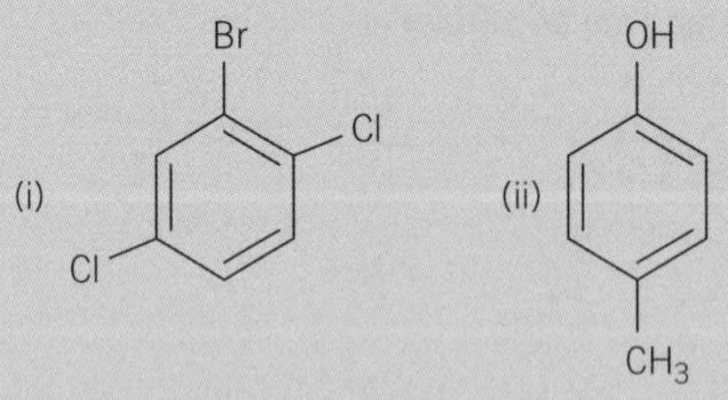

   (i)   (ii)

   b Draw out the structure of (i) 2-hydroxybenzoic acid
      (ii) 3-nitrobenzenesulfonic acid. *(2 marks)*

3 The Kekulé model of benzene is shown in Figure 1 above. Discuss to what extent this model is consistent with the observed shape of the benzene molecule. *(4 marks)*

12.5 Reactions of arenes

Specification reference: CD (g)

Electrophilic substitution reactions

The benzene ring in arenes and other aromatic compounds generally takes part in electrophilic substitution reactions, although the reactions are often quite slow.

Explaining the reactions of benzene

The reactions of the benzene ring can be explained using ideas about the delocalised structure of benzene:

- Benzene contains an area of high electron density (the ring of delocalised electrons), so it is likely to be attacked by electrophiles.
- Substitution reactions (replacing one of the H atoms from the benzene ring with a different atom or group of atoms) does not affect the delocalised system of electrons, so the product is relatively stable.
- Benzene does not usually take part in addition reactions, because addition would destroy the delocalised system of electrons.

Because the rate of reaction can be very slow, specific catalysts often need to be present, as well as certain key conditions. These are necessary in order to generate a species that is reactive enough to act as an electrophile.

Halogenation reactions

Benzene reacts with bromine or chlorine in the presence of certain catalysts to form halogenated derivatives of benzene.

▼ **Table 1** *Halogenation reactions of benzene*

Reaction	Reagent	Catalyst	Conditions	Electrophile	Product
bromination	bromine	iron (or iron(III) bromide)	reflux	Br^+	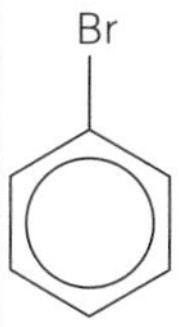 bromobenzene
chlorination	chlorine	aluminium chloride	room temperature, anhydrous conditions	Cl^+	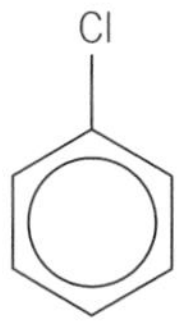 chlorobenzene

Equations for these reactions

The overall equation for the bromination of benzene is:

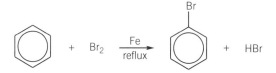

▲ **Figure 1** *Equation for the bromination of benzene*

Synoptic link

The definition of an electrophile is given in Topic 12.2, Alkenes.

Substitution reactions are described in Topic 13.2, Haloalkanes.

Revision tip

You should be able to compare the reactions of the delocalised benzene structure with those predicted by the **Kekulé** structure; this contains double bonds so addition reactions would be most likely. However it also has areas of high electron density so would also be attacked by electrophiles.

Synoptic link

The naming system for derivatives of benzene is explained in Topic 12.4, Arenes.

Common misconception: The molecular formula of the products of substitution reactions

If you are required to write the molecular formula of the products of substitution reactions, it is important to remember that the benzene ring has lost a H atom during the reaction – so the benzene ring shown in the structural formula now has formula C_6H_5, not C_6H_6.

It can sometimes be helpful to show the process that occurs with the actual electrophile (Br⁺):

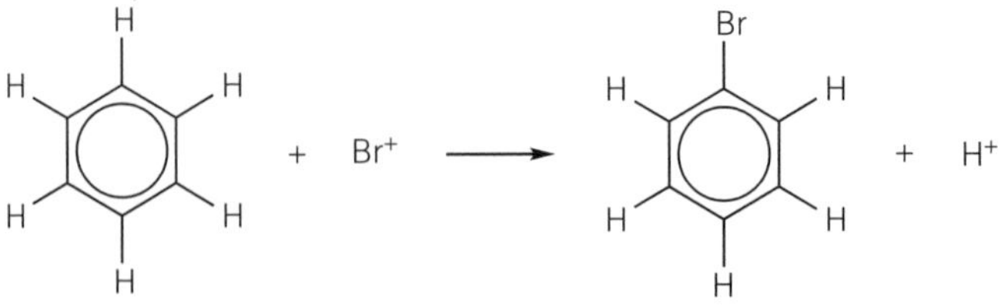

▲ **Figure 2** *The reaction that occurs with the Br⁺ electrophile*

The equations for chlorination are very similar to those for bromination.

Nitration and sulfonation reactions

Nitration and sulfonation reactions are important in modifying the properties of dye molecules based on benzene rings.

▼ **Table 2** *Nitration and sulfonation reactions*

Reaction	Reagent	Catalyst	Conditions	Electrophile	Product
nitration	conc. nitric acid	conc. sulfuric acid	below 55 °C (to avoid multiple nitrations)	NO_2^+	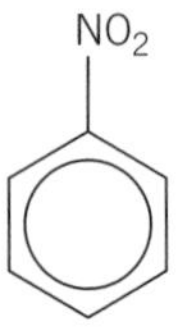 NO_2 nitrobenzene
sulfonation	conc. sulfuric acid	none required	reflux	SO_3	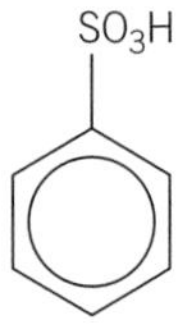 SO_3H benzene sulfonic acid

Equations for these reactions

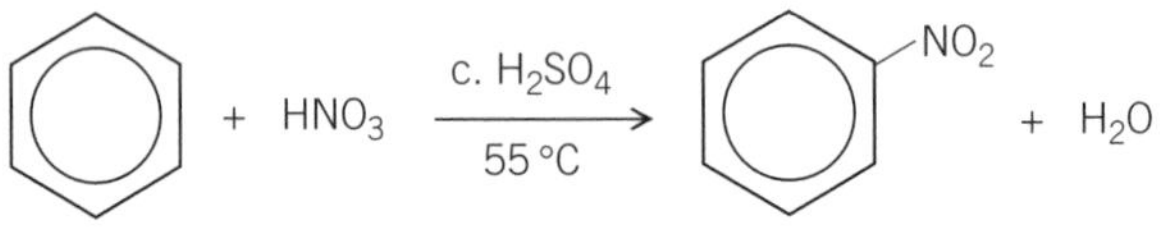

▲ **Figure 3** *The equation for nitration of benzene*

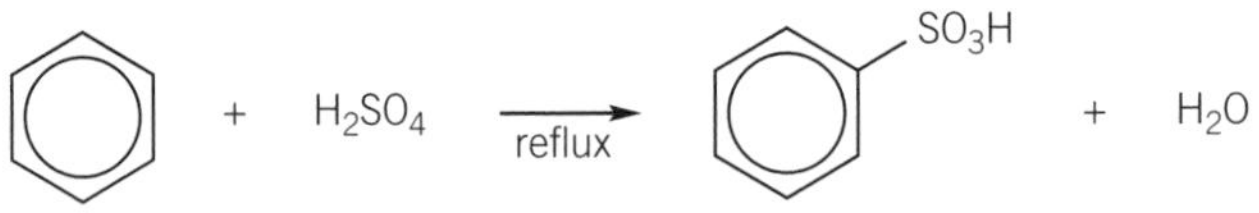

▲ **Figure 4** *The equation for sulfonation of benzene*

Mechanism for the nitration reaction

The NO_2^+ is generated by the reaction between the sulfuric acid catalyst and nitric acid.

This NO_2^+ ion acts as an electrophile, accepting a pair of electrons from the benzene ring to form a positively charged intermediate.

This intermediate is unstable. A pair of electrons from a C–H bond becomes part of the delocalised system and an H⁺ ion is lost from the intermediate.

Finally, the H⁺ ion causes the regeneration of the sulfuric acid catalyst.

The mechanism is shown in Figure 5.

Step 1: $HNO_3 + H_2SO_4 \rightarrow NO_2^+ + HSO_4^- + H_2O$

Step 2:

NO_2^+

intermediate

$+ \ H^+$

Step 3: $H^+ + HSO_4^- \rightarrow + H_2SO_4$

▲ **Figure 5** *The mechanism of nitration*

Friedel–Crafts reactions

Alkyl groups (such as ethyl groups –CH_2CH_3) or acyl groups (such as ethanoyl groups –$COCH_3$) can be attached to a benzene ring using Friedel–Crafts reactions (named after the two 19th century chemists who developed these processes).

▼ **Table 3** *Products of Friedel–Crafts reactions*

Reaction	Reagent	Catalyst	Conditions	Electrophile	Product
alkylation	haloalkane e.g. CH_3CH_2Cl	aluminium chloride, $AlCl_3$	anhydrous conditions, heat under reflux	$CH_3CH_2^+$	CH_2CH_3 ethylbenzene
acylation	acyl chloride e.g. CH_3COCl	aluminium chloride, $AlCl_3$	reflux	SO_3	SO_3H benzene sulfonic acid

Summary questions

1 Explain why benzene takes part in substitution reactions, rather than addition reactions. *(3 marks)*

2 Give the reagents and conditions necessary for these reactions: *(4 marks)*

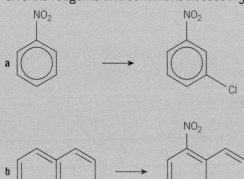

3 Benzene can react with propanoyl chloride (CH_3CH_2COCl) in a Friedel–Crafts reaction:
   a State what other substance needs to be present and explain the role of this substance. *(3 marks)*
   b Write an equation, using structural formulae, for this reaction. *(3 marks)*

12.6 Azo dyes

Specification reference: CD (b), CD (h)

Key term

Azo group: An organic functional group with the structure $R_1–N=N–R_2$.

Key term

Diazonium compound (or ion): A compound or ion that contains a diazo functional group ($R–N^+≡N$).

Azo dyes

A dye is a coloured molecule that can attach to fabrics and other solid substance. Many dye molecules are described as azo dyes because they contain an azo group.

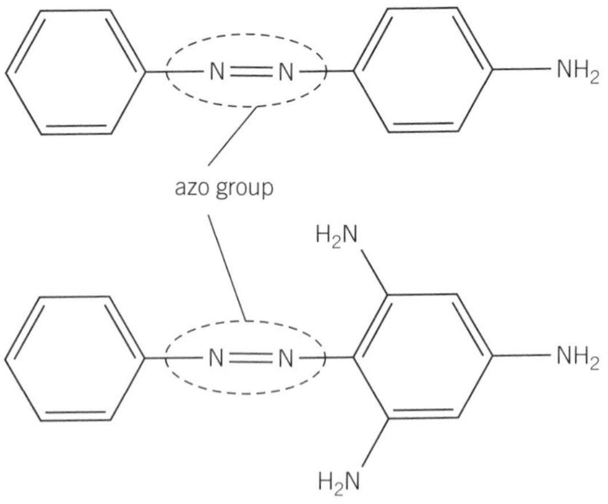

▲ **Figure 1** *The structure of some azo dyes, showing the azo functional group*

Formation of azo dyes

Azo dyes are synthesised from aromatic amines in two stages:

- A diazonium compound is formed as an intermediate.
- The diazonium compound is then coupled to a second aromatic molecule.

Forming a diazonium compound

Aromatic amines, such as phenylamine, react with nitric(III) acid, HNO_2, in the presence of hydrochloric acid.

The nitric(III) acid is usually formed in the reaction mixture by adding a solution of sodium nitrate(III) to hydrochloric acid.

This is added to a solution of phenylamine in hydrochloric acid.

The temperature must be kept below 5 °C to slow down the rate at which the diazonium compound decomposes.

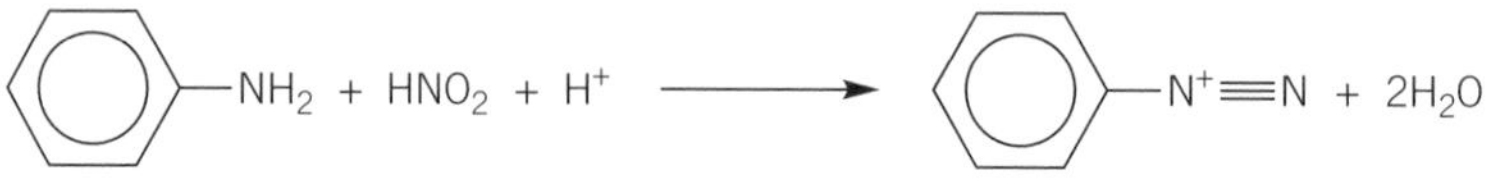

▲ **Figure 2** *The equation for the formation of a diazonium ion*

Coupling reactions

The unstable diazonium compound must be used as soon as possible in the second stage of the synthesis.

The diazonium ion reacts with aromatic compound, called a coupling agent. This coupling agent is usually a phenol, or another aromatic amine:

- A solution of the coupling agent is made up.
- An ice-cold solution of the diazonium compound is added to it.
- A coloured suspension of the azo dye forms immediately.

This reaction is an electrophilic substitution reaction because the diazonium ion acts as an electrophile and replaces one of the H atoms on the benzene ring of the coupling agent.

The presence of the phenol or amine group increases the reactivity of the coupling agent by increasing the electron density on the benzene ring. When phenols are used as coupling agents they are usually dissolved in sodium hydroxide to form a negative phenoxide ion which increases the electron density even further.

Revision tip

You may need to predict a possible structure for the azo dye formed from a given combination of aromatic amine and coupling agent, or to work backwards to deduce the structures of the starting materials needed to form a particular azo dye.

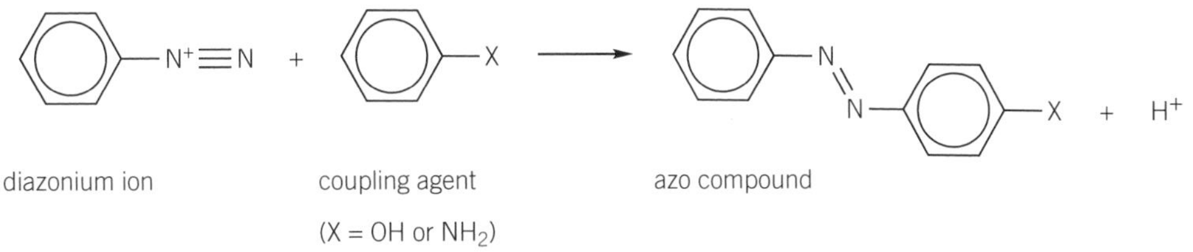

▲ **Figure 3** *The equation for a coupling reaction, forming an azo dye*

Remember that one of the nitrogen atoms in the azo group comes from the aromatic amine; the other one comes from the nitrogen atom from the nitric(III) acid. The coupling agent will retain its OH or amine group, even after the coupling reaction.

Chromophores in azo dyes

Azo dyes usually have intense orange or yellow colours because the delocalised system of the chromophore is extensive enough to absorb visible light in the blue part of the spectrum.

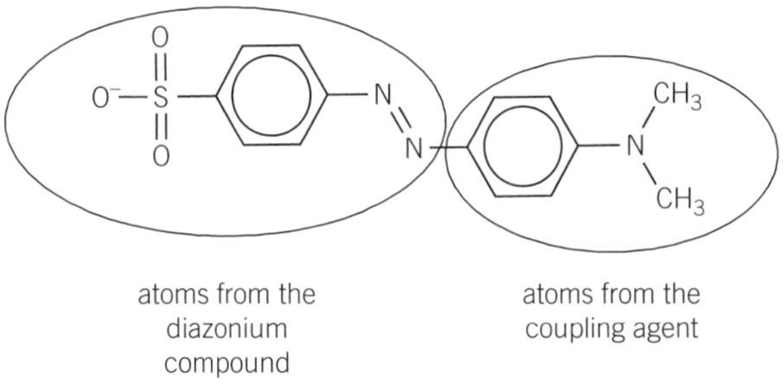

atoms from the diazonium compound — atoms from the coupling agent

▲ **Figure 4** *The structure of methyl orange, showing the parts of the molecule that originated in the diazonium compound and the coupling agent*

Modifying dye molecules

Chemists modify the structure of dyes by substituting various functional groups into the benzene rings of the azo dye.

Modifying the chromophore

The chromophore can be modified by including groups of atoms, such as nitro groups ($-NO_2$), that extend the delocalised system. This will alter the colour of the dye.

Increasing solubility

Many dyes need to be water soluble in order to be used in the dyeing process. Sulfonic acid groups ($-SO_3H$) in the dye structure increase solubility because these groups can easily be converted into the ionic sulfonate group ($-SO_3^-$).

Bonding dye molecules to fibres

Groups such as sulfonic acid (SO_3H), hydroxyl (OH), or amine (NH_2) are also introduced into dye molecules to allow them to bond more effectively to fabric.

> ## Synoptic link
>
> The concept of a chromophore was introduced in Topic 6.9, Coloured organic molecules.

> ## Synoptic link
>
> Substituting nitro- and sulfonic acid groups into benzene rings was described in Topic 12.5, Reactions of arenes.

> ## Synoptic link
>
> The bonding of dyes to fabric is described in detail in Topic 5.7, Bonding dyes to fibres.

Summary questions

1 Describe how phenylamine, $C_6H_5NH_2$, can be converted into the diazonium compound phenyldiazonium chloride, $C_6H_5N_2^+Cl^-$. (*3 marks*)

2 The structure of the azo dye acid black 1 is shown:

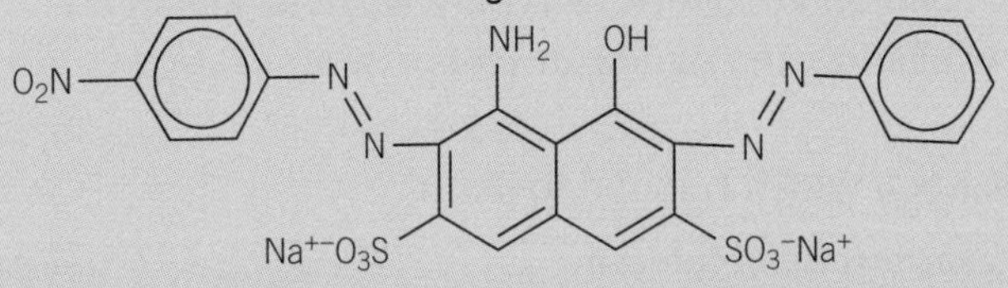

Suggest possible reasons for the inclusion of the following groups in the structure of this dye molecule:

a the $SO_3^-Na^+$ groups   c the NO_2 (O_2N) group
b the OH and NH_2 groups.   (*3 marks*)

3 The azo dye shown below is synthesised from two organic starting materials:

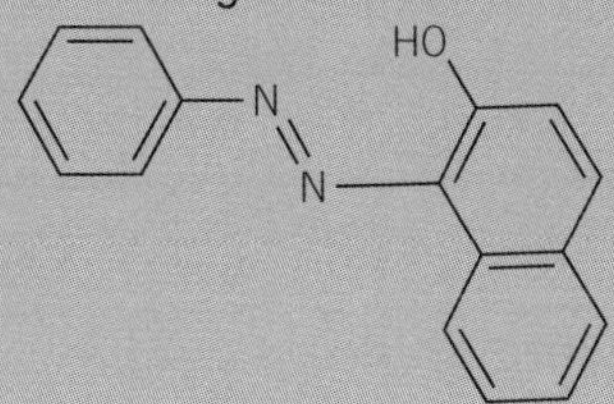

Deduce the structures of the two organic starting materials. (*2 marks*)

1 What is the name of the molecule $CH_3CH(CH_3)CH(CH_3)CH_3$?

   A hexane

   B 2,3-dimethylhexane

   C 2,3-dimethylbutane

   D butane-2,3-dimethyl   *(1 mark)*

2 Skeletal formulae

   A use straight lines to represent C–C bonds

   B show hydrogen atoms

   C cannot represent functional groups

   D accurately represent the 3D shapes of molecules   *(1 mark)*

3 Ethanol and methoxymethane

   A are examples of position isomerism

   B both have the molecular formula C_2H_5OH

   C are examples of functional group isomerism

   D are both alcohols   *(1 mark)*

4 Which molecule does **not** have the molecular formula C_5H_{10}?

   A pent-1-ene

   B pentane

   C cyclopentane

   D 2-methylbut-2-ene   *(1 mark)*

5 Which statements are true about alkanes and alkenes?

     1 Alkanes and alkenes are hydrocarbons.

     2 Alkanes and alkenes can both undergo addition polymerisation.

     3 Alkenes with one double C=C bond have the same molecular formula as the cycloalkane with the same number of carbons.

   A 1 only

   B 2 and 3 only

   C 1 and 3 only

   D 1, 2, and 3   *(1 mark)*

6 A hydrocarbon has an M_r of 58. It reacts with bromine water and can produce an addition polymer. It has E/Z isomers. Identify the hydrocarbon, explaining the significance of each of its properties.   *(4 marks)*

7 Decane is a hydrocarbon which is found in petrol.

   a Give the molecular formula of decane.   *(1 mark)*

   b Explain why all the bond angles in decane are 109°.   *(3 marks)*

   Decane can undergo catalytic cracking to produce two shorter molecules, an alkene, C_3H_6, and an alkane, C_7H_{14}.

   c Name C_3H_6 and C_7H_{14}.   *(2 marks)*

   d Name the *types* of isomerism that C_3H_6 and C_7H_{14} can show.   *(2 marks)*

   C_3H_6 was reacted with aqueous bromine.

   e Name the type of reaction that occurs.   *(1 mark)*

   f Give the name of the main product.   *(1 mark)*

   g Explain why the main product is contaminated with $CH_2BrCH(OH)CH_3$.   *(1 mark)*

8 The molecular formula of benzenesulfonic acid is:

   A $C_6H_5SO_3H$ **B** $C_6H_5SO_3$ **C** $C_6H_7SO_4$ **D** $C_6H_6SO_3$ (*1 mark*)

9 Nitro groups ($–NO_2$) are sometimes substituted into a benzene ring in a dye molecule. The most likely reason for this is:

   A To alter the wavelength at which the dye absorbs visible light.

   B To increase the solubility of the dye.

   C To improve the ability of the dye to bond to fibres.

   D To make the dye more stable.   (*1 mark*)

10 Which two substances would react in a coupling reaction to form the dye molecule shown (left)?

   A Benzene diazonium chloride and 1,3-dihydroxybenzene

   B Phenylamine and 1,3-dihydroxybenzene

   C Benzene and 1,3-dihydroxybenzene diazonium chloride

   D Phenylamine and 1,3-dihydroxybenzene   (*1 mark*)

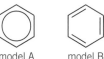

11 Which of the following intermediates are involved in the nitration of benzene by a mixture of concentrated nitric and sulfuric acids

   1 $C_6H_6NO_2^+$ **2** NO_2^+ **3** $C_6H_5^+$

   A 1,2, and 3   **B** 1 and 2   **C** 2 and 3   **D** Only 3   (*1 mark*)

12 The reaction between benzene and 1-chloropropane to produce propyl benzene can be described as:

   1 Electrophilic substitution   **2** A Friedel-Crafts reaction   **3** Acylation

   A 1,2, and 3   **B** 1 and 2   **C** 2 and 3   **D** Only 3   (*1 mark*)

13 The structure and bonding in benzene can be represented by two different models, as shown (left).

   a Describe how the bonding differs in these two models.   (*2 marks*)

   b Discuss the extent to which experimental evidence is consistent with these two models.   (*6 marks*)

model A   model B

14 The following sequence of reactions shows a four-step synthesis of a dye molecule.

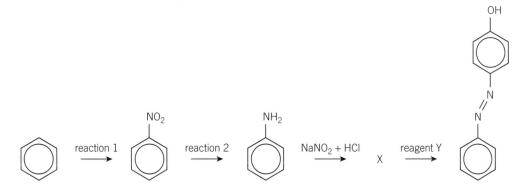

   a State the reagents that would be required to carry out:

     i Reaction 1   **ii** Reaction 2   (*2 marks*)

   b Draw out the structure of:

     i Molecule X   **ii** Reagent Y   (*2 marks*)

   c Reaction 1 and the reaction that forms X are both carried out at particular conditions of temperature. For **each** reaction, state the temperature that is required and explain why it is necessary.   (*4 marks*)

13.1 Naming organic compounds: Alkanes, haloalkanes, and alcohols

Specification reference: DF(m), OZ(j), WM(b)

Alkanes

Alkanes are named by looking for the longest chain of carbon atoms. The name ends in -ane and depends on the number of carbon atoms in the chain: *meth*ane, *eth*ane, *prop*ane, *but*ane, *pent*ane, etc. Branched molecules are named based on the parent alkane. For example, CH_3 is a *methyl* group and C_2H_5 is an *ethyl* group. Numbers are used to show the position of the group, using the lowest numbers possible. For example, $CH_3CH_2CH_2CH(CH_3)CH_3$ is 2-methylpentane.

Haloalkanes

The homologous series of the halogenoalkanes is an alkane series with hydrogen atoms substituted by one or more halogen atoms. They are often shown as R–Hal, where Hal could be F, Cl, Br, or I.

Naming haloalkanes

The alkane chain name is *prefixed* with the name of the halogen. The halogens are listed in alphabetical order, with a number indicating the position of each. For example, C_2H_5Cl is chloroethane, $CH_3CHBrCH_3$ is 2-bromopropane, and CH_2BrCH_2I is 1-bromo-2-iodoethane.

Alcohols

Alcohols all contain the –OH functional group and their names end in -ol. They may be classfied as primary, secondary, or tertiary. It is the position of the –OH group which determines the classification.

Type of alcohol	Position of –OH group	General formula	Example
primary	–OH bonded to a carbon bonded to *one* other carbon atom	RCH_2OH	butan-1-ol
secondary	–OH bonded to a carbon bonded to *two* other carbon atoms	$RCH(OH)R$	butan-2-ol
tertiary	–OH bonded to a carbon bonded to *three* other carbon atoms	$R_2C(OH)R$	2-methylpropan-2-ol

Summary questions

1 Describe the differences between:
   a 2-methylpentane and 3-methylpentane;   *(1 mark)*
   b 2,2-dimethylbutane and 2,3-dimethylbutane;   *(1 mark)*
   c 3-methylhexane and 3-methylheptane.   *(1 mark)*

2 Explain why these alkanes are incorrectly named:
   a 1-methylpropane   *(1 mark)*
   b 2-ethylbutane   *(1 mark)*
   c 2,3-methylbutane.   *(1 mark)*

3 Draw and name primary, secondary, and tertiary alcohols each containing four carbon atoms.   *(3 marks)*

4 For each pair of haloalkanes, choose the correct name:
   a 1-bromo-2-chloroethane or 1-chloro-2-bromoethane   *(1 mark)*
   b 2-chloro-2-iodopropane or 2-iodo-2-chloropropane   *(1 mark)*
   c 1,2,3-iodopropane, triiodopropane, or 1,2,3-triiodopropane.   *(1 mark)*

13.2 Haloalkanes

Specification reference: OZ(d), OZ(k), OZ(l), OZ(m), OZ(n), OZ(q)

Physical properties of haloalkanes

The boiling points increase with a heavier halogen atom (R–I > R–F) or with increasing numbers of halogen atoms (CCl_4 > CH_2Cl_2). As the halogen introduced is larger or the number of halogen atoms increases, the overall number of electrons increases. This increases the instantaneous dipole–induced dipole bonds. With stronger intermolecular bonds, more energy is needed to pull the molecules apart from each other, so the boiling point is higher.

Bond enthalpies and reactivity of haloalkanes

Bond within the molecule	Bond strength	Reactivity
C–F	decreasing strength	increasing reactivity
C–Cl		
C–Br		
C–I		

The C–Hal bond becomes weaker as the size of the halogen atom increases. This makes the bond easier to break and the compounds become more reactive. Although the C–F bond is the most polar, fluoroalkanes are very unreactive. This shows that it is bond strength rather than bond polarity that has the greatest effect on the reactivity of haloalkanes.

- Fluoro- compounds are very unreactive.
- Chloro- compounds are reasonably stable in the troposphere and can react to produce chlorine radicals that deplete ozone.
- Bromo- and iodo-compounds are reactive and so are useful as intermediates in chemical synthesis.

Reactions of haloalkanes

Homolytic fission (forming radicals):

$$R–Hal \rightarrow R\bullet + Hal\bullet$$

The C–I bond most easily undergoes homolytic fission, as it has the lowest bond enthalpy.

Heterolytic fission

$$R–Hal \rightarrow R^+ + Hal^-$$

The carbon–halogen bond breaks to give ions. If the polar C–Hal bond is broken completely, a negative halide ion moves away, leaving the C group positively charged. This is now a **carbocation**.

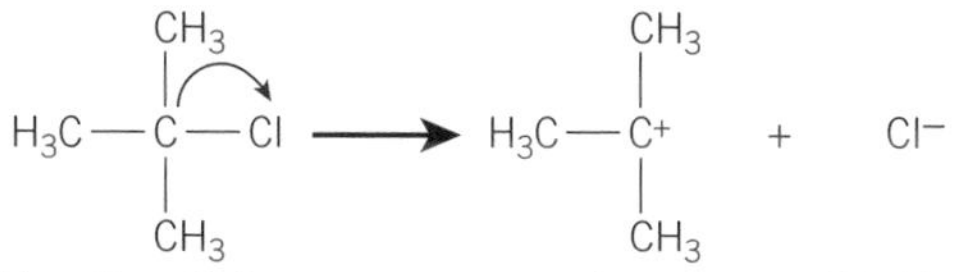

2-chloro-2-methylpropane   carbocation   chloride ion

▲ **Figure 1** *Heterolytic fission in 2-chloro-2-methylpropane*

Substitution reactions:

$$R–Hal + X^- \rightarrow R–X + Hal^-$$

Synoptic link

You learned about instantaneous dipole–induced dipole bonds in Topic 5.3, Bonds between molecules.

Revision tip

Bond polarity suggests that the C–F bond should be the easiest to break as it is the most polar, and that the C–I bond should be the hardest to break as it is least polar.

Revision tip

Bond strength indicates that the C–F bond should be the hardest to break as it is the strongest, and that the C–I bond should be the easiest to break as it is weakest.

Revision tip

As iodoalkanes are the most reactive the C–I bond is the easiest to break and therefore bond strength, not bond polarity, is the determining factor.

Key term

Homolytic fission: A type of bond breaking where each atom in a covalent bond gains one electron, forming radicals.

Key term

Radical: A species with an unpaired electron.

Common misconception: Radicals

When describing radicals, avoid saying they contain a lone electron.

In this case, the C–Hal bond breaks and the halogen atom is replaced by another functional group. Since the halogen is replaced by a nucleophile these reactions are called **nucleophilic substitution** reactions.

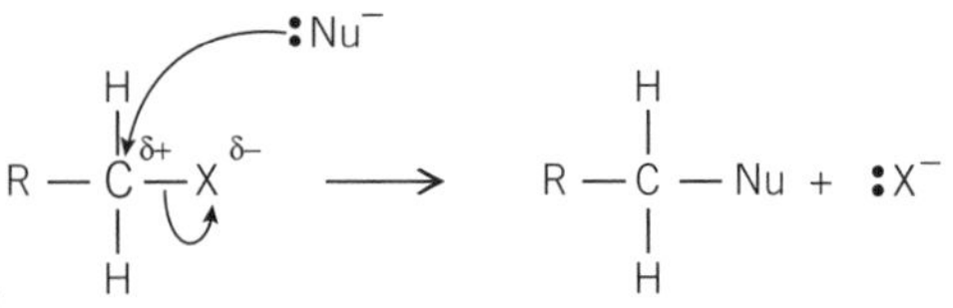

The mechanism of nucleophilic substitution is shown in Figure 2.

▲ **Figure 2** *General reaction mechanism for nucleophilic substitution*

Nucleophile	Equation	Product	Reaction conditions
H_2O	R–Hal + H_2O → R–OH + H^+ + Hal^-	alcohol	heat under reflux: this is sometimes called hydrolysis
OH^-	R–Hal + OH^- → R–OH + Hal^-	alcohol	heated under reflux with NaOH(aq), with ethanol as solvent
$:NH_3$	R–Hal + NH_3 → R–NH_2 + Hal^- + H^+	amine	the haloalkane is heated with concentrated ammonia solution in a sealed tube

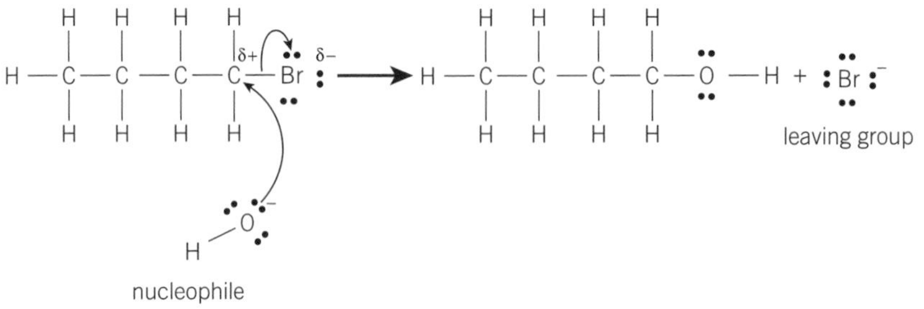

▲ **Figure 3** *Nucleophilic substitution reaction between hydroxide ions and 1-bromobutane*

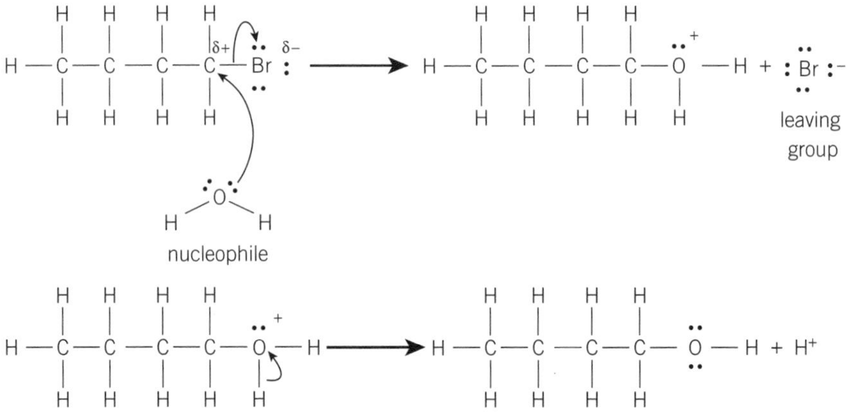

▲ **Figure 4** *Nucleophilic substitution reaction between water molecules and 1-bromobutane – this occurs in two steps. The ion resulting from step 1 loses H^+ to form an alcohol. 'This is known as a hydrolysis reaction.*

Revision tip

Conditions for homolytic fission are gas phase with high temperatures or the presence of UV radiation (e.g. in the stratosphere).

Key term

Heterolytic fission: A type of bond breaking where both electrons in a covalent bond go to one atom, forming ions.

Revision tip

Conditions for heterolytic fission are: dissolved in a polar solvent such as an ethanol / water mixture.

Key term

Carbocation: an ion with a positively-charged carbon atom.

Revision tip

A full headed arrow indicates the movement of a pair of electrons. A half-headed arrow indicates the movement of one electron.

Key term

Nucleophile: A molecule, or negatively charged ion, with a lone pair of electrons that it can donate to a positively charged atom to form a covalent bond.

Summary questions

1 Draw the skeletal formula for $CH_3CH_2CHBrCCl_2CHICH_3$ and name the compound. *(1 mark)*

2 Which in each of the following pairs of molecules has the higher boiling point? Explain your reasoning.

  a CH_3Br and CH_3F

  b CH_3Br and CBr_4. *(4 marks)*

3 Explain why iodoethane is more reactive than fluoroethane. *(1 mark)*

4 Explain the term *nucleophile* and give two examples. *(3 marks)*

5 Give the conditions needed for ammonia to react with a haloalkane. *(1 mark)*

6 Draw the mechanism for the nucleophilic substitution reaction between the hydroxide ion and 1-iodopropane. *(4 marks)*

13.3 Alcohols

Specification reference: WM(b), WM(d)

Primary, secondary, and tertiary alcohols

Primary alcohols have an –OH group bonded to a carbon bonded to one other carbon atom. Secondary alcohols have an –OH group bonded to a carbon bonded to two other carbon atoms. Tertiary alcohols have an –OH group bonded to a carbon bonded to three other carbon atoms.

Oxidation of alcohols

The –OH group can be oxidised using acidified potassium dichromate(VI), $K_2Cr_2O_7$. The –OH group is oxidised to a carbonyl group. At the same time, the $Cr_2O_7^{2-}$ ion, which is orange, is reduced to $Cr^{3+}(aq)$, which is green. During the oxidation reaction, two hydrogen atoms are removed – one from the oxygen atom and one from the carbon atom. The reaction conditions for the oxidation of alcohols are heating the alcohol under reflux with excess acidified potassium (or sodium) dichromate(VI) solution.

condenser

water out

water in

reaction mixture

anti-bumping granules

heat

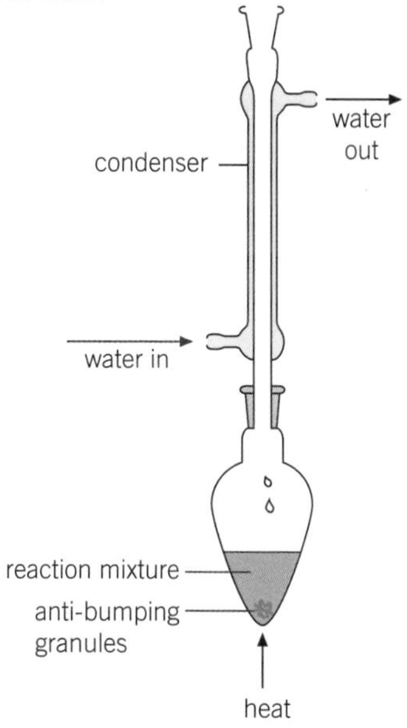

▲ **Figure 1** *Heating under reflux*

The products of oxidation

The product depends on the type of alcohol used. With a primary alcohol, an aldehyde is produced, which can oxidise further to give a carboxylic acid. The colour of the reaction mixture changes from orange to green as the dichromate(VI) ion is reduced. If the aldehyde is required it can be distilled out of the reaction mixture as it is produced, in order to prevent further oxidation. If the carboxylic acid is required, the mixture is heated under reflux with excess potassium dichromate(VI) solution. With a secondary alcohol, a ketone is produced and no further oxidation occurs. The colour of the reaction mixture changes from orange to green. Tertiary alcohols do not undergo oxidation with acidified potassium dichromate(VI), because they do not have a hydrogen atom on the carbon to which –OH is attached. The colour of the reaction mixture does not change, but remains orange.

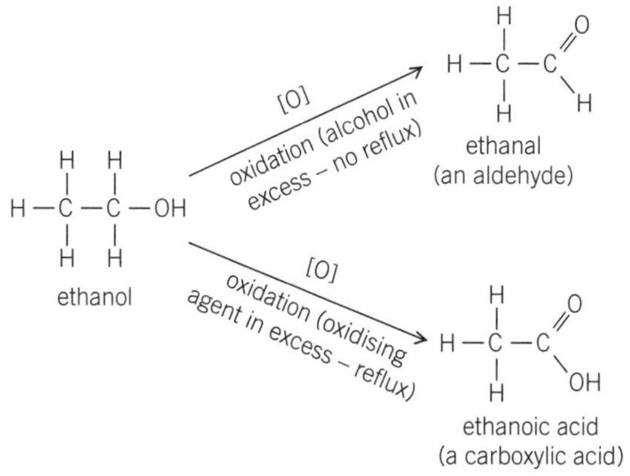

[O]
oxidation (alcohol in excess – no reflux)

ethanol

ethanal
(an aldehyde)

[O]
oxidation (oxidising agent in excess – reflux)

ethanoic acid
(a carboxylic acid)

▲ **Figure 2** *Oxidation of a primary alcohol*

Dehydration of alcohols

Alcohols can lose a molecule of water to produce an alkene. This is known as dehydration and is an example of an elimination reaction.

Synoptic link

There is more detail on the structure and nomenclature of alcohols in Topic 13.1, Naming organic compounds.

Revision tip

You need to learn the reaction conditions for the oxidation of alcohols.

Key term

Reflux: Boiling a volatile liquid with a condenser attached vertically to the flask, to prevent escape of reactants and / or products while the reaction is in progress.

Key term

Carbonyl compound: The C=O bond is called the carbonyl bond, so compounds containing a C=O group (e.g. aldehydes and ketones) are often referred to as carbonyl compounds.

Revision tip

Oxidation of alcohols:

Primary alcohol → Aldehyde (distillation) → Carboxylic acid (reflux)

Secondary alcohol → Ketone (reflux)

Tertiary alcohol – no reaction.

Key term

Elimination reaction: Reactions in which a small molecule is removed from a larger molecule, leaving an unsaturated molecule.

Synoptic link

There is more about carboxylic acids in Topic 13.4, Carboxylic acids and phenols.

Synoptic link

There is more detail on nucleophilic substitution and haloalkanes in Topic 13.2, Haloalkanes.

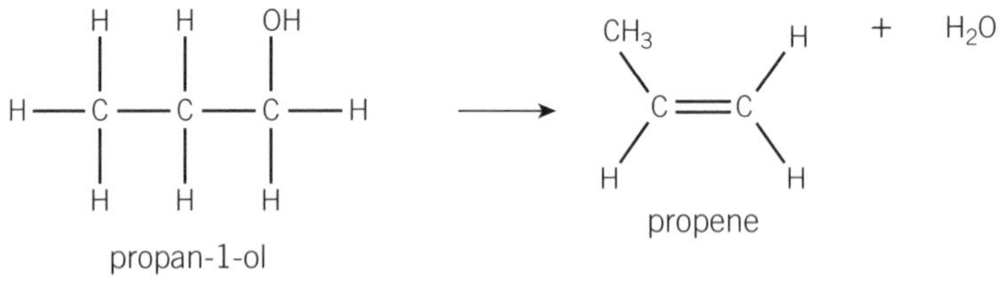

▲ **Figure 3** *Dehydration of an alcohol*

Typical reaction conditions would be using an Al_20_3 catalyst at 300 °C and 1 atm or refluxing with concentrated sulfuric acid.

Formation of esters

Carboxylic acids react with alcohols in the presence of a strong acid catalyst (concentrated sulfuric acid or concentrated hydrochloric acid) when heated under reflux. This reaction (called esterification) is reversible and reaches equilibrium during refluxing:

alcohol + Carboxylic acid ⇌ ester + water

$$CH_3COOH(l) + CH_3CH_2OH(l) \rightleftharpoons CH_3COOCH_2CH_3(l) + H_2O(l)$$

This is shown using the skeletal formulae below:

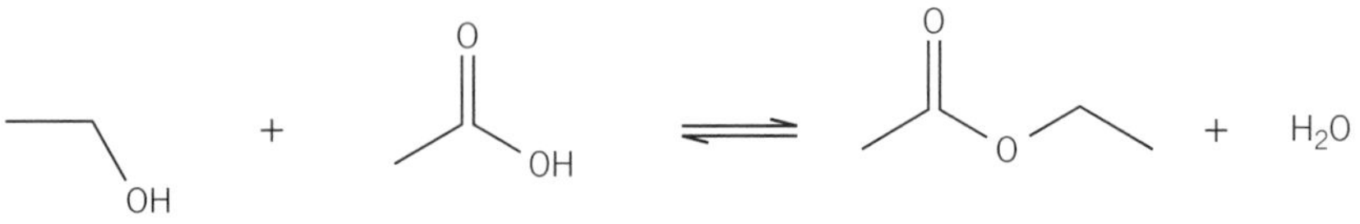

▲ **Figure 4** *Formation of ethyl ethanoate*

Acid anhydrides can also react with alcohols to form esters. Acid anhydrides are more reactive than carboxylic acids, so the reaction goes to completion and does not require a catalyst:

$$(CH_3CO)_2O + CH_3CH_2OH \rightarrow \underset{ester}{CH_3COOC_2H_5} + CH_3COOH$$

Formation of haloalkanes

Nucleophilic substitution reactions can be used to make haloalkanes from alcohols. The reactants are a halide salt (X^-) in the presence of a strong acid catalyst. The H^+ protonates the oxygen in the alcohol:

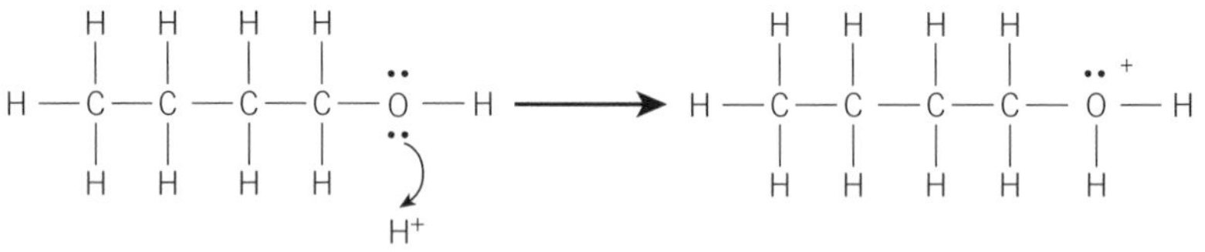

The X^- acts as the nucleophile. The 'leaving group' is a water molecule:

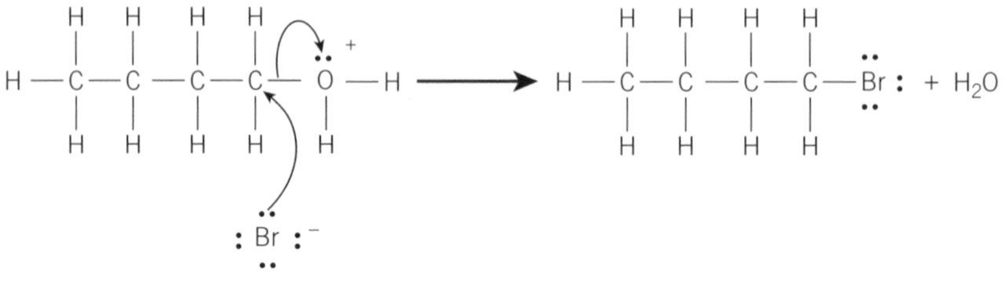

▲ **Figure 5** *Formation of a haloalkane by nucleophilic substitution*

Summary questions

1 Name the homologous series of the following molecules:
  a butanal
  b propanone
  c propanoic acid
  d butan-2-ol
  e ethyl butanoate. *(5 marks)*

2 Name the two oxidation products that could be obtained from the oxidation of butan-1-ol. *(2 marks)*

3 What colour change would you observe when the following alcohols react with acidified potassium dichromate(VI) solution? Explain your answer.
  a 2-methylpropan-2-ol
  b 2-methylpropan-1-ol. *(5 marks)*

4 What reagents and conditions would you use to turn propan-1-ol into propanal? *(3 marks)*

13.4 Carboxylic acids and phenols

Specification reference: WM(a), WM(c), WM(d)

Carboxylic acids contain the **carboxyl** functional group, abbreviated to –COOH. Their systematic names all end with **-oic acid**.

When two carboxyl groups are present, the ending **-dioic acid** is used. Note how the *e* in the name of the alkane is left, for example *ethanedioic* acid.

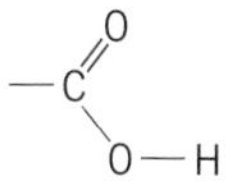

▲ **Figure 1** *Carboxyl group*

Reactions of carboxylic acids

Reactions with alkalis

Carboxylic acids react with alkalis to produce salts:

$$CH_3COOH(aq) + NaOH(aq) \rightarrow CH_3COONa(aq) + H_2O(l)$$

> **Revision tip**
> Remember: acid + base →
> salt + water.

Esterification reactions

In the presence of a strong acid, carboxylic acids react with alcohols as follows:

$$\text{Carboxylic acid + alcohol} \rightarrow \text{ester + water}$$

Esters contain an R-COO-R functional group:

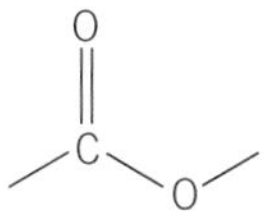

▲ **Figure 2** *An ester group*

> **Revision tip**
> CH$_3$COONa is the salt, sodium ethanoate. It is an ionic compound: CH$_3$COO$^-$ Na$^+$.

Tests for carboxylic acids

Although carboxylic acids are weak acids, they will react with carbonates to produce carbon dioxide. Sodium carbonate or sodium hydrogencarbonate solutions are commonly used to test for acids:

$$2CH_3COOH(aq) + Na_2CO_3(aq) \rightarrow 2CH_3COONa(aq) + H_2O(l) + CO_2(g)$$

$$CH_3COOH(aq) + NaHCO_3(aq) \rightarrow CH_3COONa(aq) + H_2O(l) + CO_2(g)$$

> **Revision tip**
> Remember: carbonate + acid → salt + water + carbon dioxide.

The reaction will produce bubbles of carbon dioxide gas, which are readily seen and can be confirmed by testing the gas with limewater, which turns milky.

Phenols

Phenols are compounds that have one or more –OH groups attached directly to a benzene ring:

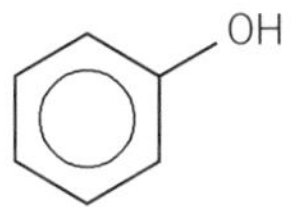

▲ **Figure 3** *Phenol*

Although phenols look similar to alcohols, their chemical reactions are different.

Synoptic link

For more details on esterification reactions, see Topic 13.3, Alcohols.

Acidic properties of phenols

Phenols react with alkalis:

$$C_6H_5OH(aq) + NaOH(aq) \rightarrow C_6H_5ONa(aq) + H_2O(l)$$

But they do not react with carbonates.

Test for phenols

When **neutral iron(III) chloride solution** is added to phenol, a **purple** complex is formed. This test allows us to distinguish between phenols and alcohols.

Formation of esters

To produce an ester from a phenol, an acid anhydride is used under alkaline conditions, as phenols do not react with carboxylic acids:

$$C_6H_5OH(aq) + (CH_3CO)_2O \rightarrow CH_3COOC_6H_5 + CH_3COOH$$

Revision tip

Carboxylic acids are more acidic than phenols.

Revision tip

Carboxylic acids react with alkalis and carbonates.

Phenols react with alkalis, but not carbonates.

Alcohols do not react with alkalis or carbonates.

Revision tip

C_6H_5ONa is the salt, sodium phenoxide. It is an ionic compound: $C_6H_5O^- \, Na^+$.

Revision tip

Remember that alcohols can form esters from carboxylic acids or acid anhydrides. With a phenol, you must use an acid anhydride.

Summary questions

1   Name the following carboxylic acids:

   a $CH_3CH_2CH_2CH_2CH_2COOH$

   b $HOOCCH_2CH_2CH_2COOH$

   c $CH_3CH(CH_3)CH_2COOH$.   *(3 marks)*

2   Write the equation for the reaction of propanoic acid with potassium hydroxide.   *(2 marks)*

3   State the names and formulae of the reactants needed to make propyl butanoate.   *(4 marks)*

4   A student had unlabelled samples of propan-1-ol, 4-methylphenol, and ethanoic acid. Which of the samples would:

   a give a purple colour with neutral iron(III) chloride

   b react with sodium carbonate solution, producing carbon dioxide gas

   c react with acidified potassium dichromate(VI)

   d react with sodium hydroxide solution?   *(4 marks)*

5   Write an equation for the reaction of phenol with sodium carbonate solution.   *(1 mark)*

13.5 Carboxylic acids

Specification reference: PL(h), PL(p)

The structure of carboxylic acids

As you saw in Topic 13.4, Carboxylic acids and phenols, carboxylic acids contain the –COOH functional group, known as the carboxyl group. When counting carbons in the longest chain to name a carboxylic acid, the carbon in the carboxylic acid group is included. For example, CH_3CH_2COOH is propanoic acid. When two carboxyl groups are present the ending is **-dioic acid**. In this case the e in the name of the alkane is kept, e.g. $HOOCCH_2COOH$ is propanedioic acid.

Reactions of carboxylic acids

In aqueous solution, carboxylic acids exist in equilibrium with their dissociated ions:

$$RCOOH(aq) + H_2O(l) \rightleftharpoons RCOO^-(aq) + H_3O^+(aq)$$

Carboxylic acids are weak acids. They are only partially dissociated in solution.

The reaction with metals

Carboxylic acids react with metals to form a salt and hydrogen. The general reaction is:

$$RCOOH + M \rightarrow RCOO^- M^+ + \frac{1}{2}H_2$$

For example, magnesium reacts with benzoic acid as follows:

$$2C_6H_5COOH(aq) + Mg(s) \rightarrow (C_6H_5COO)_2Mg(aq) + H_2(g)$$

The reaction with bases

Carboxylic acids react with bases to form a salt and water. Many useful derivatives can be made from the salts of carboxylic acids. The general reaction is:

$$RCOOH + MOH \rightarrow RCOO^- M^+ + H_2O$$

For example, potassium hydroxide reacts with propanoic acid as follows:

$$C_2H_5COOH(aq) + KOH(aq) \rightarrow C_2H_5COOK(aq) + H_2O(l)$$

The reaction with carbonates

Carboxylic acids react with carbonates to form a salt, carbon dioxide, and water. This is the basis of a test for carboxylic acids – they fizz when added to sodium carbonate. The reaction produces bubbles of carbon dioxide gas, which are readily seen and can be confirmed by testing the gas with limewater, which forms a cloudy precipitate.

The general equation for the reaction of carboxylic acids with carbonates is:

$$RCOOH + M_2CO_3 \rightarrow RCOO^- M^+ + CO_2 + H_2O$$

For example, sodium carbonate reacts with ethanoic acid as follows:

$$2CH_3COOH(aq) + Na_2CO_3(s) \rightarrow 2CH_3COONa(aq) + CO_2(g) + H_2O(l)$$

> **Synoptic link**
>
> You learned about weak acids in Topic 8.2, Strong and weak acids and pH.

> **Revision tip**
>
> Remember, a magnesium ion is Mg^{2+}, and carboxylate ions are $RCOO^-$.

> **Revision tip**
>
> The test for carbon dioxide involves the reaction of limewater, $Ca(OH)_2$, with CO_2:
>
> $$Ca(OH)_2(aq) + CO_2(g) \rightarrow CaCO_3(s) + H_2O(l)$$
>
> $CaCO_3$ forms a cloudy precipitate.

> **Revision tip**
>
> In this example, M is an ion with a single + charge, such as Na^+.

Summary questions

1 Name the following carboxylic acids:
     a $CH_3CH_2CH_2COOH$ b $HOOCCH_2CH_2COOH$ c C_6H_5COOH (3 marks)

2 Write an equation for the reaction of ethanoic acid with:
     a potassium   b potassium hydroxide   c potassium carbonate.   (3 marks)

3 A student reacts propanoic acid with two substances, X and Y. The reaction with X produces a colourless gas that burns with a squeaky pop, and the reaction with Y produces a colourless gas that turns limewater cloudy. What type of substances are X and Y?   (2 marks)

13.6 Amines

Specification reference: PL(j), PL(k), PL(l), PL(n)

Revision tip

The NH_2 functional group is called the amino group.

Revision tip

Remember, R stands for an alkyl or aryl group.

Revision tip

You will sometimes see the prefix 'amino' used instead, as in aminobutane for $C_4H_9NH_2$.

Key term

Base: A hydrogen ion (proton) acceptor.

Revision tip

The lone pair on the nitrogen atom forms a dative covalent bond with the hydrogen ion.

Synoptic link

You learned about bases in Topic 8.1, Acids, bases, alkalis, and neutralisation.

Revision tip

The $OH^-(aq)$ ion is responsible for the alkaline characteristics.

Revision tip

The longer the hydrocarbon chain, the less soluble the amine, as hydrocarbon chains are hydrophobic.

Revision tip

When amines dissolve, H_2O acts as an acid, donating a proton to the amine.

Revision tip

Do not confuse am*ine* ($-NH_2$) with am*ide* ($-CONH-$).

Formulae and nomenclature of amines

Amines are organic compounds, derived from ammonia, NH_3. In amines, one or more of the hydrogens in ammonia are substituted by alkyl groups. The formulae of amines are therefore RNH_2, R_2NH, or R_3N.

Primary, secondary, and tertiary amines

In primary amines, there is one R group. In secondary amines, there are two R groups, and in tertiary amines, there are three R groups.

Naming primary amines

Primary amines are named by taking the name of the R group and adding the suffix 'amine'. For example, CH_3NH_2 is methylamine, $CH_3CH_2NH_2$ is ethylamine, $CH_3CH_2CH_2NH_2$ is propylamine, and $C_6H_5NH_2$ is phenylamine.

Where the amino group is attached to the middle of a chain, numbers are required to indicate its position. For example, $CH_3CH(NH_2)CH_3$ is 2-propylamine.

Diamines

If there are two $-NH_2$ groups, the prefix 'diamino' is used. For example, 1,6-diaminohexane is $H_2NCH_2CH_2CH_2CH_2CH_2CH_2NH_2$. It is one of the monomers for the production of nylon-6,6.

Amines as bases

The nitrogen atom on the NH_2 group has a lone pair of electrons. This is responsible for much of the chemistry of amines. Amines act as bases, as they can accept protons to form a dative covalent bond.

When amines dissolve in water, they form alkaline solutions:

$$RNH_2 + H_2O \rightarrow RNH_3^+(aq) + OH^-(aq)$$

Amines can accept a hydrogen ion from an acid. The general reaction is:

$$RNH_2 + H^+ \rightarrow RNH_3^+(aq)$$

For example, ethylamine reacts with hydrochloric acid to form a chloride salt:

$$CH_3CH_2NH_2(aq) + HCl(aq) \rightarrow CH_3CH_2NH_3^+(aq) + Cl^-(aq)$$

The formation of amides

Amides can be formed from amines by reaction with carbonyl compounds.

Amides contain the –CONH– functional group. Primary amides have this group at the end of the hydrocarbon chain – for example **ethanamide** is CH_3CONH_2.

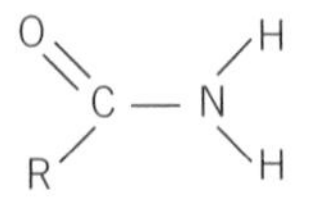

▲ **Figure 1** *A primary amide*

Primary amides have the structure in Figure 1.

Secondary amides have the functional group in the middle of the chain. The nitrogen has one hydrogen and one alkyl group attached to it.

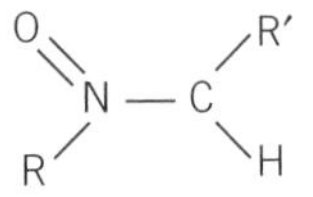

▲ **Figure 2** *A secondary amide*

Secondary amides have the structure in Figure 2.

The polymer nylon is an example of a polyamide.

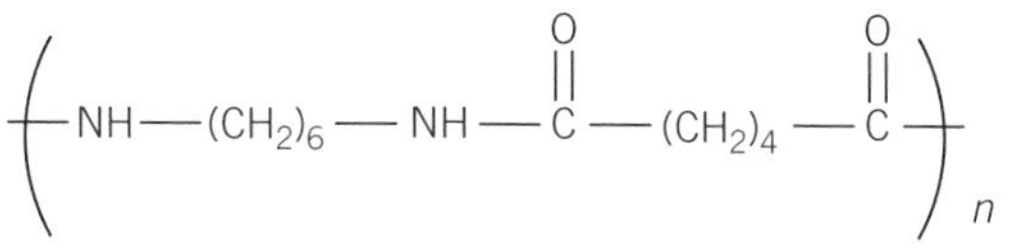

The repeating unit of nylon-6,6

▲ **Figure 3** *The repeating unit of nylon-6,6*

The reactions of amines with acyl chlorides

Acyl chlorides are derivatives of carboxylic acids, with the COCl functional group.

They are reactive forms of carboxylic acids, because the chlorine and oxygen atoms are both significantly more electronegative than carbon.

Acyl chlorides react with ammonia to form primary amides. The general reaction is:

$$RCOCl + NH_3 \rightarrow RCONH_2 + HCl$$

Acyl chlorides react with amines to form secondary amides. The general reaction is:

$$RCOCl + R'NH_2 \rightarrow RCONHR' + HCl$$

For example, ethanoyl chloride (CH_3COCl) reacts with ethylamine as follows:

$$CH_3COCl + C_2H_5NH_2 \rightarrow CH_3CONHC_2H_5 + HCl$$

Acyl chlorides also react with alcohols. In this case, an ester is formed:

$$RCOCl + R'OH \rightarrow RCOOR' + HCl$$

> **Revision tip**
>
> The carbon in the amide group is counted when determining the name of the amide.

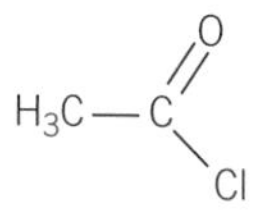

ethanoyl chloride

▲ **Figure 4** *Ethanoyl chloride*

> **Revision tip**
>
> The HCl from these reactions produces white fumes in moist air.

Summary questions

1 Write the structural formulae of:
   a butylamine
   b 2-aminopentane
   c 1,3-diaminopropane. *(3 marks)*

2 Write an equation for the reaction of ethylamine with sulfuric acid. *(2 marks)*

3 Write equations for the reaction of benzoyl chloride, C_6H_5COCl, with:
   a ammonia
   b methylamine
   c methanol. *(3 marks)*

13.7 Hydrolysis of amides and esters

Specification reference: PL(m)

Key term

Hydrolysis: The breaking of a C–O bond in an ester or a C–N bond in an amide by reaction with water.

Key term

Carboxylate salt: The product of the reaction of a carboxylic acid with an alkali. A hydrogen ion is lost from the carboxylic acid to leave an $RCOO^-$ ion.

Revision tip

Alkaline hydrolysis of esters is preferred as the yield is better.

Revision tip

The salt is produced because the amine reacts with H^+ ions.

Synoptic link

You learned about the basic nature of amines in Topic 13.6, Amines.

Revision tip

Proteins, which also contain –CONH– groups, can be hydrolysed this way too.

Revision tip

The carboxylate salt is formed because the carboxylic acid reacts with the hydroxide ions.

Synoptic link

You learned about the reactions of carboxylic acids with alkalis in Topic 13.4, Carboxylic acids and phenols, and Topic 13.5, Carboxylic acids.

What is hydrolysis?

'Hydro' means water and 'lysis' means to break. Therefore *hydrolysis* involves the breaking of a chemical bond through a reaction with water. Hydrolysis reactions can be considered as the reverse of a condensation reaction.

Hydrolysis of esters

Esters are in equilibrium with their constituent carboxylic acid and alcohol:

$$CH_3CH_2OOCCH_3 + H_2O \rightleftharpoons C_2H_5OH + CH_3COOH$$

Hydrolysis brings about the forward reaction, breaking the C–O bond in the ester and forming the carboxylic acid and alcohol.

Acid hydrolysis of esters

Typically sulfuric acid is used as a catalyst.

Alkaline hydrolysis of esters

In this case, OH^- ions are used as a catalyst. The hydrolysis reaction occurs as shown above, but the carboxylic acid reacts with the OH^- ions, so a **carboxylate salt** is produced instead:

$$CH_3CH_2OOCCH_3 + OH^- \rightarrow C_2H_5OH + CH_3COO^-$$

This has the effect of removing the carboxylic acid from the equilibrium above, and consequently the position of equilibrium moves to the right. Therefore the yield of products increases as the reaction goes to completion.

Hydrolysis of amides

The C–N bond breaks and the products are a carboxylic acid and an amine. The reaction is catalysed by either acid or alkali, resulting in the formation of salts.

Acid hydrolysis of amides

The amide is heated with concentrated sulfuric or hydrochloric acid. The products are a carboxylic acid and the salt of an amine:

$$CH_3CONHCH_3 + H_2O + H^+ \rightarrow CH_3COOH + CH_3NH_3^+$$

Alkaline hydrolysis of amides

The amide is heated with moderately concentrated alkali, typically sodium hydroxide. The products are a carboxylate salt and an amine:

$$CH_3CONHCH_3 + OH^- \rightarrow CH_3COO^- + CH_3NH_2$$

Summary questions

1   Write an equation for the hydrolysis of ethyl ethanoate, $CH_3COOC_2H_5$, to form ethanoic acid and ethanol.   *(1 mark)*

2   Write an equation for the alkaline hydrolysis of propyl ethanoate.   *(1 mark)*

3   Write equations for the hydrolysis of $CH_3CH_2CONHCH_2CH_3$ under acid and alkaline conditions.   *(2 marks)*

13.8 Amino acids, peptides, and proteins

Specification reference: PL(a), PL(b), PL(i), PL(q)

The structure of amino acids

Amino acids contain two different functional groups: an amine (–NH_2 group), and a carboxylic acid group (–COOH group). Naturally occurring amino acids have these two groups joined to the same carbon atom. This gives amino acids important characteristics:

- They can form **peptide bonds** by condensation polymerisation to make proteins.
- They act as acids *and* as bases.
- They display a type of isomerism known as optical isomerism as they contain **chiral** carbon atoms.

The general formula of an amino acid is H_2N–CH(R)–COOH. The R-group is a side chain, and each amino acid has a unique R-group. Some examples are shown in Table 1.

▼ **Table 1** *Some amino acids*

R-group	Name and abbreviation of amino acid	Structure
H	Glycine, Gly	H_2NCHCO_2H | H
CH_3	Alanine, Ala	H_2NCHCO_2H | CH_3
CH_2OH	Serine, Ser	H_2NCHCO_2H | CH_2OH
CH_2SH	Cysteine, Cys	H_2NCHCO_2H | CH_2SH
$CH(CH_3)_2$	Valine, Val	H_2NCHCO_2H | $CHCH_3$ | CH_3

There are 20 different naturally occurring amino acids, all with unique R-groups.

Optical isomerism

All amino acids, except glycine, have four different groups attached to the central carbon atom. This gives rise to **optical isomers**.

Chirality

Amino acids, except glycine, contain a chiral centre. Four groups can be arranged in different ways around a carbon atom. This is a type of stereoisomerism, and it is known as optical isomerism.

Figure 1 shows the two optical isomers of alanine. They are mirror images of each other, but the mirror images are non-superimposable.

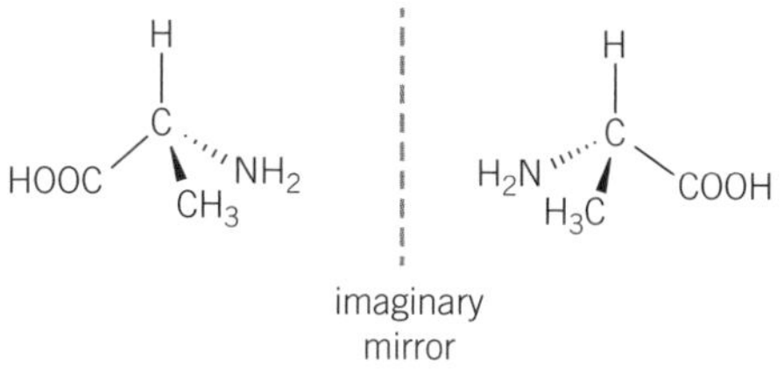

▲ **Figure 1** *The optical isomers of alanine*

Revision tip

Naturally occurring amino acids are known as α-amino acids. They have the –NH_2 and –COOH groups attached to the same carbon.

Revision tip

You do not need to learn the R-groups or abbreviations of amino acids.

Key term

Optical isomer: A stereoisomer with a non-superimposable mirror image.

Key term

Chiral centre: A carbon atom which is attached to four different groups.

Synoptic link

Stereoisomerism was introduced in Topic 12.3, Structural isomerism and E/Z isomerism.

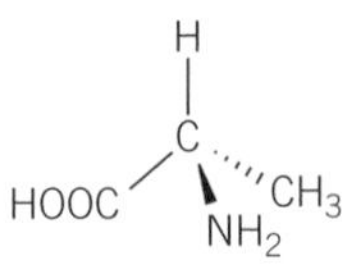

▲ **Figure 2** *Rotating the molecule*

If you rotate the right-hand image you can see that it is not the same as the left-hand image, see Figure 2. The $-CH_3$ group and the $-NH_2$ group are not in the same position. They are separate molecules. The only way to make them superimpose would be to break the bonds, but this would require a chemical reaction to make the new molecule.

Whenever a molecule contains a chiral carbon, with four different groups attached, it will have optical isomers. The optical isomers are sometimes known as **enantiomers**. Enantiomers are separate molecules to each other, although most of their physical properties, such as melting point, are the same. However their biological activity can be very different, due to the way they may fit into the active site of an enzyme, and they sometimes smell or taste different to each other.

Key term

Enantiomer: Either one of a pair of optical isomers.

Acid–base chemistry of amino acids

In alkaline conditions, the –COOH group of an amino acid can lose a proton, leaving a $-COO^-$ group:

$$H_2N\text{–}CHR\text{–}COOH \rightarrow H_2N\text{–}CHR\text{–}COO^- + H^+$$

In acidic conditions, the $-NH_2$ group can accept a proton as the nitrogen has a lone pair and can form a dative covalent bond with H^+:

$$H_2N\text{–}CHR\text{–}COOH + H^+ \rightarrow H_3N^+\text{–}CHR\text{–}COOH$$

Amino acids can also exist as zwitterions, where the NH_2 group is protonated and the COOH is deprotonated. This happens in neutral solution.

Synoptic link

The acid–base chemistry of amines and carboxylic acids are covered in Topic 13.5, Carboxylic acids and 13.6, Amines.

Amino acids are soluble in water due to their ionic nature. Adding small quantities of acid or alkali does not have much effect on the pH as the zwitterions act like buffers, accepting or donating H^+ ions.

Key term

Zwitterion: An amino acid in which the NH_2 group is protonated to form NH_3^+, and the COOH group is deprotonated to form COO^-.

The ionic forms of alanine are shown below.

In acid solution	In neutral solution	In alkaline solution
$H_3N^+\text{–}CH(CH_3)\text{–}COOH$	$H_3N^+\text{–}CH(CH_3)\text{– }COO^-$	$H_2N\text{–}CH(CH_3)\text{–}COO^-$
NH_2 group is protonated due to high concentration of H^+.	Zwitterion.	COOH group deprotonated due to high concentration of OH^-.

Proteins – condensation polymers of amino acids

Amino acids are bifunctional compounds, and the –COOH group can react with the $-NH_2$ group of a neighbouring molecule in a condensation reaction:

$$H_2N\text{–}CHR\text{–}COOH + H_2N\text{–}CHR'\text{–}COOH \rightarrow H_2N\text{–}CHR\text{–}CONH\text{–}CHR'\text{–}COOH + H_2O$$

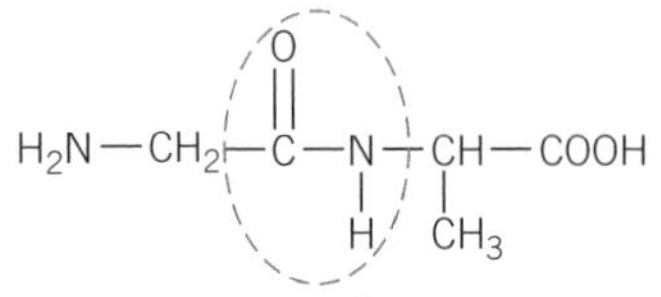

▲ **Figure 3** *A peptide link between glycine and alanine*

The –CONH– group is known as a **peptide link**, and the product above is known as a **dipeptide**. Figure 3 shows the dipeptide formed from glycine and alanine.

Revision tip
A peptide link is a secondary amide.

A polypeptide, or protein, is a molecule containing many amino acid residues joined together with peptide links.

Synoptic link

You learned about amides in Topic 13.7, Hydrolysis of esters and amides, and 13.8, Amino acids, peptides, and proteins.

The primary structure of proteins

Insulin is a protein made from 51 amino acid monomers. The precise order in which the amino acids are joined together is known as the **primary structure** of the protein. For example, the primary structure of insulin begins Gly-Ile-Val-Glu-Gln-Cys..., where each three-letter code refers to an amino acid.

Revision tip
A protein is said to contain amino acid 'residues' as the amino acid lost the elements of water when the peptide bond was formed.

All proteins have different primary structures as they are made of amino acids joined together in different orders.

Hydrolysis of peptide links

Peptides are secondary amides, and hydrolysis of the C–N bond produces the amino acids that the protein is composed of. The protein is hydrolysed by heating with moderated concentrated acid or alkali.

Paper chromatography

Paper chromatography can be used to identify the amino acids that are present in a protein. Once the protein has been hydrolysed, a spot of the product mixture can be placed on a piece of chromatography paper.

Using a suitable solvent, the chromatogram is allowed to run, and the different amino acids in the mixture will separate as the solvent rises up the paper. To reveal the spots, a locating agent, or ultraviolet light, may be needed.

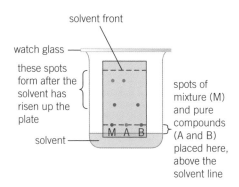

▲ **Figure 4** *Paper chromatography*

R_f values can be calculated: $Rf = \dfrac{\text{distance moved by spot}}{\text{distance moved by the solvent}}$. R_f values are always between 0 and 1, and have no units as the units of distance cancel out.

By comparing the R_f values with known data, or by repeating the chromatography with known samples of amino acids under the same conditions, the identity of the amino acids from the hydrolysed protein can be determined.

Summary questions

1   Draw the structure of valine under:
    a   alkaline conditions
    b   neutral conditions
    c   acidic conditions.   (*3 marks*)

2   Draw the two different dipeptides that can be formed by serine and alanine.   (*2 marks*)

3   Draw the products of the hydrolysis of the dipeptide $H_2NCH_2CONHCH(CH_3)COOH$ when hydrolysed with:
    a   hydrochloric acid
    b   sodium hydroxide.   (*2 marks*)

13.9 Oils and fats

Specification reference: CD(c)

Synoptic link

You learned about esters in Topic 13.3, Alcohols, and Topic 13.4, Carboxylic acids and phenols.

Revision tip

The only difference between oils and fats is that oils are liquid at room temperature, and fats are solid.

Revision tip

Triesters are sometimes called **triglycerides** because they are based on glycerol.

Key term

Fatty acid: A long-chain carboxylic acid, found as an ester in an oil or fat.

Revision tip

Unsaturated fats are considered healthier than saturated fats.

Revision tip

Hydrolysing an oil or fat with an alkali tends to make the reaction go to completion (see Topic 13.7, Hydrolysis of esters and amides).

Revision tip

Hydrolysis of a triester with NaOH produces sodium salts of the fatty acids. The free fatty acids can be released by treating the sodium salt with dilute HCl.

Revision tip

Remember that three moles of NaOH are required to hydrolyse a triester.

Oils and fats are naturally occurring **triesters** of propane-1,2,3-triol (known as glycerol) and long-chain carboxylic acids (known as fatty acids). The three ester groups can be identical, from the same fatty acid, or they can be different. Naturally occurring oils and fats tend to have different fatty acid groups.

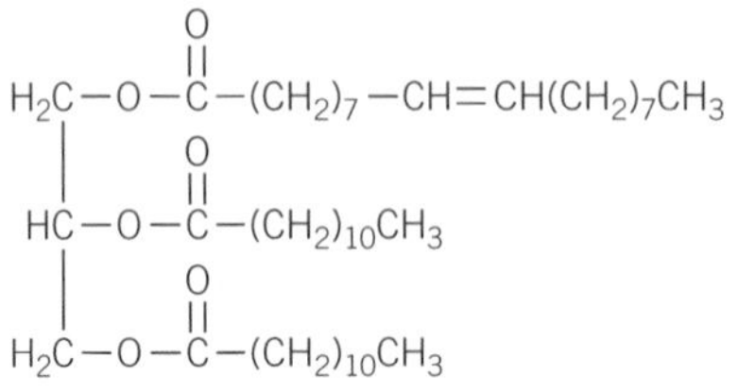

2,3-dilauryl-1-oleoylglycerol

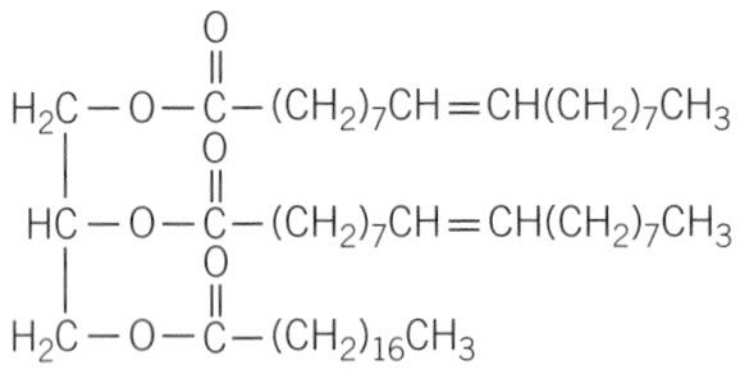

1,2-dioleoyl-3-stearoylglycerol 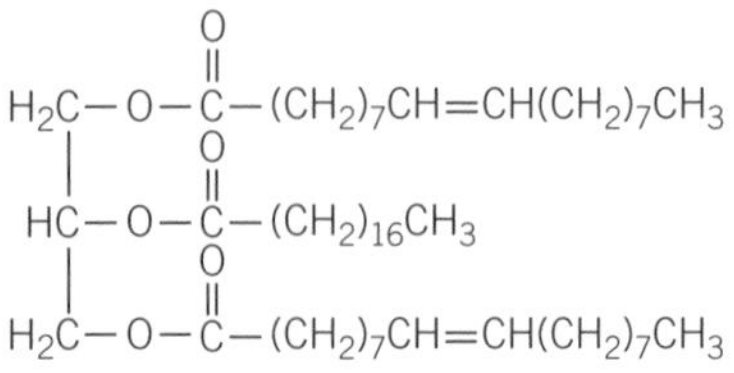   1,3-dioleoyl-2-stearoylglycerol

▲ **Figure 1** *Three different triesters*

Fatty acids

The carboxylic acids in oils and fats are known as 'fatty acids'. They are long hydrocarbon chains, and are usually unbranched. They have an even number of carbon atoms in their chains, up to 24, but typically contain 16 or 18. The presence of these long hydrocarbon chains, which are non-polar, means that oils and fats do not mix with water.

The fatty acid chains can be saturated, containing only single C–C bonds, or unsaturated, with one or more C=C double bonds.

Unsaturated fatty acid chains can be hydrogenated to make the fat more spreadable. In this process C=C bonds react with H_2 to produce C–C bonds.

Hydrolysis of triesters

Hydrolysing a triester involves heating it with concentrated acid or alkali. This produces the three fatty acids and propane-1,2,3-triol.

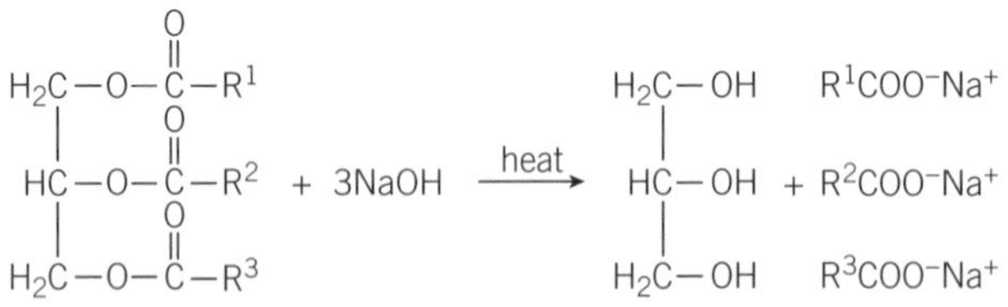

▲ **Figure 2** *The hydrolysis of a triester*

Summary questions

1   What is meant by the term 'mixed triester'?   (*1 mark*)

2   Draw the products of hydrolysis of a triester under alkaline conditions. Use R^1, R^2, and R^3 to represent the fatty acid chains.   (*1 mark*)

3   A mixed triester was hydrogenated. 0.15 mol of the triester required 7.2 dm³ of hydrogen at room temperature and pressure for complete hydrogenation. How many C=C bonds did the triester contain? (*3 marks*)

13.10 Aldehydes and ketones

Specification reference: CD(i), CD(k)

Carbonyl compounds

Many functional groups, such as carboxylic acids and amides, contain C=O bonds. This section considers two homologous series containing the carbonyl group, C=O, aldehydes and ketones.

Aldehydes and ketones

Aldehydes have a carbonyl group at the end of a carbon chain. They have the general formula R–CHO. Their names end in –al.

Ketones have a carbonyl group in the middle of a chain. They have the general formula R^1–CO–R^2. Their names end in –one.

Although they contain the same functional group, aldehydes are more reactive than ketones as they can be oxidised to form carboxylic acids.

Oxidation of aldehydes

Aldehydes can be oxidised using acidified potassium dichromate(VI) to form the corresponding carboxylic acid:

$$RCHO + [O] \rightarrow RCOOH$$

For example, methanal reacts to form methanoic acid:

$$CH_2O + [O] \rightarrow HCOOH$$

Ketones **cannot** be oxidised because they do not contain a hydrogen attached to the carbonyl group.

Distinguishing between aldehydes and ketones

Because aldehydes can be oxidised, they can be distinguished from ketones by refluxing with acidified potassium dichromate. Aldehydes cause a colour change from orange to green, whereas the colour remains orange with ketones.

Fehling's solution

The test substance is warmed with Fehling's solution. Fehling's solution contains Cu^{2+} ions, which oxidise aldehydes. This causes the Cu^{2+} ions to be reduced and a precipitate of copper(I) oxide, Cu_2O, is formed. The colour changes from blue to red. This is a positive test for aldehydes. Ketones do not cause the colour change.

Tollens' reagent

Tollens' reagent contains Ag^+ ions, which oxidise aldehydes. When an aldehyde is warmed with Tollens' reagent, the Ag^+ ions are reduced to metallic Ag, which appears as a silvery layer on the inside of the test tube. Ketones do not produce a silver mirror.

Reaction with hydrogen cyanide, HCN

Aldehydes and ketones can react with HCN to produce a cyanohydrin. This is a nucleophilic addition reaction.

Hydrogen cyanide is too hazardous to use in the laboratory, so acidified KCN is often used instead.

> **Revision tip**
> The aldehyde functional group is written as CHO, not COH, in order to distinguish it from alcohols.

> **Synoptic link**
> In Topic 13.3, Alcohols, you learned that aldehydes are produced by the oxidation of primary alcohols, and ketones are produced by the oxidation of secondary alcohols.

> **Revision tip**
> [O] represents an oxygen from the oxidising agent.

> **Revision tip**
> Because acidified dichromate solution also oxidises alcohols, this is not a conclusive test for aldehydes.

> **Key term**
> Cyanohydrin: A molecule containing an OH and a CN group.

> **Revision tip**
> The CN^- ion has a lone pair which acts as a nucleophile, to form a covalent bond.

> **Revision tip**
> This is an addition reaction because the HCN molecule is added across the C=O bond.

Nucleophilic addition

- The cyanide ion is attracted to the partial positive charge of the carbon atom in the C=O bond.
- A new carbon–carbon bond is formed.
- A pair of electrons from the C=O bond moves to the oxygen atom, which then becomes negatively charged.
- The negatively charged ion then picks up a hydrogen ion, H^+, from the solvent water.

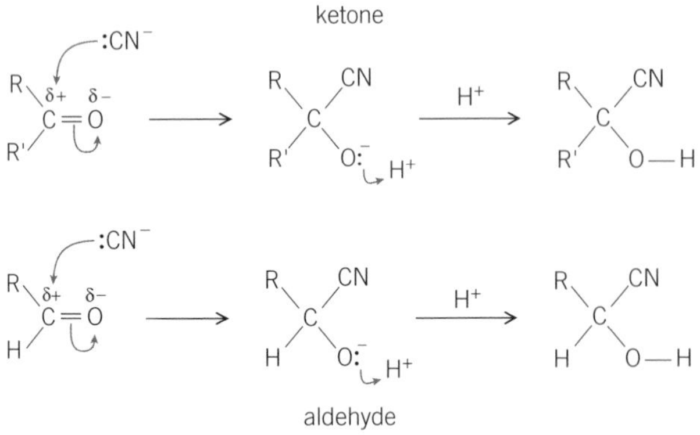

▲ **Figure 1** *The mechanism of nucleophilic addition of a ketone (top) and an aldehyde (bottom)*

This sort of reaction is valuable in organic synthesis as it creates a new carbon–carbon bond. Therefore it is a way of increasing the carbon chain length.

> **Revision tip**
> The C=O bond is polarised because oxygen is more electronegative than carbon.

> **Revision tip**
> Remember, a curly arrow represents the movement of a pair of electrons.

Summary questions

1 A student has mixed up two bottles, one of which contains an aldehyde and one of which contains a ketone. Suggest how the student could identify them using a chemical test. *(2 marks)*

2 Write an equation for the oxidation of :
   a propanal
   b propanone. *(2 marks)*

3 Draw the mechanism of the nucleophilic addition of H–CN to ethanal. *(3 marks)*

Experimental techniques

Specification reference: WM (f)

Heating under reflux

This technique is used to heat volatile reactants together in order to allow a chemical reaction to take place without any of the mixture escaping. This increases yield and reduces the risk of fires.

The reactants and products evaporate on heating but change back into liquids in the condenser and fall back into the reaction mixture in the flask.

1 Put the reactants in a pear-shaped or round-bottomed flask. Add a few anti-bumping granules. Attach a condenser vertically to the flask as shown in the diagram. Do NOT stopper the condenser.

2 Connect the condenser to the water supply.

3 Heat so that the liquid boils gently, using a Bunsen burner or heating mantle.

> ### Common misconceptions: Reflux
>
> It is a common mistake to say that refluxing 'prevents substances evaporating'. Refluxing prevents the evaporated substances from *escaping from the reaction mixture*.

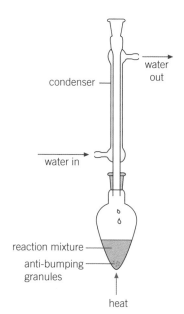

▲ **Figure 1** *Heating under reflux*

Simple distillation

This technique is used to separate two or more miscible liquids with different boiling points. For details of the apparatus used see Figure 2. When the mixture is heated the liquid with the lowest boiling point evaporates first.

The mixture is heated until the thermometer shows the vapour temperature to be at the boiling point of the desired liquid. A clean receiver beaker is placed below the condenser to collect the condensed liquid. When the temperature on the thermometer rises above the boiling point of the desired liquid, the distillation process is stopped.

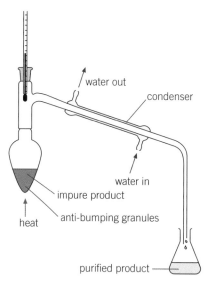

▲ **Figure 2** *Simple distillation*

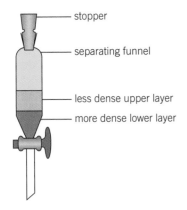

▲ **Figure 1** *A separating funnel*

Revision tip

The lower layer is the denser liquid.

Revision tip

If you are not sure which layer is the aqueous layer, you can add a few drops of water to the separating funnel and watch to see which layer they join.

Purifying an organic liquid product

1 Two immiscible liquids can be separated using a separating funnel. Often an organic product (hydrophobic) is mixed with an aqueous solution. The two layers will separate and can be run off through the seaparating funnel. The aqueous layer can be disposed of.

2 Acidic impurities can be neutralised by adding sodium hydrogen carbonate solution. Alkaline impurities can be neutralised by adding a dilute acid.

3 The crude product can be dried using an anhydrous salt such as anhydrous sodium sulfate or calcium chloride.

4 The pure product can be separated by distillation.

Recrystallisation

Recrystallisation is used to purify solid organic products, by dissolving them in a hot solvent. When cooled, the pure compound will form (recrystallise), with soluble impurities remaining in solution.

1 A suitable solvent is one in which the organic product is very soluble at higher temperatures and insoluble (or virtually insoluble) at lower temperatures.

2 Dissolve the impure solid in the minimum quantity of hot solvent.

3 Filter to remove impurities.

4 Leave the filtrate to cool until crystals form.

5 Collect the crystals by vacuum filtration, and dry them in an oven or in the open, covered by an upside-down filter funnel.

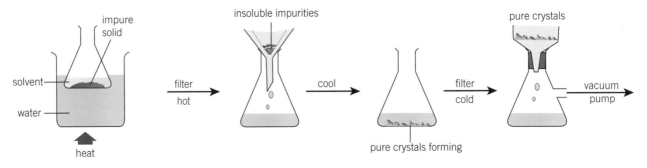

▲ **Figure 2** *Recrystallisation of an impure solid*

Revision tip

You will need to remove the stopper when you are running the liquids off from the separating funnel.

Revision tip

It is important to use the minimum quantity of solvent, as this will give a better yield of product.

Revision tip

It is best to preheat the filter funnel and conical flask to prevent early recrystallisation.

Vacuum filtration

Vacuum filtration enables you to separate a solid rapidly from a filtrate.

1 Connect a side-arm conical flask to a vacuum pump.

2 Place a damp piece of filter paper in the Buchner funnel. The paper should be placed flat.

3 Switch on the vacuum pump and carefully pour in the mixture to be filtered. The partial vacuum ensures that the filtration happens quickly.

Determining melting points

The melting point of an organic solid is evidence of the identity of the substance and of its purity.

1 Heat a glass melting point tube in a Bunsen flame to seal the end.

2 Put a small amount of the solid into the tube by tapping the open end of the tube into the solid, so that a small quantity goes in. Tap the tube so that the solid falls to the bottom of the sealed end.

3 Place the tube in the melting point apparatus and begin heating.

4 Note the temperature range over which the solid melts – when it starts melting, and when it finishes melting.

Thin-layer and paper chromatography

Chromatography involves a mobile phase (a solvent) and a stationary phase (paper or, in thin-layer chromatography, a silica plate). It relies on the fact that different compounds have different affinities for different mobile phases, and so are carried up the stationary phase at different rates. Therefore chromatography can be used to identify the components of a mixture.

Paper chromatography and thin-layer chromatography (TLC) use very similar procedures:

1 Draw a pencil line 1 cm from the base of the chromatography paper or plate.

2 Spot the test mixture and reference samples along the line.

3 Suspend the paper or plate in a covered beaker containing the solvent.

4 Wait for the solvent to reach near the top. Remove the plate, mark the solvent front, and allow it to dry.

5 Locate the spots using iodine, ninhydrin, or ultraviolet light.

6 Match the heights reached, or R_f values, with known data using the same solvent.

> **Revision tip**
> The experimental melting point range can be compared with published values.

> **Revision tip**
> A pure compound will melt within 0.5°C of the true melting point.

> **Revision tip**
> An impure compound will have a wider melting range.

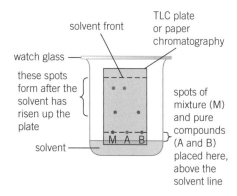

▲ **Figure 3** *Thin-layer or paper chromatography*

1 What is the correct name for $CH_3CH_2CH_2CH_2CH(CH_3)CH_3$?

A heptane

B 5-methylhexane

C 2-methylhexane

D 1-methylheptane (*1 mark*)

2 Which of these homologous series does not contain oxygen?

A carboxylic acid

B phenol

C ketone

D alkane (*1 mark*)

3 $CH_3C(CH_3)(OH)CH_3$ is

A a tertiary alcohol

B an aldehyde

C a carboxylic acid

D a ketone (*1 mark*)

4 Which of these reactions do primary alcohols undergo?

1 Nucleophilic substitution   **2** Oxidation   **3** Dehydration

A 1 only

B 1 and 2 only

C 2 and 3 only

D 1, 2, and 3 (*1 mark*)

5 Which of the following is the formula of an ester?

A C_6H_5COOH

B $C_6H_5COOC_2H_5$

C $C_6H_5COOCOC_6H_5$

D $C_6H_5CH_2OH$ (*1 mark*)

6 Which statement correctly describes the acid-base properties of carboxylic acids and phenols?

A Carboxylic acids react with sodium carbonate but not sodium hydroxide.

B Carboxylic acids react with sodium hydroxide but not sodium carbonate.

C Phenols react with sodium carbonate but not sodium hydroxide.

D Phenols react with sodium hydroxide but not sodium carbonate. (*1 mark*)

7 What are the correct conditions for converting C_2H_5Br into C_2H_5OH?

A Heat under reflux with NaOH(aq) with ethanol as solvent.

B Heat under reflux with acidified dichromate(VI).

C Distil with acidified dichromate(VI).

D React with a carboxylic acid in the presence of a strong acid catalyst. (*1 mark*)

8 Which of the following haloalkanes has the highest boiling point?

A CH_3F

B C_2H_5F

C C_2H_5Cl

D $C_2H_4Cl_2$ (*1 mark*)

9 A student has an unlabelled sample of a substance which is suspected to be a phenol.

The student adds a range of reagents to the separate samples of the substance.

State what the student would expect to see in each case if the substance is a phenol.

   a Adding neutral iron(III) chloride. *(1 mark)*

   b Adding sodium hydrogen carbonate. *(1 mark)*

   c Esterifying the sample with a carboxylic acid and a strong acid catalyst. *(1 mark)*

10 A substance X has the molecular formula $C_4H_{10}O$. When heated under reflux with acidified sodium dichromate(VI) there is no colour change.

   a State the colour of the reaction mixture. *(1 mark)*

   b Explain what this result tells you about substance X. *(2 marks)*

   c Give the structure and name of substance X. *(2 marks)*

   d There are two isomers of X that contain an –OH group at the end of the carbon chain. Give the structure of these isomers. *(2 marks)*

   e Give the products that would be obtained when these isomers are separately oxidised with acidified sodium dichromate(VI). *(4 marks)*

11 The physical properties and reactions of haloalkanes depend on the fact that the C−Hal bond is polar.

   a Describe and explain the polarity of the C−Hal bond. *(2 marks)*

   b Give an example of a physical property that depends on the polarity of the C−Hal bond. *(1 mark)*

   c Nucleophiles are attracted to the polar bond and can substitute the halogen atom. Describe the mechanism of this reaction. *(3 marks)*

12 Which statement is true about carboxylic acids? *(1 mark)*

   A Carboxylic acids react with metals to produce carbon dioxide.

   B The formula of calcium ethanoate is CH_3COOCa.

   C Carboxylic acids are strong acids.

   D Carboxylic acids react with alcohols to form esters.

13 Which of the following reactions of amines is **incorrect**? *(1 mark)*

   A $CH_3NH_2 + H^+ \rightarrow CH_3NH_3^+$

   B $CH_3NH_2 + OH^- \rightarrow CH_3NH^- + H_2O$

   C $CH_3NH_2 + HCl \rightarrow CH_3NH_3^+ Cl^-$

   D $CH_3NH_2 + CH_3COCl \rightarrow CH_3NHOCCH_3 + HCl$

14 What are the products of alkaline hydrolysis of $C_6H_5COOC_2H_5$? *(1 mark)*

   A $C_6H_5OH + C_2H_5COO^-$ **C** $C_6H_5COO^- + C_2H_5OH$

   B $C_6H_5O^- + C_2H_5COOH$ **D** $C_6H_5COOH + C_2H_5O^-$

15 Which statement describes a property of the amino acid alanine, $H_2NCH(CH_3)COOH$? *(1 mark)*

   A Alanine can form two different dipeptides with glycine, H_2NCH_2COOH.

   B Alanine is insoluble in water.

   C Alanine has no isomers.

   D Alanine is deprotonated at low pH.

16 Pentanoic acid occurs naturally in the flowering plant valerian. It has a distinctive, unpleasant odour.

  a Give the molecular formula of pentanoic acid. (*1 mark*)

  b Write an equation for the reaction of pentanoic acid with sodium carbonate. Explain how this is used as a test for carboxylic acids. (*2 marks*)

  c Pentanoic acid reacts with ammonia. Write an equation for this reaction and classify the product as a primary or secondary amide. (*2 marks*)

17 The tripeptide Gly-Ala-Ser was hydrolysed by a student.

  a Draw the structure of the tripeptide before hydrolysis. (Assume that the NH_2 group has not formed a peptide link and is the 'N terminal'.) (*1 mark*)

  b Explain why different products are produced, depending on whether acid or alkaline hydrolysis was used. (*1 mark*)

  c Explain why only two of the products display optical isomerism. Identify the product that does not display optical isomerism. (*3 marks*)

  d Describe how paper chromatography could be used to identify the products of the hydrolysis. (*3 marks*)

18 The naturally occurring compounds heptan-2-one, $C_7H_{14}O$, found in blue cheese, and benzaldehyde, C_6H_5CHO, found in almonds, are both colourless liquids at room temperature. A student carried out a range of reactions on both substances.

  a Which homologous series do these compounds belong to? (*1 mark*)

  b Draw the structure of 2-heptanone. (*1 mark*)

  c Which of the compounds will react with Fehling's solution? What will be observed during the reaction? (*2 marks*)

  d Which of the compounds will react with Tollens' reagent? What will be observed during the reaction? (*2 marks*)

  e Which of the compounds will react with HCN? What product(s) will be formed? (*3 marks*)

14.1 Risk and benefits of chlorine
Specification reference: ES(n), WM(g)

Greener industry

Chemical manufacturers pay careful attention to the area of green chemistry, which involves several different considerations.

Batch processes and continuous processes

During a batch process, products are removed at the end of the reaction and the vessel is cleaned for the next batch. This is cost-effective for small quantities but is time-consuming and may require a larger workforce.

In a continuous process, the starting materials are fed in at one end of the plant and the product emerges at the other end. This is more easily automated, operating for months at a time, but is usually designed to make one product.

Raw materials

Raw materials are usually obtained from the ground or the atmosphere. They are converted into feedstocks – the reactants which are fed in at the start of the process.

Costs and efficiency

The efficiency of a chemical process depends on many factors. High temperatures increase the rate of a reaction, but may affect the position of equilibrium and reduce yields. High pressure can improve the yield, but requires more expensive reaction vessels and has safety implications. Recycling unreacted feedstock is an important way of reducing costs. Efficient use of energy is also essential.

Plant location

Chemical plants were traditionally sited near sources of raw materials. Nowadays, factors which affect the location of a chemical plant include availability of good transport links, skilled labour, cheap energy, and plentiful water.

Health and safety

Safety is a major consideration, and companies have to abide by legislation. Every stage of a manufacturing process is checked to reduce exposure of employees to hazardous chemicals or procedures.

Waste disposal

There are strict limits on the amount of hazardous chemicals that can be released. All chemical waste must be treated before disposal.

Revision tip

Most bulk chemicals are made by continuous processes.

Revision tip

Raw materials can be obtained from rocks, crude oil, seawater, or the atmosphere.

Key term

Feedstock: Feedstocks are chemicals produced from raw materials. They are the reactants in the chemical reactions in the process.

Synoptic link

You learned about rate of reaction in Topic 10.1, Factors affecting reaction rates, and about equilibrium in Topic 7.1, Chemical equilibrium.

Revision tip

You should link your knowledge of green chemistry to any industrial process you study.

Synoptic link

You learned about the halogens in Topic 11.3, The p-block: Group 7.

Chlorine

Chlorine is produced by electrolysis of the feedstock brine, from the raw material rock salt. This is a continuous process and requires large quantities of electricity.

Chlorine is transported under pressure as a liquid in special containers. They are lined with steel and designed to vent small quantities of chlorine if the pressure or temperature gets too high, rather than risk a catastrophic explosion.

These precautions are necessary, as chlorine is a toxic gas. Workers using chlorine must follow strict regulations. About 50 million tonnes of chlorine are used every year in water purification and the production of bleach.

Summary questions

1 Explain the difference between a batch process and a continuous process. *(1 mark)*

2 Suggest if the following processes are batch or continuous:
   a catalytic cracking of gas oil in the petrol industry
   b fermentation of grapes to make wine
   c the manufacture of ammonia in the Haber process. *(3 marks)*

3 Suggest why locating a new chemical plant on an existing site of chemical manufacture can lower costs. *(3 marks)*

14.2 Atom economy

Specification reference: ES(a)

Percentage yield and atom economy

The percentage yield for a reaction is calculated using the equation:

$$\% \text{ yield} = \frac{\text{actual mass of product}}{\text{theoretical maximum mass of product}} \times 100$$

The atom economy for a reaction is calculated using the equation:

$$\% \text{ atom economy} = \frac{\text{relative formula mass of useful product}}{\text{relative formula mass of reactants used}} \times 100$$

 Worked example: Calculating percentage yield and atom economy

Ethanol can be converted into ethene in the following reaction:

$$C_2H_5OH \rightarrow C_2H_4 + H_2O$$

In one reaction, 23.0 g of ethanol produced 6.00 g of ethene. Calculate the percentage yield and the atom economy of this reaction.

Step 1: moles of ethanol used $= \dfrac{\text{mass}}{M_r} = \dfrac{23.0}{46.0} = 0.50 \text{ mol}$

Step 2: We can see from the equation that 0.50 mole of ethanol should, in theory, produce 0.50 mole of ethene, or 14.0 g (28.0 × 0.50).

So % yield $= \dfrac{6.0}{14.0} = 43\%$

Step 3: % atom economy $= \dfrac{28.0}{46.0} = 61.0\%$

Percentage yield used to be a significant factor in deciding if a chemical process was economically viable. With green chemistry high on the agenda, atom economy is just as important.

Atom economy is a way of measuring the efficient use of atoms in a chemical reaction. The higher the atom economy, the more atoms from the reactant molecules end up in the useful product, and the fewer end up in waste products. If the atom economy is 100% there are no waste products, and this has environmental benefits.

Revision tip

Add up all formula masses of all the reactants when calculating % atom economy.

Revision tip

Learn the equations for % yield and % atom economy.

Synoptic link

The theoretical yield is calculated using your knowledge of stoichiometric relationships, which you covered in Topic 1.3, Using equations to work out reacting masses.

Synoptic link

You learned how to calculate formula masses in Topic 1.1, Amount of substance.

Revision tip

If there is only one product, the % atom economy is 100%, as there are no waste products.

Summary questions

1. Epoxyethane, C_2H_4O is an important feedstock in the chemical industry. It is made by the following reaction:

$$2C_2H_4 + O_2 \rightarrow 2C_2H_4O$$

 a Calculate the atom economy of the reaction. (*1 mark*)

 b Suggest what happens to the unreacted ethene in the reaction. (*1 mark*)

2. Place these types of chemical reaction in order from highest to lowest atom economy. Explain your answer. (*2 marks*)
 - addition
 - elimination
 - substitution.

3. Chloroethene, C_2H_3Cl is the monomer for poly(chloroethene) manufacture. It can be produced in a two-step process:

 Step 1: $C_2H_4 + Cl_2 \rightarrow C_2H_4Cl_2$

 Step 2: $C_2H_4Cl_2 \rightarrow C_2H_3Cl + HCl$

 Calculate the atom economies for step 1 and step 2. (*2 marks*)

14.3 Operation of a chemical manufacturing process

Specification reference: CI (i), CI(k)

Chemical reactions in industry

Industry uses a wide range of reaction types. Examples you encounter in this course include:

- Direct combination of elements to make simple molecules

$$N_2 + 3H_2 \rightleftharpoons 2NH_3$$

- Simple oxidation processes

$$2SO_2 + O_2 \rightarrow 2SO_3$$

- Reduction of the oxides in metal ores

$$Fe_2O_3 + 3CO \rightarrow 2Fe + 3CO$$

- Substitution reactions involving organic molecules

$$C_6H_6 + HNO_3 \rightarrow C_6H_5NO_2 + H_2O$$

- Polymerisation reactions

$$nCH_2CH_2 \rightarrow (CH_2CH_2)_n$$

Choices in chemical industry

Chemists need to take decisions about how to manufacture a particular product. These decisions are taken to attempt to ensure that the process is as cost-effective as possible whilst minimising the environmental impact.

Factors that need to be considered include:

- Choice of starting materials (feedstocks) for the reaction
- Conditions needed for the reaction (pressure, temperature)
- The possible use of a catalyst
- The atom economy of the reaction used
- The yield of the reaction
- The rate at which products are formed
- The formation of any waste products that need to be disposed of
- The hazards associated with feedstocks or products.

Most of these factors are described in more detail in other chapters of this guide.

Costs of industrial processes

There are several components to the cost of an industrial process:

- The cost of the raw materials and any costs associated with converting them into feedstocks that can be added to the reactor.

  For example, crude oil is a relatively cheap raw material, although there are considerable costs associated with the fractional distillation and cracking needed to extract ethene molecules which are necessary for polymerisation reactions.

- Energy costs. Fuel needs to be burnt, or electricity used to create high temperatures and pressures.

- Costs associated with plant. Many chemical processes require expensive specialist apparatus – for example thick-walled reactors to enable reactions to take place at high pressures.

Revision tip

You will not be required to recall the details of particular industrial processes, but if you are given information about a process you will need to be prepared to use and analyse the information provided.

Key term

Raw materials: The naturally occurring resource from which the feedstock is obtained. Examples include crude oil, natural gas, metal ores, the air, sea water, and biomass such as plant material.

Synoptic link

The factors affecting the rate of reaction are described and explained in Topic 10.1, Factors affecting reaction rates, and Topic 10.2, The effect of temperature on rate.

Synoptic link

The factors affecting equilibrium position of industrial reactions are covered in Topic 7.5, Equilibrium constant K_c, temperature, and pressure.

Synoptic link

Atom economy is explained in Topic 14.2, Atom economy.

Discussing the conditions used in the operation of a chemical process

An example of how to do this is provided in the model answer below:

Model answer: Manufacture of ammonia

Ammonia, NH$_3$ is manufactured from nitrogen and hydrogen. The conditions used are 200 atmospheres and 450 °C with the use of an iron catalyst. Discuss why these are the most cost-effective conditions to use.

High pressure increases the rate of the reaction and the yield at equilibrium (because the equilibrium position favours the side of the equation with the fewest moles). However, high pressure requires a lot of energy to maintain and requires expensive reaction vessels.

High temperature increases the rate but decreases the yield (because the equilibrium position favours the endothermic direction). However, high temperature requires a lot of energy to maintain.

A very high pressure is chosen, but above 200 atmospheres the extra costs outweigh the increase in rate and yield. A moderately high temperature is a compromise, to produce an acceptable yield without decreasing the rate too much. The use of a catalyst helps to increase the rate without the need for high temperature.

> The answer discusses the effect of the high pressure on rate and yield.

> The answer discusses the effect of the moderately high temperature on rate and yield.

> The answer explains how the chosen conditions are the most cost effective, using ideas from the previous two paragraphs.

Co-products and by-products

The overall cost of operating an industrial process is also affected by the formation of any additional products formed alongside the main product. Additional products will need to be separated from the main product which increases the cost of the process. They may also be hazardous or require special handling. However, if the additional products can be sold then this can help to make the overall process more profitable.

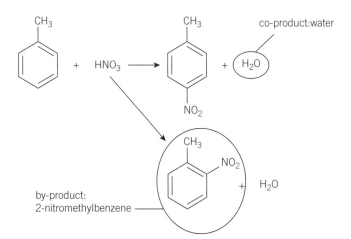

▲ **Figure 1** *In the nitration of methylbenzene, water is a co-product because it is formed alongside the product in the intended reaction. 2-nitromethylbenzene is a by-product because it is formed in an unintended reaction*

Choices for society

If a product is particularly hazardous, then society may need to take an ethical decision about whether its manufacture and use should be permitted.

These decisions may be based on:

- the benefit that the product may bring to individuals or society.

Revision tip

You may be required to discuss the advantages and disadvantages of the use and manufacture of a particular product, and reach a reasoned conclusion about whether its use and manufacture is justified.

This is weighed up against:

- the risks or hazards associated with the use or manufacture.

An example of how to do this is provided in the model answer.

Model answer: Discuss the risks and benefits in the use of chlorine in water treatment

Chlorine kills bacteria that may cause serious or fatal diseases such as cholera.

However chlorine can be harmful to humans if the concentration in water is too high.

There are also risks associated with leaks of chlorine gas during manufacture or transport.

If suitable safety precautions are taken during transport, and if the concentration of the chlorine is kept to as low a level as possible, then the benefits of the use and manufacture of chlorine outweigh the risks.

> Describe a benefit of using chlorine.

> Describe some of the hazards and risks of handling or using chlorine.

> Balance the benefits against the risks and reach a justified conclusion about the use if chlorine.

Summary questions

1 2-chlorobutane can be converted to but-2-ene in an elimination reaction:

A small amount of but-1-ene is also formed.
Classify HCl and but-1-ene as either by-product or co-product and explain your answer.   (3 marks)

2 Many important products can be prepared in two (or more) ways. For this pair of processes, select the process which is likely to be the greenest (has the lowest environmental impact). Give a reason for your answer.
   Chloroethane can be synthesised in industry by two different processes:
   A From the reaction of ethene and hydrogen chloride at 130 °C with an aluminium chloride catalyst
   $C_2H_4 + HCl \rightarrow C_2H_5Cl$
   B From the reaction of ethanol and hydrochloric acid over a zinc chloride / aluminium oxide catalyst at 325 °C
   $C_2H_5OH + HCl \rightarrow C_2H_5Cl + H_2O$
   Suggest and explain three reasons why reaction A is currently more cost effective than reaction B   (6 marks)

3 Hydrcarbons, such as methane, can be formed from CO and water. The exothermic reaction is carried out at 400 °C and atmospheric pressure, using a catalyst that is a mixture of iron(III) oxide and chromium(III) oxide:
   $CO + H_2O \rightleftharpoons CO_2 + H_2$
   a Explain the reasons for the choice of these conditions.   (7 marks)
   b Discuss whether the use of this reaction will be an overall benefit to society, bearing in mind the risks and benefits of the substances involved.   (5 marks)

1   Which of the following chemicals are produced by batch processes?

   A Ammonia in the Haber Process

   B Ethanol by reacting ethene and steam

   C Pharmaceuticals to be tested for biological effect

   D Sodium hydroxide by electrolysis of brine.   (*1 mark*)

2   Which of the following considerations are taken into account when deciding the location of a chemical plant?

   1 Availability of labour

   2 Transport links

   3 Availability of raw materials.

   A 1 only   **B** 2 only

   C 2 and 3 only   **D** 1, 2, and 3   (*1 mark*)

3   What is the correct formula for calculating atom economy?

   A $\dfrac{\text{relative formula mass of useful product}}{\text{relative formula mass of reactants used}} \times 100$

   B $\dfrac{\text{relative formula mass of reactants used}}{\text{relative formula mass of useful product}} \times 100$

   C $\dfrac{\text{actual mass of product}}{\text{theoretical maximum mass of product}} \times 100$

   D $\dfrac{\text{theoretical maximum mass of product}}{\text{actual mass of product}} \times 100$ (*1 mark*)

4   The equation for the formation of ethene from ethanol is $C_2H_5OH \rightarrow C_2H_4 + H_2O$. What is the atom economy of this reaction?

   A 1.64%   **B** 28%

   C 39%   **D** 61%   (*1 mark*)

5   A student carried out a reaction to produce propanone from propan-2-ol:

$$CH_3CH(OH)CH_3 + [O] \rightarrow CH_3COCH_3 + H_2O$$

   [O] represents an oxygen from the oxidising agent.

   a What reagent and conditions would the student use for this reaction?   (*3 marks*)

   b What functional groups do propanone and propan-2-ol contain? (*2 marks*)

   c Calculate the atom economy of the reaction.   (*1 mark*)

   d The student started with 6.0 g of propan-2-ol. Calculate the amount in moles that the student used.   (*1 mark*)

   e What amount (in moles) of propanone would the student expect to make?   (*1 mark*)

   f What is the theoretical maximum mass of propanone in this reaction?   (*1 mark*)

   g The student's actual yield was 2.61 g. Calculate the percentage yield.   (*1 mark*)

6   A student carried out a reaction to produce ethanol from bromoethane:

$$C_2H_5Br + H_2O \rightarrow C_2H_5OH + HBr$$

   a Calculate the atom economy of the reaction.   (*1 mark*)

   The student reacted 5.445 g of C_2H_5Br and produced 1.426 g of C_2H_5OH.

   b Calculate the percentage yield.   (*4 marks*)

7 Nitrogen oxide is converted into nitrogen dioxide in industry:

$2NO(g) + O_2(g) \rightleftharpoons 2NO_2(g) \Delta_rH = -115\,kJ\,mol^{-1}$

Which changes will produce an increase in both rate and yield?

A Using a catalyst and increasing the pressure.

B Increasing the temperature and decreasing the pressure.

C Using a catalyst and increasing the temperature.

D Decreasing the pressure and decreasing the temperature. *(1 mark)*

8 Titanium(IV) oxide can be formed from titanium(IV) chloride by oxidation at high temperature. $TiCl_4(l) + O_2(g) \rightarrow TiO_2(s) + 2Cl_2(g)$

Which aspect is likely to be a significant environmental issue?

1 The reaction must be carried at high pressure, increasing the risk of explosions.

2 The raw materials used to produce oxygen are not renewable.

3 The atom economy of the process is very low.

A 1,2, and 3   B 1 and 2   C 2 and 3   D Only 3   *(1 mark)*

9 Nylon 6,6 can be formed in the laboratory from hexanedioyl dichloride and 1,6-diaminohexane. However in industry hexanedioic acid is used in place of hexanedioyl dichloride. The main reason for this is that:

A Hexanedioyl dichloride must be formed from crude oil which is non-renewable.

B Using hexanedioyl dichloride creates a toxic co-product during the reaction.

C Hexanedioic acid reacts more rapidly than hexanedioyl dichloride.

D No catalyst is necessary when using hexanedioic acid. *(1 mark)*

10 Which of these types of reaction will have the highest atom economy?

A Substitution   B Elimination   C Addition   D Condensation *(1 mark)*

11 Ammonium nitrate can be made from the reaction of ammonia and concentrated nitric acid: $NH_3(g) + HNO_3(aq) \rightarrow NH_4NO_3(aq)$

Which of these are likely to be significant hazards of this process:

1 Ammonia is flammable.

2 Nitric acid is corrosive.

3 The reaction is likely to be highly exothermic.

A 1,2, and 3   B 1 and 2   C 2 and 3   D only 3   *(1 mark)*

12 As part of the process that results in the manufacture of nitric acid, ammonia is oxidised to nitrogen oxide (NO).

The equation for the reaction is:

$4NH_3(g) + 5O_2(g) \rightleftharpoons 4NO(g) + 6H_2O(g) \Delta_rH = -905.2\,kJ\,mol^{-1}$

Under some conditions, some nitrogen dioxide, NO_2 may also be formed.

a Calculate the atom economy of this process, and comment on your answer. *(2 marks)*

b Identify a **co-product** and a **by-product** of the reaction to form nitrogen oxide. Explain your answer. *(4 marks)*

   i The reaction is carried out industrially at a temperature of 500 K. Explain why a higher temperature is not used. *(2 marks)*

   ii A pressure of about 5 atmospheres is often chosen for the reaction. Discuss the reasons for the choice of 5 atmospheres. *(6 marks)*

15.1 Atmospheric pollutants

Specification reference: DF(k)

Production and effects of pollutants from petrol

Combustion of hydrocarbons in car engines results in a wide range of pollutants being produced. These atmospheric pollutants can have significant environmental implications.

Particulates

Particulates are small carbon particles below 2.5×10^{-12} m formed from burning fossil fuels. They can penetrate into the human body causing lung cancer and heart attacks. They also make surfaces dirty.

Unburnt hydrocarbons

These are formed from incomplete combustion or leakage from fuel tanks. They can go on to form photochemical smog.

Carbon monoxide

CO is formed from incomplete combustion of hydrocarbons. It is a toxic gas which is directly harmful to humans. Catalytic converters can oxidise CO into CO_2.

> **Revision tip**
> Particulates and carbon monoxide harm humans directly.

Carbon dioxide

CO_2 is formed from the complete combustion of hydrocarbons. It is a greenhouse gas and is responsible for global warming.

Nitrogen oxides

Nitrogen monoxide is formed from the reaction of oxygen and nitrogen (from the air) at high temperatures inside car engines. Nitrogen monoxide is readily oxidised to nitrogen dioxide by oxygen in the atmosphere. Nitrogen oxides (often written as NO_x) dissolve in rainwater to cause acid rain. Catalytic converters can reduce nitrogen monoxide to nitrogen, N_2.

> **Revision tip**
> Exposure to NO_2 has been linked to respiratory illnesses such as asthma or bronchitis.

Sulfur oxides

Sulfur oxides, SO_x, are formed from the oxidation of sulfur impurities in fuels. Sulfur oxides form acid rain. It is possible to remove the sulfur from fuel before burning – this is called desulfurisation.

> **Revision tip**
> Global warming and acid rain harm humans indirectly.

Summary questions

1. Write an equation for the formation of nitrogen monoxide in a car engine. Describe one polluting effect of nitrogen monoxide. *(2 marks)*

2. Describe the harmful effects of carbon monoxide. Suggest an equation for the formation of carbon monoxide from methane. *(2 marks)*

3. A lean burn engine uses a higher ratio of air to petrol vapour than other engines. Explain why less unburnt hydrocarbons are produced. *(1 mark)*

15.2 Alternatives to fossil fuels

Key terms

Carbon neutral: On burning, a carbon-neutral fuel only releases as much CO_2 as the plant it came from absorbed when it was growing.

Sustainable: Able to be used without negative impact on future generations.

Renewable: Renewable fuels will not run out, because they come from sources such as plants, which can be regrown.

Finite resource: A finite resource will run out. Crude oil is a finite resource.

Revision tip
A fuel can only be considered truly carbon neutral if its production, transportation, and so on are carbon neutral as well.

Alternative fuels for a car engine

People need to make choices, both now and in the future, regarding the use of fuels. We need to weigh up the risks, benefits, and sustainability for each fuel.

A variety of fuel choices for the future are discussed in the table below.

Alternative	Is it sustainable?	Benefits	Risks
Diesel	No; crude oil is running out	Less CO_2 produced than from a petrol engine; already sold at petrol stations	Produces more NO_x and particulates than a petrol engine; particulates can irritate lungs
LPG or autogas	No; crude oil is running out	Less CO, CO_2, C_xH_y, and NO_x than from a petrol engine; petrol engines easily converted	Needs to be stored under pressure, so that it is a liquid
Ethanol	Possibly not; large amounts of energy needed for cultivating sugar cane for fermentation	Less CO, SO_2, and NO_x than from a car engine; ethanol has a high octane number; sugar cane absorbs CO_2 in growth	Highly flammable
Biodiesel	Can be made from waste plant and animal oils and fats, so renewable, but fossil fuels may be used as an energy source in production	Living things have absorbed CO_2; it is biodegradable; less CO, C_xH_y, SO_2, and particulates than from a diesel engine	NO_x emissions higher than a diesel engine
Hydrogen	Only if the electricity needed for electrolysis of water is from a renewable source such as solar cells	Water is the only product of combustion	Highly flammable; high-pressure fuel tank needed to store it as a liquid

Summary questions

1 What problems may arise in the storage of hydrogen in a fuel tank? *(1 mark)*

2 Biodiesel can be manufactured from soya beans. Give one possible advantage and one disadvantage of such a fuel. *(2 marks)*

3 Suggest why ethanol is only likely to be a suitable choice in certain parts of the world. *(1 mark)*

The greenhouse effect

The temperature of the Earth's surface depends on the balance between energy absorbed from the Sun and energy emitted by the Earth's surface, which is then lost into space. If there were no atmosphere, the average temperature of the Earth's surface would be about 254 K (−19 °C).

The presence of greenhouse gases in the troposphere causes a greenhouse effect that increases the average temperature of the Earth's surface to about 287 K (14 °C).

Human activities have increased the concentration of greenhouse gases in the troposphere. This is resulting in an enhanced greenhouse effect, raising temperatures even higher. Current models suggest that unless emissions of greenhouse gases are reduced significantly, temperatures may rise a further 3 °C or 4 °C higher by the end of the century.

Emissions from the Sun and the Earth

The Sun and the Earth both emit electromagnetic radiation. However, because of the different temperatures at the surface of these bodies, they emit different types of electromagnetic radiation.

▼ **Table 1** *Radiation emitted from the Sun and the Earth*

Body	Surface temperature	Main type(s) of radiation emitted	Approximate range of frequencies emitted	Approximate range of wavelengths emitted
Sun	6000 K	Ultraviolet Visible Some infrared	$10^{14}–10^{15}$ Hz	3000 to 300 nm
Earth	287 K	Infrared	$10^{12}–10^{13}$ Hz	188 30 000 nm to 300 000 nm

Effect on molecules in the Earth's atmosphere

- Ultraviolet radiation can be absorbed by molecules such as ozone in the atmosphere and cause bond breaking.
- Most of the visible radiation from the Sun passes through the atmosphere without being absorbed.
- Infrared radiation from the Earth can be absorbed by the bonds in molecules, causing them to vibrate with greater energy.

The infrared window

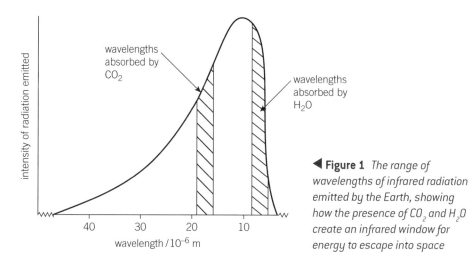

wavelengths absorbed by CO_2

wavelengths absorbed by H_2O

intensity of radiation emitted

wavelength / 10^{-6} m

◀ **Figure 1** *The range of wavelengths of infrared radiation emitted by the Earth, showing how the presence of CO_2 and H_2O create an infrared window for energy to escape into space*

The most abundant greenhouse gas in the troposphere is water vapour. This absorbs frequencies of infrared radiation that fall within the range of wavelengths emitted by the Earth's surface, leaving a 'window' of infrared wavelengths that can pass through.

Increasing the concentration of CO_2 or other greenhouse gases will increase the amount of total infrared radiation absorbed, and decrease the amount of radiation escaping into space.

Other greenhouse gases absorbing radiation in the 'window' include any molecules containing C–F bonds, for example chlorofluorocarbons (CFCs).

Greenhouse gases

▼ **Table 2** *Main sources of greenhouse gases*

Gas	Main source from human activity
carbon dioxide	burning of fossil fuels
methane	cattle farming
dinitrogen oxide	decomposition of nitrogen-based fertilisers
haloalkanes	previously used as refrigerants, solvents etc.

* Although water vapour is an abundant greenhouse gas, the emissions of water vapour by human activities have only a negligible impact on the concentration in the troposphere.

The heating effect of infrared absorption

When bonds in the molecules of greenhouse gases absorb infrared radiation from the Earth, they vibrate with greater energy.

This absorbed energy can be re-emitted in the form of infrared radiation. However, when this happens, the infrared radiation is emitted in all directions. The overall effect of this is that some of the radiation that would have been lost into space is now returned to Earth, causing heating of the surface.

Some of the vibrational energy is also transferred to other molecules in the atmosphere by collisions. This increases the average kinetic energy of molecules in the atmosphere and the temperature of the atmosphere increases.

Putting it all together

Figure 3 helps visualise the greenhouse effect.

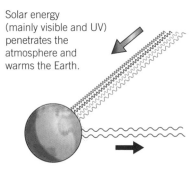

Solar energy (mainly visible and UV) penetrates the atmosphere and warms the Earth.

The Earth absorbs this energy, heats up and radiates IR Greenhouse gases in the troposphere absorb some of this IR

▲ **Figure 3** *The greenhouse effect*

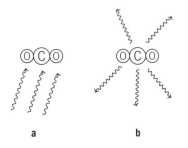

a   b

▲ **Figure 2** **a** *Greenhouse gases absorb infrared radiation emitted from the Earth* **b** *the infrared radiation is re-emitted in all directions*

Model answer: The greenhouse effect

Describe the greenhouse effect and how an increase in carbon dioxide concentration in the troposphere is responsible for an enhanced greenhouse effect.

- The Sun emits mostly visible and ultraviolet radiation.
- Most of the ultraviolet radiation is absorbed by the Earth's atmosphere, but visible light reaches the surface of the Earth.
- This visible light is absorbed by the Earth's surface and warms it up.
- The Earth's surface emits infrared radiation.
- Some frequencies of this infrared radiation are absorbed by carbon dioxide molecules in the troposphere, increasing the vibrational energy of the bonds; the remaining frequencies (in the infrared window) escape into space.
- The absorbed energy is re-emitted as infrared radiation in all directions .
- Some of this is emitted in the direction of the Earth's surface, heating it up.
- Some energy is also transferred (by collisions) to other molecules in the atmosphere, increasing the temperature of the atmosphere.
- If carbon dioxide concentration increases, more infrared radiation is absorbed and less energy is lost into space through the infrared window.

> Always start by describing the ultimate source of all the energy reaching the Earth's atmosphere – the Sun.

> Describe the processes causing the Earth to emit infrared radiation.

> Describe how molecules of greenhouse gas absorb infrared radiation.

> Describe how the absorbed energy causes heating of the atmosphere and the Earth's surface.

> Make a clear link between increased concentrations of greenhouse gases and an increase in energy absorbed by these gases.

Summary questions

1 Describe the differences in the electromagnetic radiation emitted by the Sun and the Earth. In your answer include reference to wavelength and frequency. *(3 marks)*

2 Sulfur hexafluoride (SF_6) is a powerful greenhouse gas because it absorbs in the infrared window. Explain the meaning of this statement. *(2 marks)*

3 If the temperature of the Earth rises, more water will evaporate to form water vapour. Describe and explain the further effect this will have on the temperature of the Earth. *(5 marks)*

Revision tip

When describing the greenhouse effect you should begin by describing the radiation emitted by the Sun.

1 Which of the following pollutants is produced during complete combustion of hydrocarbons?

  A carbon monoxide   **B** particulates

  C carbon dioxide   **D** unburnt hydrocarbons   *(1 mark)*

2 Which of the following statements best explains why nitrogen oxides are formed in internal combustion engines?

  A Nitrogen from the fuel reacts with oxygen from the air.

  B Nitrogen from the air reacts with oxygen from the fuel.

  C Nitrogen and oxygen from the fuel react together.

  D Nitrogen and oxygen from the air react together.   *(1 mark)*

3 Which of the following statements best explains why sulfur oxides are formed in internal combustion engines?

  A Sulfur from the fuel reacts with oxygen from the air.

  B Sulfur from the air reacts with oxygen from the fuel.

  C Sulfur and oxygen from the fuel react together.

  D Sulfur and oxygen from the air react together.   *(1 mark)*

4 What is an advantage of hydrogen fuel?

  A It is produced by electrolysis.

  B Water is the only product of combustion.

  C It is carbon neutral.

  D It can be easily stored.   *(1 mark)*

5 Why do some fuels produce less nitrogen oxide pollution than petrol?

  A They do not contain nitrogen atoms in their molecules.

  B They are more likely to undergo complete combustion.

  C They have a greater energy density.

  D They burn at lower temperatures.   *(1 mark)*

6 What is the best definition of a *sustainable* fuel?

  A It will never run out.

  B It can be used without negative impact on future generations.

  C It lasts a long time.

  D It does not contribute to global warming.   *(1 mark)*

7 Write equations for the following reactions which produce different pollutants.

  a The complete combustion of decane, $C_{10}H_{22}$.   *(1 mark)*

  b The oxidation of nitrogen monoxide to nitrogen dioxide by oxygen in the air.   *(1 mark)*

  c The oxidation of sulfur dioxide to sulfur trioxide by oxygen in the air.   *(1 mark)*

  d The formation of acid rain from sulfur trioxide.   *(1 mark)*

8 Explain why biodiesel is considered a sustainable fuel.   *(4 marks)*

9 The infrared window is the range of wavelengths of infrared radiation that are:

    A Emitted by the Earth's surface

    B Absorbed by carbon dioxide

    C Absorbed the Earth's surface

    D Not absorbed by water vapour   *(1 mark)*

10 After molecules of a greenhouse gas absorb infrared radiation, the molecules:

    A Collide more frequently

    B Reflect the radiation back to Earth

    C Pass on the energy by colliding with other molecules

    D Have electrons that have been excited to higher energy levels   *(1 mark)*

11 Dinitrogen oxide, N_2O is a powerful greenhouse gas. The best explanation of this is likely to be:

    A It has very polar bonds so absorbs infrared radiation very efficiently.

    B Its concentration is increasing rapidly due to human activities.

    C It absorbs radiation in the infrared window.

    D It is formed from reactions in the soil so is present in high concentrations close to the Earth's surface.   *(1 mark)*

12 Which of the following statements is true about radiation from the Earth and the Sun?

    1 Radiation from the Sun has a longer wavelength than radiation from the Earth.

    2 Radiation from the Sun is mostly in the form of visible and infrared radiation.

    3 The amount of energy radiated by the Earth is similar to the amount of energy that it absorbs from the Sun.

    A 1,2, and 3   **C** 2 and 3

    B 1 and 2   **D** Only 3   *(1 mark)*

13 The enhanced greenhouse effect describes:

    A The extra warming of the Earth due to human activities

    B The extra warming of the Earth due to gases other than carbon dioxide

    C The warming of the Earth that has occurred in the last century

    D The warming of the Earth due to carbon dioxide in the atmosphere   *(1 mark)*

14 The temperature of the Earth has increased by about 1°C since 1850. The concentration of carbon dioxide has increased from 270 ppm to 400 ppm. Scientists believe that the increase in carbon dioxide has caused most of this warming. Explain how an increase in carbon dioxide causes warming of the Earth's surface and atmosphere.   *(6 marks)*

Nucleic acids

Nucleic acids are condensation polymers formed from monomers called **nucleotides**.

The two types of nucleic acids found in living cells are DNA (deoxyribonucleic acid) and RNA (ribonucleic acid).

Nucleotides

Nucleotide molecules have three components:

- a phosphate group
- a sugar: deoxyribose (in DNA) or ribose (in RNA)
- a 'base': cytosine, adenine, guanine, or thymine (thymine is replaced by uracil in RNA).

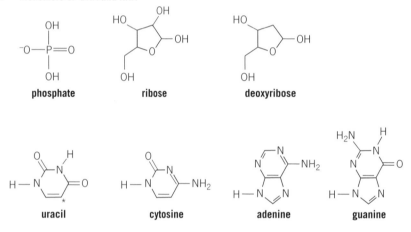

a Monomers of DNA and RNA

phosphate   ribose   deoxyribose

uracil   cytosine   adenine   guanine

(thymine has a CH₃ at position*)

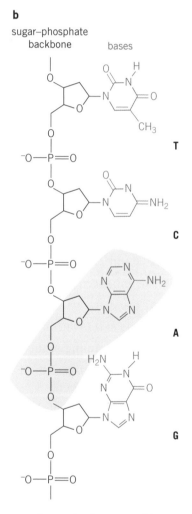

b

sugar–phosphate backbone   bases

T

C

A

G

▲ **Figure 1** **a** *the components of DNA and RNA nucleotides* **b** *A chain of nucletotides in a DNA strand. The shaded section shows the arrangement of phosphate, sugar, and base in an individual nucleotide*

Formation of nucleotides from subunits

Nucleotides are formed from the three subunits by condensation reactions in which a molecule of water is lost.

Joining a sugar to a base

A nitrogen atom from the base bonds to a carbon atom in the sugar molecule. The hydroxyl group from the sugar and a hydrogen atom from the base combine to form a water molecule.

▲ **Figure 2** *A water molecule is lost when a sugar and a base bond together*

Joining a sugar to a phosphate

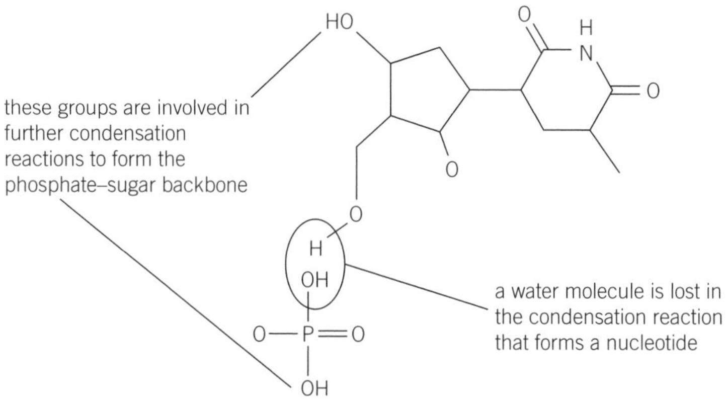

these groups are involved in further condensation reactions to form the phosphate–sugar backbone

a water molecule is lost in the condensation reaction that forms a nucleotide

▲ **Figure 3** *Water molecules are lost when a sugar and phosphate groups bond together*

An oxygen atom from the sugar molecule bonds to the phosphorus atom in the phosphate group to form a nucleotide, The hydrogen atom from the sugar and a hydroxyl group from the phosphate combine to form a water molecule.

The remaining OH groups on the phosphate and sugar then take part in a further condensation reaction to form the phosphate–sugar backbone.

The structure of nucleic acids

Nucleotides bond together to produce nucleic acids by forming a phosphate–sugar backbone.

The base remains attached to the sugar molecule in the phosphate–sugar backbone.

RNA usually consists of a single nucleic acid strand, often twisted into the shape of a helix.

DNA consists of two nucleic acid strands bonded together into a double helix by complementary base pairing.

Complementary base-pairing

The bases in a DNA molecule are in the centre of the double helix structure of the molecule.

Hydrogen bonds form between particular pairs of bases. These hydrogen bonds hold the two strands of DNA together.

Adenine (A) always bonds to thymine (T) by two hydrogen bonds.

Guanine (G) always bonds to cytosine (C) by three hydrogen bonds.

Genetic information

The sequence of bases in a DNA molecule determines the structure of the proteins that are synthesised by cells. Information stored in DNA is described as genetic information. Each codon in the DNA molecule codes for a specific amino acid in the protein.

Copying (replication) of genetic information

A DNA molecule (and the genetic information stored within it) can be copied (a process known as **replication**):

- The DNA molecule 'unzips'; the hydrogen bonds between the complementary base pairs are broken and the two strands separate.
- New DNA nucleotides bond to the exposed bases forming new complementary base pairs.

> **Key term**
>
> **Phosphate–sugar backbone:** The pattern of alternating phosphate and sugar molecules that joins nucleotides together in nucleic acids.

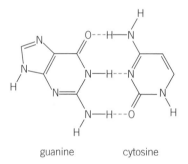

adenine   thymine

guanine   cytosine

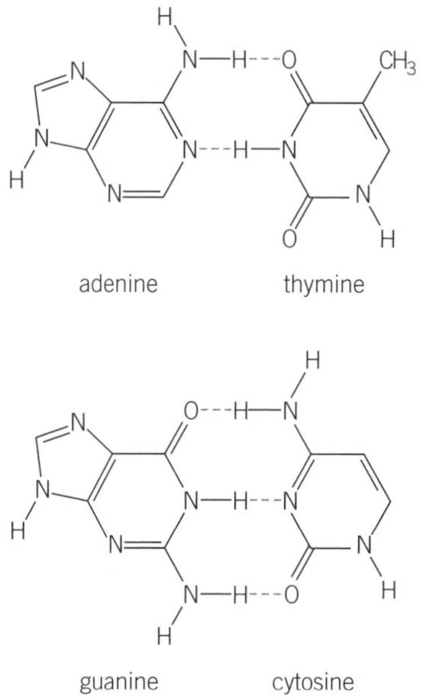

▲ **Figure 4** *Complementary base-pairs formed between bases in DNA*

> **Revision tip**
>
> You need to be able to see how the atoms in the complementary bases are arranged in exactly the right places to allow hydrogen bonds to form. Make sure that you can identify the atoms in the two bases that are involved in the hydrogen bonds.

> **Key term**
>
> **Codon:** A series of three nucleotide bases (e.g. GCC) in a DNA or RNA molecule that codes for an amino acid in a protein.

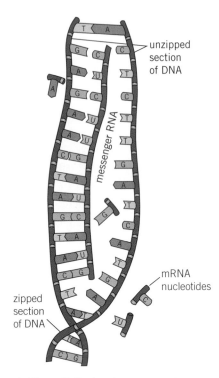

▲ **Figure 5** *During the process of transcription a mRNA molecule forms on one strand of a DNA molecule*

- The newly attached nucleotides are joined together, creating a new strand of DNA.
- The end result is two identical copies of the original DNA molecule.

How DNA encodes for RNA (transcription)

In order for proteins to be synthesised, a molecule of RNA, called messenger RNA (mRNA), is formed from a sequence of bases on one of the DNA strands:

- The DNA molecule 'unzips'; the hydrogen bonds between the complementary base pairs are broken and the two strands separate.
- New RNA nucleotides bond to the exposed bases forming new complementary base pairs.
- The newly attached nucleotides are joined together, creating a new strand of RNA.
- The RNA molecule detaches from the DNA strand.
- The DNA molecule 'zips' back up.

How RNA encodes for proteins (translation)

The sequence of codons in a messenger RNA molecule determines the sequence of amino acids in a protein. This process involves a second type of RNA molecule known as a transfer RNA (tRNA). One end of the tRNA molecule bonds to a specific amino acid, and the other end has a sequence of three bases called an **anticodon**.

- A mRNA molecule diffuses out of the nucleus.
- It attaches to a binding site in a structure in known as a ribosome.
- A tRNA molecule delivers a specific amino acid to the binding site of the ribosome.
- The anticodon on the tRNA bonds to the complementary codon on the mRNA molecule.
- A second tRNA delivers an amino acid to the second binding site of the ribosome.
- A peptide bond forms between the two adjacent amino acids.
- The tRNA molecules detach from the growing protein chain.

  The ribosome moves along the mRNA molecule and the process repeats.

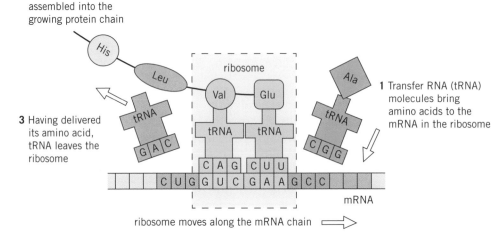

▲ **Figure 6** *The sequence of codons on the mRNA molecule are translated into a sequence of amino acids in a protein chain*

The genetic code

The four bases in RNA (or DNA) can be combined in 64 different ways to make a triplet codon. These codons must code for 20 different amino acids, so one amino acid may be coded for by 3 or even 4 codons.

Go further

The genetic code is based on triplet codons. There are enough possible combinations of bases to code for all 20 naturally occurring amino acids.

A system based on duplet codons (consisting of 2 bases) could not do this.

Calculate the number of combinations of the four bases in RNA that are possible in a duplet codon system.

Summary questions

1 **a** Describe the structure of DNA.   (*4 marks*)
   b State the three differences between the structure of DNA
          and RNA.   (*3 marks*)

2 Deduce the amino acid sequence that would be formed in a protein
   synthesised from the following mRNA base sequence:
   GCCCUGCAAGUC   (*1 mark*)

3 Discuss the roles that mRNA and tRNA play in the synthesis
   of proteins.   (*4 marks*)

1 A molecule of DNA:

   A Contains four bases described by the letters C,G,T,U

   B Consists of a backbone of sugar molecules to which bases are bonded

   C Has hydrogen bonds holding sugar molecules to phosphate groups

   D Has an overall negative charge due to the presence of phosphate groups

(1 mark)

2 During the process by which genetic information is replicated:

   A Free DNA nucleotides hydrogen bond to bases on a DNA strand

   B Free bases bond to sugar molecules on a DNA strand

   C Covalent bonds between bases break, forming a single strand of DNA

   D A DNA strand with a complementary base sequence forms hydrogen bonds to the bases on a single DNA strand *(1 mark)*

3 In which of these processes do condensation reactions occur?

   1 The formation of the phosphate–sugar backbone.

   2 The attaching of bases to deoxyribose–sugar molecules.

   3 The formation of complementary base pairs. *(1 mark)*

   A 1,2, and 3   **C** 2 and 3

   B 1 and 2   **D** Only 3   *(1 mark)*

4 RNA has several structural similarities to DNA. However, differences in the structure of RNA include:

   1 There is only one hydrogen bond formed per base pair in RNA.

   2 RNA has a uracil base in place of a cytosine base.

   3 RNA has a ribose sugar in place of a deoxyribose sugar. *(1 mark)*

   A 1,2, and 3   **C** 2 and 3

   B 1 and 2   **D** Only 3

5 Complementary base pairing occurs between pairs of bases, for example between guanine and cytosine.

   The most important reason for this is:

   A The two molecules have a similar shape.

   B Several hydrogen bonds can form between this pair of bases.

   C Ionic bonds can form between an acidic group in one molecule and a basic group in the other.

   D A condensation reaction can occur between these two molecules. *(1 mark)*

6 Part of the amino acid sequence in a protein is:

   Glycine-serine-valine

   The sequence of **DNA** codons that would encode for this sequence is:

   A CCAAGCCAG   **C** GGTTCGGTC

   B GGUUCGGUC   **D** CCTTGCCTG

7 DNA and RNA are nucleic acids, found in living cells.

   A student finds this description of the function of DNA and RNA in protein synthesis:

   "DNA encodes for RNA, which codes for an amino acid sequence in a protein"

   a Describe the process by which DNA encodes for RNA. *(3 marks)*

   b The diagram below illustrates how RNA encodes for an amino acid sequence in a protein:

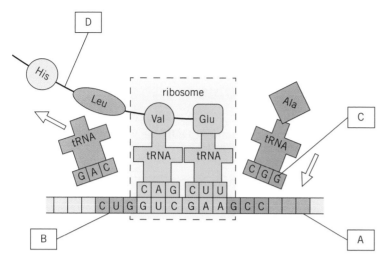

Identify the following structural features:

i The molecule labelled A. *(1 mark)*

ii The set of three bases CUG labelled B. *(1 mark)*

iii The set of three bases CGG labelled C. *(1 mark)*

iv The functional group, D that forms when amino acids are joined in this process. *(1 mark)*

c Explain how the RNA molecule, which consists of a sequence of just four different bases, is able to encode for a protein which contains 20 different amino acids. *(2 marks)*

17.1 Functional group reactions

Specification reference: CD(f), CD(j)

Key term

Polyfunctional molecule: A molecule with more than one functional group.

Synoptic link

You learned about amino acids in Topic 13.8, Amino acids, peptides, and proteins.

Polyfunctional molecules

An example of a polyfunctional molecule is an amino acid. It contains two different functional groups, $-NH_2$ and –COOH.

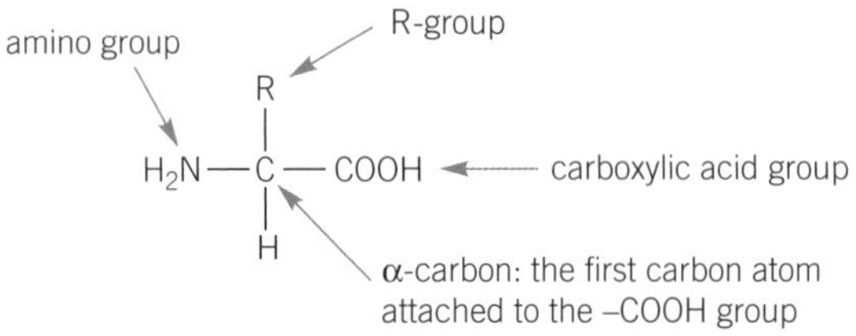

▲ **Figure 1** *An amino acid*

To predict the properties of a polyfunctional molecule, you need to be able to identify the functional groups in the molecule and predict their properties. Generally, you can assume that the properties of the functional groups in a polyfunctional molecule is the same as they are in individual molecules.

Functional groups

This is a list of various functional groups, and their chemical behaviour.

Carbon-based

- Alkenes, C=C: undergo addition reactions. See Topic 12.2, Alkenes.

Halogen-based

- Haloalkanes, R–X: undergo substitution reactions with nucleophiles. See Topic 13.2, Haloalkanes.

Oxygen-based

- Alcohols, R–OH: undergo oxidation to carbonyl compounds and/or carboxylic acids; undergo esterification; undergo nucleophilic substitution reactions with halides; undergo dehydration to form alkenes. See Topic 13.3, Alcohols.
- Carboxylic acids, R–COOH: undergo acid reactions with metals, hydroxides, and carbonates; undergo esterification; undergo conversion to acyl chlorides with $SOCl_2$. See Topic 13.4, Carboxylic acids and phenols, and 13.5, Carboxylic acids.
- Phenols, C_6H_5–OH: undergo acid reactions with hydroxides; undergo esterification with acyl chlorides. See Topic 13.4, Carboxylic acids and phenols.
- Esters, R–COOR: undergo hydrolysis to form alcohols and carboxylic acids. See Topic 13.5, Carboxylic acids, and 13.7, Hydrolysis of esters and amides.
- Acyl chlorides, R–COCl: undergo esterficiation with phenols and alcohols; undergo amide formation with amines. See Topic 13.6, Amines.
- Aldehydes, R–CHO, and ketones, R–CO–R: undergo nucleophilic addition with HCN; undergo oxidation to carboxylic acids (aldehydes only). See Topic 13.10, Aldehydes and ketones.

Nitrogen-based

- Amines, R–NH_2: undergo base reactions with acids; act as nucleophiles with acyl chlorides. See Topic 13.6, Amines.

- Nitro group, R–NO$_2$: undergoes reduction to amines with tin and concentrated hydrochloric acid.
- Nitriles, R–CN: undergo hydrolysis to carboxylic acids.
- Amides, R–CONH–R: undergo hydrolysis to carboxylic acids and amines.

Examples of polyfunctional molecules

Indigo (see Figure 2) has the following functional groups:

- **Ketone**, R–CO–R, so it could be reduced to a secondary alcohol.
- **Secondary amine**, R–NH–R, so it acts as a base by accepting H$^+$ ions, and it could act as a nucleophile.
- **Alkene**, C=C, so it will undergo electrophilic addition reactions.

3-nitrobenzoic acid has the following functional groups:

- **Nitro group**, R–NO$_2$, so it can be reduced to an amine.
- **Carboxylic acid**, R–COOH, so it can donate protons and react with metals and bases, form esters by reacting with alcohols, and be converted to an acyl chloride with SOCl$_2$.

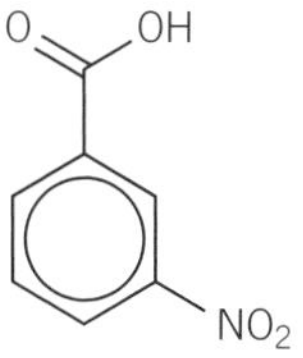

▲ **Figure 3** *3-nitrobenzoic acid*

Synthetic routes

Using your knowledge of the reactions of the functional groups above, it is possible to plan a synthesis of a target molecule. This involves identifying the functional groups in the target and considering how they can be formed from precursors.

Generally a number of intermediates will be produced. For a polyfunctional molecule, it is important to check that the proposed reagents to carry out one conversion do not react with the other functional groups present. If so, a different approach may be required, or the steps may need to be done in a different order.

Starting molecule → intermediate(s) → target molecule

▲ **Figure 2** *Indigo*

Summary questions

1. Define the term 'polyfunctional molecule'. *(1 mark)*

2. List the functional groups which can undergo:
   a addition reactions
   b substitution reactions
   c oxidation reactions. *(3 marks)*

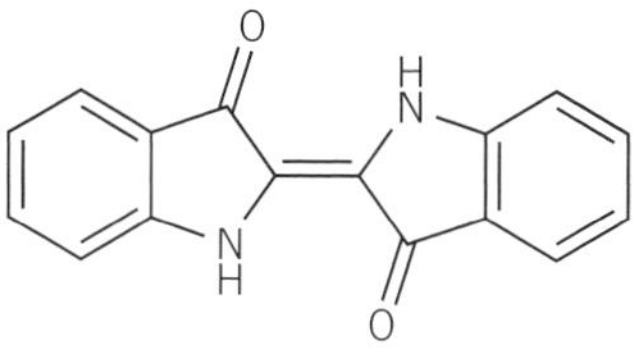

3. Draw a molecule of 3-aminobenzoic acid, identify its functional groups, and suggest the properties of the molecule. *(4 marks)*

17.2 Classification of organic reactions

Specification reference: CD(j), CD(l)

Organic mechanisms

You have learnt about a wide range of organic reactions during this course. They are summarised below.

Addition reactions

In an addition reaction, two molecules react to form a single product. The atom economy is 100% as there is no waste product. Examples include electrophilic addition of bromine to an alkene (Reaction 1), or nucleophilic addition of HCN to a carbonyl compound (Reaction 2):

Reaction 1: $C_2H_4 + Br_2 \rightarrow C_2H_4Br_2$ see Topic 12.2

Reaction 2: $CH_3CHO + HCN \rightarrow CH_3CH(OH)CN$ see Topic 13.10

Condensation reactions

Two molecules react to form a larger molecule, and a small molecule, such as H_2O or HCl, is removed. The atom economy is less than 100%. Examples include the formation of esters (Reaction 3), and the formation of amides (Reaction 4):

Reaction 3: $CH_3COOH + C_2H_5OH \rightarrow CH_3COOC_2H_5 + H_2O$ see Topic 13.3

Reaction 4: $C_6H_5COCl + CH_3NH_2 \rightarrow C_6H_5CONHCH_3 + HCl$ see Topic 13.6

Elimination reactions

A small molecule, such as HCl or H_2O, is removed from a larger molecule, leaving an unsaturated molecule. The atom economy is less than 100%. Examples include elimination of hydrogen halides from haloalkanes (Reaction 5) or dehydration of alcohols (Reaction 6):

Reaction 5: $C_2H_5Br \rightarrow C_2H_4 + HBr$ see Topic 13.2

Reaction 6: $C_2H_5OH \rightarrow C_2H_4 + H_2O$ see Topic 13.3

Substitution reactions

A group of atoms takes the place of another group in a molecule. The atom economy is less than 100% unless both products are useful. Examples include nucleophilic substitution of haloalkanes (Reaction 7), the reaction of alcohols with halides (Reaction 8), and electrophilic substitution of arenes (Reaction 9):

Reaction 7: $C_3H_7Br + OH^- \rightarrow C_3H_7OH + Br^-$ see Topic 13.2

Reaction 8: $C_4H_9OH + HBr \rightarrow C_4H_9Br + H_2O$ see Topic 13.3

Reaction 9: $C_6H_6 + C_2H_5Cl \rightarrow C_6H_5(C_2H_5) + HCl$ see Topic 12.4

Oxidation reactions

Oxygen atoms are gained and/or hydrogen atoms are lost. Examples include the oxidation of primary alcohols to aldehydes (Reaction 10), and the oxidation of aldehydes to carboxylic acids (Reaction 11):

Reaction 10: $C_3H_7OH + [O] \rightarrow C_2H_5CHO + H_2O$ see Topic 13.3

Reaction 11: $C_2H_5CHO + [O] \rightarrow C_2H_5COOH$ see Topic 13.10

Reduction reactions

Oxygen atoms are lost and/or hydrogen atoms are gained. An example is the reduction of carbonyl compounds to alcohols (Reaction 12):

Reaction 12: $CH_3COCH_3 + 2[H] \rightarrow CH_3CH(OH)CH_3$

Synoptic link

You covered atom economy in Topic 14.2, Atom economy.

Key term

Electrophile: A species that accepts a pair of electrons to form a covalent bond. Examples include Br^+ or $Br^{\delta+}$.

Revision tip

C_2H_4 is ethene, an alkene, so it has a C=C bond, and is unsaturated.

Key term

A nucleophile is a species with a lone pair that can form a covalent bond. Examples include Br^- and H_2O.

Revision tip

$C_6H_5(C_2H_5)$ is ethylbenzene.

Revision tip

Remember that [O] represents an oxygen from the oxidising agent.

Revision tip

[H] represents a hydrogen from the reducing agent.

Hydrolysis reactions

Bonds are broken by reaction with water. Examples include the hydrolysis of esters (Reaction 13), and the hydrolysis of amides (Reaction 14), which can be catalysed by acids or alkalis:

Reaction 13: $C_2H_5COOCH_3 + H_2O \rightarrow C_2H_5COOH + CH_3OH$ see Topic 13.8

Reaction 14: $CH_3CONHCH_3 + H_2O \rightarrow CH_3COOH + CH_3NH_2$ see Topic 13.8

A summary of organic reactions

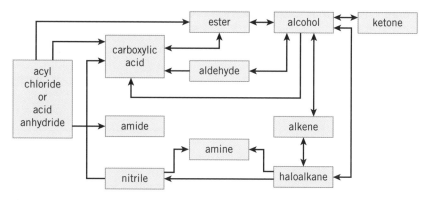

▲ **Figure 6** *Some important functional group interconversions*

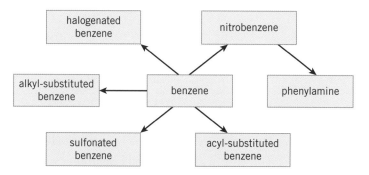

▲ **Figure 7** *Functional group interconversions for aromatic compounds*

Synthetic routes

Using the charts above it is possible to plan potential synthesis reactions.

 Worked example: Planning a synthetic route

Suggest a synthetic route to produce ethylamine from ethanol.

Step 1:   Look at the chart to find a route between an alcohol and an amine:

alcohol → alkene → haloalkane → amine

Step 2:   Identify the specific compounds involved:

ethanol → ethene → bromoethane → ethylamine

Step 3:   Consider the reagents and conditions needed:

alcohol → alkene: dehydration with Al_2O_3 at 300 °C

alkene → haloalkane: reaction with HBr

haloalkane → amine: reaction with NH_3

Summary questions

1 Write equations for the following reactions of propan-1-ol: **a** oxidation to an aldehyde **b** oxidation to a carboxylic acid **c** dehydration to an alkene **d** substitution to form a chloroalkane **e** esterification with methanoic acid. (*5 marks*)

2 Suggest a synthetic route for the formation of butanoic acid, C_3H_7COOH, from but-1-ene. Give the steps required (*2 marks*), and the reagents and conditions.   (*2 marks*)

3 Suggest a synthetic route for the formation of the cyanohydrin $C_2H_5C(OH)(CN)CH_3$ from a haloalkane. Give the three steps required (*3 marks*), name the compounds involved (*3 marks*), and give the reagents and conditions required.   (*3 marks*)

Chapter 17 Practice questions

1 Which of the following is a polyfunctional molecule?

A 2-methylpentane

B pentanedioic acid

C pentyl ethanoate

D glycine *(1 mark)*

2 Which of the following undergoes nucleophilic addition reactions?

A carbonyl compounds

B arenes

C haloalkanes

D alkenes *(1 mark)*

3 Which of the following reacts with NaOH?

A phenol

B ethanol

C propene

D nitrobenzene *(1 mark)*

4 Which of the following can react to form a carboxylic acid in one step?

1 nitriles

2 ketones

3 alkenes

A 1 only

B 1 and 2

C 2 and 3

D 1, 2, and 3 *(1 mark)*

5 Which of the following reacts with HBr in an electrophilic addition reaction?

A propene

B propan-1-ol

C propylbenzene

D propanone *(1 mark)*

6 Identify the functional groups in Disperse Red 60 dye, and suggest the chemical properties of the molecule. *(7 marks)*

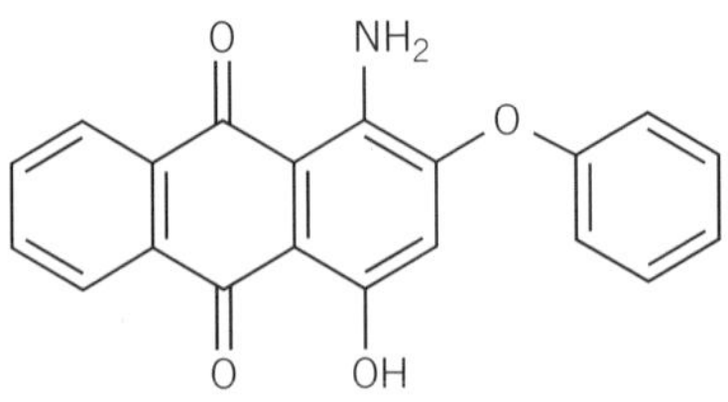

▲ **Figure 8** *Disperse Red 60 dye*

7 Suggest a synthetic route to make $CH_3CH_2CH_2CN$ from CH_3CH_2CHO, giving the intermediates, and reagents and conditions. *(3 marks)*

8 Suggest a synthetic route to make phenylamine from benzene in two steps, giving the intermediate, and reagents and conditions. *(2 marks)*

Answers to summary questions

1.1

1 a 342

b 128 *[1 for each]*

2 92.3% *[1]*

3 Relative isotopic mass is the relative mass of a single isotope *[1]*; relative atomic mass may be the (weighted) average mass of several isotopes *[1]*

4 C and O have different masses *[1]*. An atom of C has a smaller mass than an atom of O *[1]*.

5 a 79.99 *[1]*

b 20% ^{11}B and 80% ^{10}B *[1]*

6 % of O = 16.33% *[1]*. Empirical formula of the compound = $C_6H_{10}O$ *[1]*. 2 empirical formula units are needed to make a M_r of 196, so molecular formula = $C_{12}H_{20}O_2$ *[1]*.

1.2

1 a S^{2-}

b Rb^+ *[1 mark each]*

2 a $Ba(OH)_2$

b Al_2O_3

c Na_2SO_4 *[1 mark each]*

3 a $CaO(s) + 2HCl(aq) \rightarrow CaCl_2(aq) + H_2O(l)$. *[1 mark formulae, 1 mark balancing, 1 mark state symbols]*

b $Ba^{2+}(aq) + SO_4^{2-}(aq) \rightarrow BaSO_4(s)$ *[1 mark formulae, 1 mark state symbols]*

1.3

1 36.7 g *[1]*

2 M_r of UF_6 = 352.1, M_r of ClF_3 = 109 *[1]*. Moles of UF_6 formed = 100/352.1 = 0.284 mol, so moles of ClF_3 needed = 0.568 *[1]*. Mass of ClF_3 needed = 0.568 × 109 = 61.9 g *[1]*.

3 Moles U = 1000/238.1 = 4.20 mol; mol ClF_3 = 1000/109 = 9.17 *[1]*. ClF_3 is in excess and U is limiting reagent *[1]*. Expected yield of UF_6 = 352.1 × 4.20 = 1478 g *[1]* % yield = (1.36/1.478) × 100 = 92.0% *[1]*

1.4

Go further: dilution factor = $\frac{250}{(3.57)}$ = 70.0. Original concentration = 70.0 × 0.045 = 3.15 mol dm^{-3}

Summary questions

1 a 0.800 mol dm^{-3}

b 0.040 mol dm^{-3}

c 0.125 mol dm^{-3} *[1 mark each]*

2 Moles of HCl = 28.6/36.5 = 0.7836 mol *[1]*. Concentration = 0.7836/0.250 = 3.13 mol dm^{-3} *[1]*. Dilution factor = 10/250, so concentration of diluted solution = (10/250) × 3.13 = 0.125 mol dm^{-3} *[1]*

3 Moles H_2SO_4 = 2.09 × 10^{-4} mol *[1]*. Moles KOH = 4.18 × 10^{-4} mol *[1]*. Concentration = $\frac{4.18 \times 10^{-4}}{0.0200}$ = 0.0209 mol dm^{-3} *[1]*. M_r of KOH = 56, so concentration = 0.0209 × 56 = 1.17 g dm^{-3} *[1]*.

1.5

1 a 0.00182%

b 400 ppm [1 mark each]

2 M_r (H_2O_2) = 34, so 2.40 g H_2O_2 = 0.07059 mol *[1]*. Moles of O_2 formed = 0.07059/2 = 0.03429 mol *[1]*. Volume of O_2 at RTP = 0.03429 × 24.0 = 0.847 dm^3 *[1]*

3 $n = \frac{pV}{RT}$ *[1]*. 360 dm^3 = 0.36 m^3 *[1]*, so $n = \frac{(200\,000 \times 0.36)}{(8,314 \times 400)}$ = 21.65 mol *[1]*.

2.1

1 It was realised that the atom was able to be divided into smaller particles OR consisted of positive and negative parts. *[1]*

2 a 1 proton + 2 neutrons

b 20 protons and 27 neutrons

c 11 protons + 12 neutrons *[1 mark for each part question]*

3 The atom is mostly empty space (so most passed through); the atom contains a positive, dense / small nucleus (so some particles bounce back). *[2]*

2.2

1 To overcome the repulsion *[1]* between positive nuclei *[1]*

2 a $^1_0 n$ *[1]*

b $^{16}_8 O$ *[1]*

3 a $^3_2 He \rightarrow {}^4_2 He + 2{}^1_1 H$ *[1]*

2.3

1

↑↓	↑	↑

4 electrons in total arranged as above. *[2]*

2 a $1s^2\ 2s^2\ 2p^6\ 3s^2\ 3p^1$ *[1]*

b $1s^2\ 2s^2\ 2p^6$ *[1]*

c $1s^2 2s^2 2p^6 3s^2 3p^6$ *[1]*

3 a $4s^2\ 3d^3$ (or reversed)

b $6p^4$ *[1 mark each*

3.1

1 a ionic *[1]*

b covalent *[1]*

2

$$\left[Mg \right]^{2+} \left[\begin{smallmatrix} & \times\times & \\ \times & O & \times \\ & \bullet\times & \end{smallmatrix} \right]^{2-}$$

[1 mark: correct structures of Mg and O, 1 mark correct charges on ions]

3

: N ⦂ N ⦂ O :

[1 mark: correct number of electrons on nitrogen atoms (5) and oxygen atom (6); 1 mark: triple bond between N and N, single bond between N and O; 1 mark: single bond shown as a dative bond.]

3.2

1 a octahedral, 90° [2]

b trigonal planar, 120° *[1 mark for shape, 1 mark for bond angle in each case]*

2 3 bonding pairs and 1 lone pair around P atom *[1]*. 3 lone pairs around each. Cl *[1]*.

There are 4 pairs of electrons around the central P atom *[1]*. These repel and get as far apart as possible (to minimise repulsion) *[1]*. Electrons arranged tetrahedrally *[1]*. One lone pair and 3 bonding pairs so shape of molecule is pyramidal *[1]*. Extra repulsion from lone pair reduces tetrahedral bond angle (109.5°) to 107° *[1]*.

3 a 2 double bonds from S to O atoms *[1]*. 2 single bonds from S to O atoms *[1]*. Extra electron on the single-bonded O atoms shown clearly *[1]*.

b

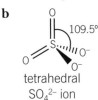

tetrahedral
SO_4^{2-} ion

Shape [1], bond angle. [1].

4.1

1 a i Negative   **ii** Temperature increases [2]

b Energy is released by forming bonds AND energy is required to break bonds *[1]* Energy released is greater than energy required *[1]*

2 Energy released to the water = 25 707 J = 25.707 kJ *[1]*. Moles of methanol burnt = (1.60)/32 = 0.0500 mol *[1]*. Energy released by 1 mole = (25.707)/(0.05) = 514.14 *[1]*. $\Delta_c H = -514$ kJ mol^{-1} (3 s.f.) *[1]*

3 Moles of NI_3 = 1000/394.7 = 2.524 mol *[1]*. Ratio of moles reacted: moles in equation = 2.524/2 = 1.267 *[1]*. $\Delta H = -290 \times 1.267 = -367$ kJ, so 367 kJ of energy is released *[1]*.

4.2

1 $\Delta_f H$ [C_2H_4], $\Delta_f H$ [C_2H_6] [1 mark each]. Any additional substance, e.g. H_2 loses one of these 2 marks.

2 $\Delta_f H$(products) $-\Delta_f H$ (reactants) *[1]* (or correct use of the enthalpy changes in an enthalpy cycle); $\Delta_f H$(products) = 2 × −285.8 = −571.6 *[1]*. $\Delta_f H$ (reactants) = +50.6 *[1]* $\Delta_r H$ = −571.6 − 50.6 = −622.2 kJ mol^{-1}. *[1]*

3

3 a Top line of cycle shown as $3C(s) + 4H_2(g) \rightarrow C_3H_8(g)$ *[1]* Combustion products at bottom = $3CO_2(g) + 4H_2O(l)$ *[1]* + $5O_2$ included on each downward arrow. *[1]*

b $\Delta H_2 = (3 \times -394) + (4 \times -286) = -2326$ kJ mol^{-1} *[1]* $\Delta H_3 = -2219$ kJ mol^{-1} *[1]*. ΔH_1 ($\Delta_f H$ [C_3H_8] = $-2316 - (-2219) = -97$ kJ mol^{-1}. *[1]*

4.3

1 a C=C is stronger than C−C *[1]*

b C−C is longer than C=C. *[1]*

2 a Bonds broken = 4 C−H + 2O=O AND bonds formed = 2C=O + 4 O−H *[1]*. Energy changes = 2468 AND −3466 *[1]*. Enthalpy change of reaction = −818 kJ mol^{-1} *[1]*.

b Bond enthalpies are average values OR H_2O is a liquid in its standard state, bond enthalpies are for gaseous molecules *[1]*.

3 $C_3H_6 + 4\frac{1}{2}O_2 \rightarrow 3CO_2 + 3H_2O$ *[1]*

Bonds formed = 6C=O + 6O−H = 7614 *[1]*

Bonds broken = 3C−C + 6C−H + 4$\frac{1}{2}$ O=O = 3C−C + 4719 *[1]*

$\Delta_c H = -2091 = (3\text{C−C} + 4719) - 7614 = 3\text{C−C} - 2895$ *[1]*

3C−C = −2091 + 2895 = 804. *[1]*

C−C = $\frac{804}{3}$ = + 268 kJ mol^{-1} *[1]*

Significantly lower than usual C−C bond enthalpy so bond is easier to break in cyclopropane. *[1]*

4.4

1

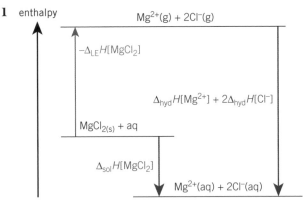

Correct substances on lines *[1 for each]*; correct positions of lines *[1 for each]*; correct labels on each arrow. *[1 for each][Total 7]*

2 For calcium chloride to dissolve, ionic bonds between calcium ions and chloride ions must break *[1]*; only weak ion–dipole bonds can form *[1]*; bonds broken would be stronger than bonds formed *[1]*; energy released by forming bonds would not compensate for energy needed to break bonds. *[1]*

3 a Sum of enthalpy changes of hydration = enthalpy change of solution + lattice enthalpy [1]

$= -155 - 2526 = -2681\,kJ\,mol^{-1}$ [1]

So $\Delta_{hyd}H\,[Mg^{2+}(aq)] + 2\,\Delta_{hyd}H\,[Cl^-(aq)] = -2681$ [1]

$2\,\Delta_{hyd}H\,[Cl^-(aq)] = -2681 - (-1926) = 755$;

$\Delta_{hyd}H\,[Cl^-(aq)] = -\dfrac{755}{2} = -377.5\,kJ\,mol^{-1}$ [1]

b Ca^{2+} ions have a smaller charge density (than Mg^{2+}) [1] weaker ion-dipole bonds are formed between Ca^{2+} and water molecules than between Mg^{2+} and water molecules. [1]

4.5

1 Negative [1]; there are fewer moles of gas on the RHS of the equation [1]; so there are fewer ways of arranging the particles after the reaction [1]; this means that entropy decreases (so ΔS is negative). [1]

2 $\Delta_{sys}S = \Sigma S$ (products) $- \Sigma S$ (reactants) [1] $= (2 \times 26.9) + (4 \times 240) + 205 - (2 \times 164) = +890.8\,J\,K^{-1}\,mol^{-1}$ [1]

$\Delta_{surr}S = -\Delta H/T$ [1] $= -510.8 \times \dfrac{1000}{298}$

$= -1714\,J\,K^{-1}\,mol^{-1}$ [1]

$\Delta_{tot}S = \Delta_{sys}S + \Delta_{surr}S = 890.8 - 1714$

$= -823.2\,J\,K^{-1}\,mol^{-1}$ [1]; reaction is not feasible / spontaneous at this temperature. [1]

3 At room temperature $\Delta_{tot}S$ is negative (because the reaction is not feasible) [1]; reaction is feasible at certain temperatures so $\Delta_{surr}S$ and $\Delta_{sys}S$ mist have opposite signs [1]; reaction becomes feasible when temperature is increased, so $\Delta_{surr}S$ must be negative [1]; $\Delta_{sys}S$ must be positive. [1]

5.1

1 a Add silver nitrate solution, [1] a cream precipitate shows presence of bromide ions. [1]

b Add barium chloride solution [1], a white precipitate shows presence of sulfate ions. [1]

2 a blue

b white

c brown

d yellow [8]

3 a iron(II) sulfate. [2]

b $Fe^{2+}(aq) + 2OH^-(aq) \rightarrow Fe(OH)_2(s)$ [1 formulae, 1 balancing and state symbols] $Ba^{2+}(aq) + SO_4^{2-}(aq) \rightarrow BaSO_4(s)$. [2]

5.2

Go further:

Octahedral

Go further:

Summary questions

Group 1 ions have a small (1+) charge and have a relatively large ionic radius (compared to other ions in the same period). These two factors reduce the electrostatic attraction between the ions and the delocalised electrons.

1 Calcium ions [1] are arranged in a regular pattern / in layers [1] to form a giant 3-dimensional lattice [1] delocalised electrons surround the ions. [1]

2 Calcium oxide: high melting (and boiling) point [1], probably soluble in water [1] conducts electricity when molten or in solution. [1]

3 Silicon dioxide has a covalent network / giant covalent structure [1] carbon dioxide has a covalent molecular structure [1] weak intermolecular bonds between carbon dioxide molecules [1] (no molecules in silicon dioxide so) all bonds are (strong) covalent bonds [1] (much) more energy is needed to break covalent bonds than intermolecular bonds. [1]

5.3

1 C–Cl: C δ^+, Cl δ^- [1]   Cl–F: Cl δ^+, F δ^- [1]

H–N: H δ^+, N δ^- [1]   Cl–S: Cl δ^-, S δ^+. [1]

2 a instantaneous dipole–induced dipole [1]

b Electrons in a molecule are in continuous random motion [1]. At a particular instant, the electrons may be distributed unevenly [1]. This creates an instantaneous dipole. [1] The dipole induces a dipole on a neighbouring molecule, creating an induced dipole [1]. There is an electrostatic attraction between the two dipoles [1].

3 CCl_2F_2 is tetrahedral [1] C–Cl and C–F bonds are polar but C–F bonds are more polar (or can be shown on diagram using appropriate δ^- δ^+ convention) [1]; centre of positive charge is not in same place as centre of negative charge / dipole do not cancel out / charges are not arranged symmetrically [1]; there is an overall dipole. [1]

5.4

1 N is a small atom with a δ^- charge / small electronegative atom [1]. H has a δ^+ charge (because it is bonded to a N atom) [1] there is a lone pair of electrons on the N atom. [1]

2

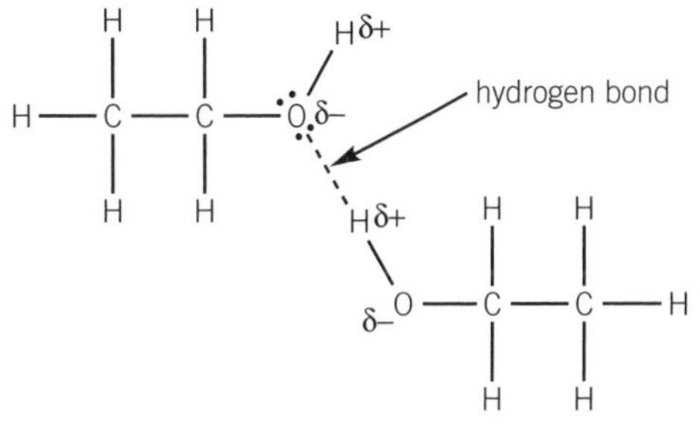

Correct charges on O and H atoms [1] Lone pair on O [1] shown pointing to H [1] H....O–H shown in a line / 180° bond angle around H atom. [1]

3 id–id bonds between methane molecules, H-bonds (and id–id) between water and between hydrogen fluoride [1]. H-bonds are stronger than id–id [1]. Water has more H-bonds per molecule than hydrogen fluoride [1], stronger and/or greater number of bonds = more energy needed to separate molecules [1].

5.5

1 Secondary structure is the folding of the chain into 3-dimensional structures (helix and sheet) *[1]*; tertiary structure is the further folding into the overall complex 3-dimensional structure of the enzyme *[1]* secondary structure is maintained by hydrogen bonds between peptide groups *[1]* tertiary structure is maintained by various types of bond between the amino acid side groups. *[1]*

2 O of C=O group shown bonding to H atom of NH group (using dotted line to denote hydrogen bond) *[1]* O shown as δ-, H as δ- *[1]* lone pair on O shown pointing in direction of hydrogen bond. *[1]*

3 Changes in primary structure means that different amino acids may be present in active site *[1]* this affects the pattern of bonding that maintains the shape of the active site *[1]* different shaped active site may no longer be complementary to the substrate of the enzyme OR different pattern of amino acids may affect ability of active site to bind to substrate. *[1]*

5.6

1 a the part of a molecule that is responsible for its pharmacological / biological activity. *[1]*

   b to increase effectiveness *[1]*; to reduce side effects *[1]*; OR give specific beneficial properties that might be improved e.g. altered solubility (in fats / blood) *[1]*; altered rate of breakdown in organism. *[1]*

2 Circle round either structure as shown below: *[1]*

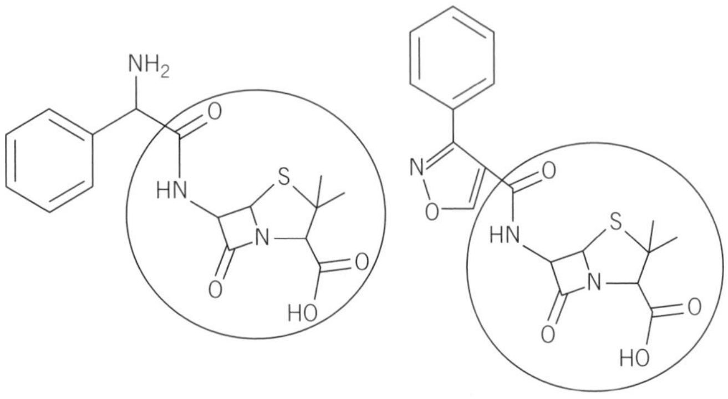

3 a The molecule contains a chiral C atom *[1]*; this is the left-hand carbon atom in the right-hand ring. *[1]*

   b The three-dimensional arrangement of the groups in the two enantiomers is different *[1]*; may have a different bonding pattern with a receptor site / active site. *[1]*

5.7

1 hydrogen bonds, instantaneous dipole-induced dipole, ionic, covalent *[1 mark each]*

2 The sulfonic acid group on the acid blue molecule can ionise to form SO_3^- *[1]*; NH_2 groups gain H^+ ions to form NH_3^+ *[1]*; ionic bonds form between SO_3^- and NH_3^+ *[1]*

3 Hydrogen bonds form between the dye and cotton *[1]*; because of the presence of OH groups in both molecules *[1]*; there are hydrogen bonds between water molecules *[1]*; a few / weak hydrogen bonds form between the dye and water *[1]*; the bonds that can be formed are weaker than the bond that need to break *[1]*; the energy released from bond formation is not sufficient to compensate for bond breaking. *[1]*

6.1

Go further:

Lines get closer at high frequency: High frequency lines arise from transitions with large ΔE values; these are produced from transitions from higher energy levels; high energy levels are much closer together than lower energy levels; transitions from adjacent energy levels will have very similar ΔE values hence produce light of similar frequencies.

Several series of lines: Lines are caused when electrons drop to a lower energy level; all transitions ending at, e.g. $n = 1$ produce one series; other series are produced by drops to $n = 2$, $n = 3$, etc.

Summary questions

1 a yellow *[1]*

   b (pale) green *[1]*

2 Differences in appearance: emission spectra consists of coloured / bright lines on a dark background *[1]* absorption spectra consists of black lines on a coloured background *[1]*. Difference in formation: emission spectra formed by electron falling from higher to lower energy levels and emitting light *[1]* absorption spectra formed by electrons being excited from lower to higher energy levels and absorbing light. *[1]*

3 Electrons drop from higher to lower energy levels *[1]* emit energy in the form of light *[1]* energy lost is related to frequency of photon / ΔE = h ν *[1]* each drop in energy levels causes one line in the spectrum *[1]* series of lines caused when electrons drop to a specific level, e.g. $n = 2$. *[1]*

6.2

1 a Infrared *[1]*

   b Visible light (accept ultraviolet) *[1]*

2 a Visible has higher energy / greater frequency / shorter wavelength (any two) *[2]*

   b visible light excites electrons to higher energy levels OR can cause bond breaking *[1]* infrared causes bonds to vibrate (with greater energy) *[2]*

3 Energy of photon absorbed when bond breaks = 7.227×10^{-19} J *[1]*. Energy needed to break 1 mole of bonds = $7.227 \times 10^{-19} \times 6.02 \times 10^{23}$ = $\times 10^3$ J *[1]*. 435 kJ mol^{-1}. *[1]*

6.3

Go further:

Reaction of methane with chlorine:

$$CH_4 + Cl \rightarrow CH_3 + HCl$$
$$CH_3 + Cl_2 \rightarrow CH_3Cl + Cl$$

overall: $CH_4 + Cl_2 \rightarrow CH_3Cl + HCl$

Depletion of ozone:

$$O_3 + Cl \rightarrow ClO + O_2$$
$$ClO + O \rightarrow Cl + O_2$$

Summary questions

1 Homolytic fission: one electron goes to each atom; heterolytic: both electrons go to a single atom. *[1]*

   Homolytic: radicals are formed; heterolytic: ions are formed. *[1]*

2 **a** Ozone is broken down to form oxygen molecule and an oxygen atom / $O_3 + h\nu \rightarrow O_2 + O$. *[1]*

   b Prevents high frequency ultraviolet radiation reaching the Earth's surface *[1]* high frequency uv can cause skin cancer / eye cataracts. *[1]*

3 $NO + O_3 \rightarrow NO_2 + O_2$ *[1]*

   $NO_2 + O \rightarrow NO + O_2$ *[1]*

6.4

1 The bonds vibrate *[1]* with greater energy / more *[1]*

2 2500–3200: O–H bond *[1]* in a carboxylic acid *[1]*; 1715: C=O bond *[1]* in a carboxylic acid *[1]*

3 $\lambda = \dfrac{c}{f}$ *[1]* $= \dfrac{3.00 \times 10^8}{9.04 \times 10^{13}} = 3.318 \times 10^{-6}\,m$ *[1]* $=$ $3.318 \times 10^{-4}\,cm$, so wavenumber $= \dfrac{1}{3.318} \times 10^{-4} = 3013\,cm^{-1}$ *[1]*

6.5

Go further:

a 3 peaks are: $^{35}Cl^{35}Cl$, $^{35}Cl^{37}Cl$ and $^{37}Cl^{37}Cl$; m/z values are 70, 72, and 74.

b Abundance of $^{35}Cl^{35}Cl = (75.78/100)2 \times 100 = 57.4\%$, $^{37}Cl^{37}Cl = (24.22/100)2 \times 100 = 6.87\%$, abundance of $^{35}Cl^{37}Cl = 2 \times (75.78/100) \times (24.22/100) \times 100 = 36.7\%$ (or by subtraction from previous 2 answers).

Summary questions

1 Number of different isotopes, mass number / isotopic mass of each isotope, relative abundance of each isotope. *[1 mark each]*

2 m/z 78 = molecular ion / M^+ ion / unfragmented molecule with a 1+ charge *[1]* $m/z = 79$ is M^+ ion containing a ^{13}C atom. *[1]*

6.6

1 High-resolution mass spectrometry measures m/z (or relative mass) of M+ peak to 4 d.p. *[1]*; individual atoms / isotopes have relative masses that are not whole numbers *[1]*; so each molecular formula has an M+ peak with a unique mass *[1]*; identify molecule by comparing with database. *[1]*

2 **a** molecular mass = 72, so 57 is formed by loss of 15 = CH_3 *[1]*; 43 is formed by loss of 29 = C_2H_5 *[1]*; 29 is formed by loss of 43 = CH_3CO *[1]*

   b Formulae of fragments: 57 = $C_3H_5O^+$ or $CH_3CH_2CO^+$ *[1]*; 43 = $C_2H_3O^+$ or $COCH_3^+$ *[1]*; 29 = $C_2H_5^+$ or $CH_3CH_2^+$ *[1]*

3 M^+ peak will be same (88) for both isomers in low resolution *[1]* and in high resolution *[1]*; some fragments will be identical because same groups are present, e.g. at 15 (CH_3^+) and 29 ($CH_3CH_2^+$) *[1]*; some fragments will be identical because different groups may have same mass, e.g. 43 ($C_3H_7^+$ in propyl ethanoate and CH_3CO^+ in ethyl ethanoate) , 45 ($HCOO^+$ for propyl methanoate, $OC_2H_5^+$ for ethyl ethanoate) *[1]*; there are no fragments that are likely to be unique to one of these isomers *[1]*; however, the heights of the fragment peaks may be different database could be used to identify molecule). *[1]*

6.7

Go further

Three protons can be aligned in four different ways:

all protons with field ↑↑↑, all protons against field ↓↓↓, 2 protons with field + 1 proton against field ↑↑↓, 2 protons against field + 1 proton with field ↓↓↑

The ratios of heights of the multiplet peaks will be 1 : 3 : 3 : 1

Summary questions

1 **a** 2 peaks *[1]*   **b** 3 peaks *[1]*

2 Quartet is produced by a CH_3 group adjacent to the CH proton environment OR there are three H atoms on the C atom that is adjacent to the CH group *[1]*; doublet is produced by the CH group adjacent to the CH_3 proton environment OR there is one H atom on the C atom that is adjacent to the CH_3 group. *[1]*

3 **a** There are 4 proton environments *[1]*; doublet at δ = 1.2 ppm suggests that there is a CH group adjacent to the CH_3 environment producing this peak *[1]*; quartet suggests there is a CH_3 group adjacent to the CH environment producing this peak *[1]*; peak at δ = 11.0 suggests COO<u>H</u> *[1]*; peak at δ = 2.8 is not split so could be an OH group. *[1]* *[any 4 points]* Structure is:

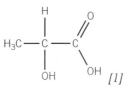

[1]

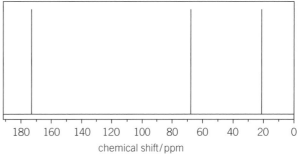

chemical shift / ppm

b Three peaks *[1]*; δ = 180–220 (for <u>C</u>OOH) *[1]*; δ = 0–45 (for <u>C</u>H₃–C) *[1]*; δ = 50–90 (for <u>C</u>H–O). *[1]*

6.8

1 Empirical formula can be deduced from percentage by mass data *[1]*; relative molecular mass can be deduced from m/z value of M^+ peak *[1]*; combining both data allows molecular formula to be deduced. *[1]*

2 a mass spectrometry *[1]*; fragmentation pattern will be different *[1]*; e.g. fragment at 43 ($C_3H_7^+$) for (i), or 29 (C_2H_5+) for (ii) *[1]*

b infrared spectrometry *[1]*; different peaks present *[1]*; e.g. C=C peak at 1620–1680 cm⁻¹ in **i**, OR C=O peak at 1720–1740 cm⁻¹ in **ii**, OR O–H peak at 3200–3600 cm⁻¹ in **i**. *[1]*

3 Points that could be made include:

Mass spectrum: empirical formula is C_5H_6O; molecular mass (from M^+ peak) is 164, hence molecular formula is $C_{10}H_{12}O_2$; Low ratio of H:C suggests an aromatic compound;

IR spectrum: peak at above 1700 suggests C=O in aldehyde, ketone or carboxylic acid; No broad peak at 3200–3600 cm⁻¹, so carboxylic acid not present;

Peak at 1620 cm⁻¹ could suggest alkene or arene C=C;

^{13}C nmr spectrum: 8 peaks so 8 types of C atom; Low-res ^1H nmr spectrum: 5 peaks so 5 different ^1H environments Ratio of H atoms in each environment = 2 : 2 : 3 : 2 : 3; 4H peak between δ =7.0 and 8.0 could suggest 4H atoms in arene; Grouping of 2:2 in arene group suggests H atoms on C atoms 2, 3, 5, and 6; High res. ^1H spectrum: triplet at δ = 1.2 suggests CH_2 adjacent to CH_3; Quartet at δ = 2.6 suggests CH_3 adjacent to CH_2; Singlet at 3.8 suggests no H atoms on adjacent carbons; could be <u>C</u>H₃O Most likely structure is:

Level of response mark scheme:

Marks 5–6: Identifies molecule correctly, linking most features of the molecule clearly to the evidence, AND uses data from all sources of information (MS, IR, ^{13}C nmr, low-res ^1H nmr, and high res ^1H nmr) in reaching conclusion.

Marks 3–4: Identifies some features of the molecule correctly, linking some of the features clearly to the evidence, AND uses data from at least three sources of information in reaching conclusion.

Marks 1–2: Makes limited progress in identifying molecule, with some attempt to link with the evidence, AND uses only one or two sources of information in reaching conclusion.

6.9

1 a the part of the molecule responsible for absorbing visible light (or uv radiation). *[1]*

b an extensive conjugated system / system of alternating double and single bonds. *[1]*

2

[1 mark for circling correct part of structure]

3 The conjugated system is more extensive in A than B *[1]*; the energy gap between energy levels in A is smaller than in B *[1]*; $\Delta E = h\nu$ so A absorbs at a lower frequency than B *[1]*; A absorbs in the visible region / in the blue part of the spectrum *[1]*; B absorbs in the uv part of the spectrum. *[1]*

7.1

1 When the concentrations of the species in the reaction are constant and the rate of the forward reaction and backward reaction are equal. *[1]*

2 If the reaction starts with the reactants from the left-hand side the rate of the forward reaction is initially greater than the rate of the reverse reaction *[1]*. If the reaction starts with the substances on the right-hand side of the equation the rate of the reverse reaction is initially greater. At equilibrium the rate of both reactions is equal. *[1]*

3 Increasing the temperature means more of the particles will be in the gaseous state *[1]*. The position of equilibrium will be further to the right. *[1]*

7.2

1 a $K_c = \dfrac{[SO_3]^2}{[SO_2]^2 [O_2]}$ *[1]*

b $K_c = \dfrac{[NO_2]^2}{[N_2O_4]}$ *[1]*

2 a $K_c = \dfrac{[C_2H_5OH]}{[C_2H_4] [H_2O]}$ *[1]*

b The position of equilibrium would move to the side with fewer moles – to the right. K_c would not change. *[1]*

c The position of equilibrium would move to the endothermic side – to the left. K_c would decrease. *[1]*

d Equilibrium is reached more quickly (but the position of equilibrium is not affected). *[1]*

7.3

1 $K_c = \dfrac{[NH_3]^2}{[N_2][H_2]^3} = \dfrac{(0.142 \text{ mol dm}^{-3})^2}{(1.36 \text{ mol dm}^{-3})(1.84 \text{ mol dm}^{-3})^3}$

$= 0.00238 \text{ mol}^{-2} \text{dm}^6$ *[2]*

2 $[C_2H_5OH] = \dfrac{[CH_3COOC_2H_5][H_2O]}{[CH_3COOH] K_c}$

$= \dfrac{(3.0 \text{ mol dm}^{-3})(3.0 \text{ mol dm}^{-3})}{(0.80 \text{ mol dm}^{-3}) \times 4.0}$ *[1]*

$= 2.8125 \text{ mol dm}^{-3}.$ *[1]*

7.4

1 a Increased concentration of reactants (C_2H_4 and H_2O) *[1]*

  b Lower temperature *[1]*

  c Higher pressure. *[1]*

2 a The blood-red colour darkens as the position of equilibrium moves right. *[1]*

  b The blood-red colour becomes paler as the position of equilibrium moves left. *[1]*

3 a Increasing temperature moves the position of equilibrium left; increasing pressure moves the position of equilibrium right. *[2]*

  b Increasing temperature moves the position of equilibrium right; increasing pressure has no effect on the position of equilibrium. *[2]*

7.5

1 a $K_c = \dfrac{[COCl_2]}{[CO][Cl_2]}$ *[1]*

  b i no effect **ii** K_c decreases **iii** no effect *[3]*

2 Increasing the temperature moves the position of equilibrium in the endothermic direction *[1]*, which is to the right *[1]*; therefore K_c increases. *[1]*

3 $K_c = \dfrac{[HI]^2}{[H_2][I_2]}$ *[1]* $=$

$\dfrac{(0.027 \text{ mol dm}^{-3})^2}{(0.004 \text{ mol dm}^{-3}) \times (0.004 \text{ mol dm}^{-3})}$ *[1]* $= 46$ *[1]* [no units – *1 mark*]

7.6

1 a K_{sp} (AgBr) $= [Ag^+][Br^-]$ *[1]*

  b K_{sp} (Ag$_2$S) $= [Ag^+]^2[S^{2-}]$ *[1]*

  c K_{sp} (Ag$_2$CrO$_4$) $= [Ag+]^2[CrO_4^{2-}]$ *[1]*

2 K_{sp} (AgBr) $= [Ag^+][Br^-]$ *[1]* $= (7.07 \times 10^{-7} \text{ mol dm}^{-3})^2$ *[1]* $= 5.00 \times 10^{-13} \text{ mol}^2 \text{dm}^{-6}$ *[1]*

3 $[Pb^{2+}][SO_4^{2-}] = (1.45 \times 10^{-4} \text{ mol dm}^{-3})^2 = 2.1 \times 10^{-8} \text{ mol}^2 \text{dm}^{-6}$ *[1]*; this is greater than K_{sp}, so a precipitate will form. *[1]*

7.7

1 The mass spectrometer will confirm the relative molecular mass of the component. *[1]*

2 Retention time would decrease. *[1]*

3 Retention time on the *x*-axis and three peaks labelled A, B, and C at 3 min, 5.5 min and 6.4 min respectively. *[1]*

8.1

1 a 0.05 dm³ *[1]*

  b 2.5 dm³ *[1]*

2 a 2 mol dm⁻³ *[1]*

  b 0.05 mol dm⁻³ *[1]*

3 a 4.8 g *[1]*

  b 6.3 g *[1]*

4 a 0.02 dm³ *[1]*

  b $0.02 \times 0.100 = 0.002$ moles *[1]*

  c 0.004 moles *[1]*

  d $0.004 \times \dfrac{1000}{25} = 0.16$ moles *[1]*

8.2

1 H_2SO_4 and HSO_4^- *[1]*; OH^- and H_2O *[1]*

2 pH = 1.00 *[1]*

3 $[H^+] = K_w / [OH^-] = 2 \times 10^{-13} \text{ mol dm}^{-3}$ *[1]*; pH = 12.7 *[1]*

4 a $K_a = \dfrac{[H^+][HCOO^-]}{[HCOOH]}$ *[1]*

  b $H^+ = \sqrt{1.6 \times 10^{-4} \times 0.001} = 4 \times 10^{-4} \text{ mol dm}^{-3}$ *[1]*; pH = 3.4 *[1]*

8.3

1 Since $\dfrac{[acid]}{[salt]} = 1$, $[H^+] = K_a = 6.3 \times 10^{-5} \text{ mol dm}^{-3}$ *[1]*; pH = 4.2 *[1]*

2 $[H^+] = 3.4 \times 10^{-6} \text{ mol dm}^{-3}$ *[1]*; pH = 5.47 *[1]*

3 $[H^+] = 1.8 \times 10^{-4} \times \dfrac{0.01}{0.006} = 3.0 \times 10^{-4} \text{ mol dm}^{-3}$ *[1]*; pH = 3.52 *[1]*

9.1

1 a K +1, Br −1

  b H +1, O −2

  c C +2, O −2

  d P +5, O −2

  e Mn +4, O −2

  f Cr +6, O −2 *[6]*

2 a $CuCl_2$ **b** Cu_2O

  c $PbCl_4$ **d** MnO_4^- *[4]*

3 $2Ca \rightarrow 2Ca^{2+} + 4e^-$ and $O_2 + 4e^- \rightarrow 2 O^{2-}$. Calcium is oxidised and oxygen is reduced. *[4]*

4 $2Br^- + 2H^+ + H_2SO_4 \rightarrow Br_2 + SO_2 + 2H_2O$. Br is oxidised from −1 to 0; S is reduced from +6 to +4. *[2]*

9.2

1 a potassium at the cathode, chlorine at the anode

  b calcium at the cathode, oxygen at the anode

  c magnesium at the cathode, iodine at the anode. *[3]*

2 a hydrogen at the cathode, chlorine at the anode

  b hydrogen at the cathode, oxygen at the anode

  c copper at the cathode, oxygen at the anode. *[3]*

3 a $Al^{3+} + 3e^- \rightarrow Al$; $2Cl^- \rightarrow Cl_2 + 2e^-$

b $Pb^{2+} + 2e^- \rightarrow Pb$; $2H_2O(l) \rightarrow O_2(g) + 4H^+(aq) + 4e^-$

c $2H_2O(l) + 2e^- \rightarrow 2OH^-(aq) + H_2(g)$; $2Cl^- \rightarrow Cl_2 + 2e^-$ *[3]*

9.3

1 Cu strip immersed in a solution of Cu^{2+} ions *[1]*; Ag strip immersed in a solution of Ag^+ ions *[1]*; both solutions $1.0\,mol\,dm^{-3}$ *[1]*; voltmeter *[1]*; salt bridge. *[1]*

2 MnO_4^- (aq) + $8H^+$ (aq) + $5Fe^{2+}$ (aq) $\rightarrow$ Mn^{2+} (aq) + $4H_2O$ (aq) + $5Fe^{3+}$ (aq) *[1 mark for correct species on left- and right-hand sides; 1 mark for balancing]*

3 $E^{\varnothing}_{cell}$ = 0.74 V *[1]*; the reverse reaction is infeasible as $E^{\varnothing}_{cell}$ would be negative. *[1]*

9.4

1 The oxygen half-reaction involves O_2 and H_2O. *[1]*

2 Iron goes from 0 to +2, so it is oxidised *[1]*; oxygen in O_2 goes from 0 to –2, so it is reduced. *[1]*

3 $E^{\varnothing}_{cell}$ = 1.16 V *[1]*; this is a greater $E^{\varnothing}_{cell}$ than the reaction with iron, so zinc corrodes first. *[1]*

10.1

1 Increase the temperature, increase the concentration of sulfuric acid, divide the magnesium more finely (into a powder). *[1 mark each]*

2 Measure the rate at which hydrogen gas is produced, measure mass changes, measure pH changes as sulfuric acid is used up. *[1 mark each]*

3 As the reaction proceeds, reactants get used up. Therefore the concentration of reactants is greatest at the beginning. *[1]* The higher the concentration the faster the reaction because there are more frequent collisions between particles. *[1]*

10.2

1 The activation enthalpy is the minimum energy required by a pair of colliding particles before a reaction will occur. *[1]*

2

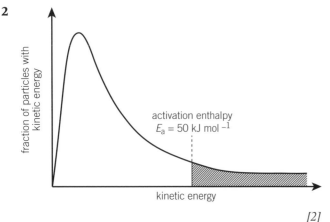

activation enthalpy
$E_a = 50$ kJ mol^{-1}

fraction of particles with kinetic energy (vertical axis)

kinetic energy (horizontal axis)

[2]

3 a The activation energy for the reaction between N_2 and O_2 is very high. However, inside a car engine the temperature is high enough for some molecules of N_2 and O_2 to collide with energies greater than the activation energy for the reaction, forming NO. *[1]*

b The activation energy for the reaction between NO and O_2 is low and many molecules have sufficient collision energy to exceed the activation energy at room temperature. *[1]*

10.3

1 a Manganese(IV) oxide is a catalyst for the reaction. *[1]*

b i manganese(IV) oxide is a heterogeneous catalyst *[1]*

ii catalase is a homogeneous catalyst. *[1]*

2 An intermediate is formed in the reaction. *[1]* This then reacts to form the product. *[1]*

3 The value of ΔH does not change if a catalyst is used. *[1]* The catalyst only affects the activation enthalpy. *[1]*

10.4

1 a) and c) are heterogeneous catalysts as they are in different states to the reactants. *[2]*

b) and d) are not heterogeneous catalysts as they are in the same state as the reactants. *[2]*

2 The reactant molecules are adsorbed onto the surface of the catalyst; bonds in the reactant break; new bonds form; the products are released from the catalyst surface and diffuse away. *[1 mark for each point]*

3 Lead is a catalyst poison for the metals in the catalytic converter. *[1]*

10.5

1 Rate of reaction is the change of concentration of a reactant or product over a given time. The unit is $mol\,dm^{-3}\,s^{-1}$. *[1]*

2 Measure the rate of production of hydrogen gas *[1]*, the change in mass as hydrogen is released *[1]*, or the rate of change of pH. *[1]*

3 If it is not possible to measure a property that continuously changes (such as colour or pH), a sample must be removed and the reaction halted by quenching. *[1]*

10.6

1 A is 2nd order; B is 1st order; the overall order is 3. *[3]*

2 If the line is horizontal, it is zero order *[1]*; if there is a straight line proportional to concentration, it is first order *[1]*; if there is a straight line proportional to concentration², it is second order. *[1]*

3 k = rate / $[X]^2$ *[1]*

$k = 4.2 \times 10^{-3}\,mol\,dm^{-3}\,s^{-1}$ / $(0.01\,mol\,dm^{-3})^2 = 42$ *[1]*

Units are $mol^{-1}\,dm^3\,s^{-1}$ *[1]*

10.7

1 Time on *x*-axis and concentration on *y*-axis *[1]*; downwards curve *[1]*; two half-lives shown – one at 50% concentration and one at 25% concentration. *[1]*

2 Rate = k [A] [H$^+$]. *[1 mark for each correct order:]* A and H$^+$ first order; B zero order.

3 Yes, because A and H appear in the rate equation and in the rate determining step. *[1]*

10.8

1 A substrate reversibly fits into the active site and reacts with the enzyme *[1]*; an inhibitor fits into the active site and blocks the substrate from entering. *[1]*

2 The active site changes shape *[1]*; at high temperatures the hydrogen-bonding of the tertiary structure is affected and at high/low pH the ionic interactions of the tertiary structure are affected. *[1]*

3 When the substrate concentration is high, the active sites are fully occupied so an increase in concentration no longer affects the rate. *[1]*

11.1

1 Around −200 °C. The actual melting point of fluorine is −219 °C. *[1]*

2 **a** fluorine **b** calcium **c** zinc *[3]*

3 **a** p-block **b** s-block **c** d-block *[3]*

4 **a** The third ionisation enthalpy of Mg is much higher than the second, so it is hard to remove the third electron and Mg forms Mg^{2+} *[1]*. For Al, it is the fourth ionisation enthalpy which is much higher so it is hard to remove the fourth electron, meaning Al forms Al^{3+} *[1]*.

  b The total energy to form Mg^{2+} is $2\,186\,kJ\,mol^{-1}$ *[1]*. The total energy to form Al^{3+} is $5\,137\,kJ\,mol^{-1}$ *[1]*. It takes less energy for Mg to form ionic compounds so Mg is more reactive. *[1]*

11.2

1 $Sr(s) + 2H_2O(l) \rightarrow Sr(OH)_2(s) + H_2(g)$ *[1]*

2 $CaCO_3(s) \rightarrow CaO(s) + CO_2(g)$ *[1]*

3 **a** $Ba(OH)_2$ is more soluble *[1]*

  b $MgCO_3$ is more soluble. *[1]*

4 $MgO(s) + 2HCl(aq) \rightarrow MgCl_2(aq) + H_2O(l)$ *[1]*

  $Mg(OH)_2(s) + H_2SO_4(aq) \rightarrow MgSO_4(aq) + H_2O(l)$ *[1]*

11.3

1 $I_2(s)$ is a grey solid; $I_2(g)$ is a purple vapour; $I_2(aq)$ is a brown solution; I_2 in organic solvents is a purple solution. *[1 mark each]*

2 $Ag^+(aq) + I^-(aq) \rightarrow AgI(s)$ *[1]*; yellow precipitate. *[1]*

3 **a** $Cl_2(aq) + 2Br^-(aq) \rightarrow Br_2(aq) + 2Cl^-(aq)$ *[1]*

  b brown colour *[1]*; bromine produced. *[1]*

  c brown layer above a yellow aqueous layer *[1]*; bromine dissolves in the cyclohexane better than in water. *[1]*

  d bromine is a weaker oxidising agent than chlorine. *[1]*

11.4

1 The total charge is $(2+) + (4 \times 0) = 2+$. *[1]*

2 Addition of ammonia causes a blue precipitate *[1]* of $Cu(OH)_2$ *[1]*. Excess ammonia gives a dark blue solution *[1]*; of $[Cu(NH_3)_4(H_2O)_2]^{2+}$. *[1]*

3 The ligands cause d-orbitals splitting *[1]*; electrons can be excited to a higher energy level *[1]*; visible light is absorbed of frequency given by $\Delta E = hv$ *[1]*; the complementary colour (orange) is transmitted. *[1]*

11.5

Go further

Colour will become darker; forward reaction is exothermic so increasing temperature shifts equilibrium to LHS (endothermic direction); more NO_2 is formed, which is darker in colour than N_2O_4.

Summary questions

1 **a** dinitrogen oxide / nitrogen(I) oxide

  b nitrate(III) **c** nitrogen dioxide / nitrogen(IV) oxide *[1 mark each]*

2 To confirm for ammonium ions: dissolve the solid in water and add NaOH *[1]*; heat gently AND moist red litmus paper (above the level of the liquid) will turn blue *[1]*; to confirm for nitrate (V): dissolve the salt in water and add a spatula measure of Devarda's alloy *[1]*; heat gently AND moist red litmus paper (above the level of the liquid) will turn blue. *[1]*

3 $NH_4^+ + 3H_2O$ *[1]* $\rightarrow NO_3^- + 10H^+$ *[1]* $+ 8e^-$ *[1]*

12.1

1 3-methylhexane has a chain of six carbons with a methyl group coming off the third carbon. 2,2,4-trimethylheptane has a chain of seven carbons with two methyl groups on the second carbon and one methyl group on the fourth. *[2]*

2 **i** $C_{10}H_{22}$

  ii $C_{10}H_{20}$ *[2]*

3 They all have the same molecular formula – C_6H_{12}. *[1]*

12.2

1 Pent-1-ene has the structure $CH_2=CHCH_2CH_2CH_3$. *[3]*

2 The product is pentane. *[1]*

3 H–Br has a dipole with Br having a partial negative charge. *[1]*

The C=C bond attacks the H–Br, causing heterolytic fission. *[1]*

A carbocation and a Br⁻ ion are formed. *[1]*

A lone pair on the Br⁻ ion attacks the carbocation forming the product 2-bromobutane. *[1]*

4 The polymer has the structure –CHCl–CHCl–CHCl–CHCl–CHCl–CHCl– *[2]*

12.3

1 The isomers are 1,2-dibromoethane and 1,1-dibromoethane. *[2]*

2 2-methylpropan-2-ol is $CH_3C(OH)(CH_3)CH_3$. 2-methylpropan-1-ol is $CH_3CH(CH_3)CH_2OH$. They are isomers. *[3]*

3 The carbons have two methyl groups or two hydrogen atoms attached to them. It does not have different groups attached to each carbon of the double bond. *[1]*

12.4

Go further

(a) aromatic (10 delocalised e⁻), (b) not aromatic (8 delocalised e⁻), (c) aromatic (6 delocalised e⁻)

Summary questions

1 Between each pair of C atoms there is one σ-bond *[1]*; each C atom has an electron in a p-orbital *[1]*; these 6 electrons are delocalised *[1]*; p-orbital overlap *[1]*; form rings of electron density above and below the plane of the C atoms. *[1]*

2 a i 1-bromo-2,5-dichlorobenzene

   ii 4-methylphenol *[1 mark each]*

b

(i) (ii)

3 Benzene has a hexagonal structure with bond angles of 120° *[1]*; this is consistent with the Kekulé model as there are three areas of electron density around each C *[1]*; the hexagon is regular with all C–C bond lengths equal *[1]*; this is not consistent with the Kekulé model as the C=C bonds would be shorter than the C–C bonds. *[1]*

12.5

Go further

a $AlCl_3 + Cl_2 \rightarrow AlCl_4^- + Cl^+$

b $H^+ + AlCl_4^- \rightarrow HCl + AlCl_3$

Summary questions

1 Benzene contains a delocalised system of electrons *[1]*; addition would destroy the delocalised system *[1]* but substitution leaves it intact *[1]*; so the product of substitution is more stable than the product of addition. *[1]*

2 a Cl_2, $AlCl_3$, reflux *[1 mark for each]*

   b conc. nitric acid, conc. sulfuric acid, temperature below 55°C *[1 mark for each]*

3 a Aluminium chloride *[1]*; acts as a catalyst *[1]*; generates a reactive electrophile / CH_3CO^+ . *[1]*

b

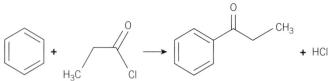

[1 mark for reactants, 2 marks for correct products. Benzene ring can be shown with circle or using Kekulé structure.]

12.6

1 Dissolve phenylamine in (ice-cold) hydrochloric acid *[1]*; make up a solution of sodium nitrate (III) in (ice-cold) hydrochloric acid *[1]*; add the two solutions together, ensuring the temperature does not rise above 5°C. *[1]*

2 a to increase solubility OR to allow dye to bond ionically to fibre *[1]*

   b to allow dye to bond (via hydrogen bonding) to fibre *[1]*

   c to modify the chromophore and hence alter the colour *[1]*

3 Structures are:

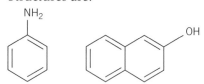

[1 mark each, Phenylamine can be shown with a circle or using Kekulé structure above.]

13.1

1 The differences are:

   a 2-methylpentane has a CH_3 group attached to the second carbon of five in the chain, whereas 3-methylpentane has a CH_3 group attached to the third carbon. *[1]*

   b 2,2-dimethylbutane has two CH_3 groups attached to the second carbon whereas 2,3-dimethylbutane has one CH_3 group on each of the second and third carbons. *[1]*

   c 3-methylhexane has a six-carbon backbone whilst 3-methylheptane has a seven-carbon backbone. Both have a CH_3 group attached to the third carbon. *[1]*

2 They are incorrectly named because:

  a 1-methylpropane has four carbons in the longest chain and should be called butane. *[1]*

  b 2-ethylbutane has five carbons in the longest chain and should be called 3-methylpentane. *[1]*

  c 2,3-methylbutane should be called 2,3-*di*methylbutane to highlight the presence of two methyl groups. *[1]*

3 Primary: butan-1-ol, secondary: butan-2-ol, tertiary: 2-methylpropan-2-ol. *[3 marks]*

4 The correct names are:

  a 1-bromo-2-chloroethane as bromo is alphabetically before chloro. *[1]*

  b 2-chloro-2-iodopropane as chloro is alphabetically before iodo. *[1]*

  c 1,2,3-triiodopropane as '1,2,3-iodopropane' does not highlight the presence of three iodine atoms, and 'triiodopropane' does not give the position of the iodine atoms in the molecule. *[1]*

13.2

1 4-bromo-3,3-dichloro-2-iodohexane – the smallest possible numbers are chosen and the halogens are listed alphabetically. *[1]*

2 The following molecules have the higher boiling point.

  a CH_3Br *[1]*; Br is bigger than F and has more electrons, so there are stronger instantaneous dipole–induced dipole bonds *[1]*

  b CBr_4 *[1]*; it has more Br atoms in the molecule, more electrons, and therefore stronger instantaneous dipole–induced dipole bonds. *[1]*

3 The C–I bond is weaker than the C–F bond, so it is easier to break. *[1]*.

4 A nucleophile has a lone pair of electrons which it can use to form a covalent bond *[1]*; examples are H_2O, NH_3, OH^-, CN^-. *[1 mark each for any two]*

5 Heat the haloalkane in a sealed tube with concentrated ammonia solution. *[1]*

6 The mechanism is similar to that between hydroxide ions and 1-bromobutane shown, except that the haloalkane involved is $CH_3CH_2CH_2I$. *[1 mark for showing the partial charges on the C-I bond; 1 mark for correct arrow from the nucleophile to the partially positive carbon atom; 1 mark for showing the C–I bond breaking; 1 mark for the correct product (propan-1-ol).*

13.3

1 The homologous series are:

  a aldehyde *[1]*

  b ketone *[1]*

  c carboxylic acid *[1]*

  d (secondary) alcohol *[1]*

  e ester. *[1]*.

2 Butanal *[1]* and butanoic acid. *[1]*

3 The colour changes are:

  a No change – remains orange *[1]* as 2-methylpropan-2-ol is a tertiary alcohol and is not oxidised. *[1]*

  b Orange to green *[1]* as 2-methylpropan-1-ol is a primary alcohol *[1]* and is oxidised, thereby reducing the dichromate ion to green Cr^{3+}. *[1]*

4 Acidified *[1]* potassium dichromate(VI) *[1]* with distillation. *[1]*

13.4

1 The names are:

  a hexanoic acid *[1]*

  b pentanedioic acid *[1]*

  c 3-methylbutanoic acid. *[1]*

2 $CH_3CH_2COOH + KOH \rightarrow CH_3CH_2COONa + H_2O$, *[1 mark for each product]*

3 Propan-1-ol *[1]*, CH_3CH_2COOH *[1]* and butanoic acid *[1]*, $CH_3CH_2CH_2OH$. *[1]*

4 The samples which would react are:

  a 4-methylphenol *[1]*

  b ethanoic acid *[1]*

  c propan-1-ol *[1]*

  d 4-methylphenol *and* ethanoic acid. *[1]*

5 Phenols are not acidic enough to react with sodium carbonate solution. *[1]*

13.5

1 a butanoic acid *[1]*   **b** butanedioic acid *[1]*

  c benzoic acid *[1]*

2 a $CH_3COOH + K \rightarrow CH_3COO^-K^+ + \frac{1}{2}H_2$ *[1] allow without charges*

  b $CH_3COOH + KOH \rightarrow CH_3COO^-K^+ + H_2O$ *[1] allow without charges*

  c $2CH_3COOH + K_2CO_3 \rightarrow 2CH_3COO^-K^+ + H_2O + CO_2$ *[1] allow without charges*

3 X is a metal as it produces hydrogen *[1]*. Y is a metal carbonate as it produces carbon dioxide. *[1]*

13.6

1 a $CH_3CH_2CH_2CH_2NH_2$ *[1]*

  b $CH_3CH(NH_2)CH_2CH_2CH_3$ *[1]*

  c $H_2NCH_2CH_2CH_2NH_2$ *[1]*

2 $2C_2H_5NH_2 + H_2SO_4 \rightarrow 2C_2H_5NH_3^+ + SO_4^{2-}$ *[1 mark for $C_2H_5NH_3^+$; 2 marks if completely correct.]*

3 a $C_6H_5COCl + NH_3 \rightarrow C_6H_5CONH_2 + HCl$ *[1]*

  b $C_6H_5COCl + CH_3NH_2 \rightarrow C_6H_5CONHCH_3 + HCl$ *[1]*

  c $C_6H_5COCl + NH_3 \rightarrow C_6H_5COOCH_3 + HCl$ *[1]*

13.7

1 $CH_3COOC_2H_5 + H_2O \rightarrow CH_3COOH + C_2H_5OH$ *[1]*

2 $CH_3COOC_3H_7 + OH^- \rightarrow CH_3COO^- + C_3H_7OH$ *[1]*

3 Acid conditions: $CH_3CH_2CONHCH_2CH_3 + H^+ + H_2O$
$\rightarrow CH_3CH_2COOH + CH_3CH_2NH_3^+$ *[1]*

Alkaline conditions: $CH_3CH_2CONHCH_2CH_3 + OH^- \rightarrow$
$CH_3CH_2COO^- + CH_3CH_2NH_2$ *[1]*

13.8

1 a $H_2NCH(CH(CH_3)_2)COO^-$ *[1]*

  b $H_3N^+CH(CH(CH_3)_2)COO^-$ *[1]*

  c $H_3N^+CH(CH(CH_3)_2)COOH$ *[1]*

2 $H_2NCH(CH_2OH)CONHCH(CH_3)COOH$ *[1]* and
$H_2NCH(CH_3)CONHCH(CH_2OH)COOH$ *[1]*

3 a Acidic conditions give $H_3N^+CH_2COOH$ Cl^- +
$H_3N^+CH(CH_3)COOH$ Cl^- ; a chloride salt is
produced because HCl is the protonating agent
(the amine group becomes protonated after the
hydrolysis). *[1]*

  b Alkaline conditions give a sodium salt of the
carboxylic acid group:

$H_2NCH_2COO^-Na^+ + H_2NCH(CH_3)COO^-Na^+$ *[1]*

13.9

1 An ester of propane-1,2,3-triol containing different
R groups from three different fatty acids. *[1]*

2

[1]

3 $7.2\,dm^3$ of hydrogen is $0.3\,mol$ *[1]*. Therefore the
triester reacts with hydrogen in a 1:2 ratio *[1]*, so
there are two C=C bonds. *[1]*

13.10

1 Warm the substances separately with Fehling's
solution or with Tollens' reagent. The aldehyde will
give a red precipitate with Fehling's solution, or a
silver mirror with Tollens' reagent *[1]*. The ketone
will not react with either. *[1]*

2 a $CH_3CH_2CHO + [O] \rightarrow CH_3CH_2COOH.$ *[1]*

  b propanone cannot be oxidised. *[1]*

3

aldehyde

The R group is CH_3. Your diagram must show curly
arrows and dipoles *[1]*, the intermediate *[1]* and the
final product $CH_3CH(OH)CN$. *[1]*

14.1

1 A batch process produces small quantities at a
time whilst a continuous process produces large
quantities and runs without interruption. *[1]*

2 a continuous

  b batch

  c continuous. *[3]*

3 There may already be good transport links / could use
products from the existing site / cheap energy
supply / good water supply / skilled labour / may be
near existing raw materials – any three suggestions.
[3]

14.2

1 a % atom economy

$$= \frac{\text{relative formula mass of useful product}}{\text{relative formula mass of reactions used}} \times 100$$

$$= \frac{2 \times 44}{(2 \times 28) + 32} \times 100 = 100\%$$ *[1]*

  b It is reused by separating it from the product and
putting it back in at the start of the plant. *[1]*

2 Generally addition > substitution > elimination. *[1]*
Addition reactions have atom economies of 100%;
elimination reactions have low atom economies
as there is always a waste product; substitution
reactions tend to be in the middle depending on
which atoms are removed from the reactant. If a
heavy atom such as Br is substituted and the Br-
containing product is a waste product, the atom
economy could be lower than for the elimination
reaction. *[1]*

3 atom economy for step 1 $= \dfrac{99}{28 + 71} \times 100 = 100\%$ *[1]*

atom economy for step 2 $= \dfrac{62.5}{99} \times 100 = 63.1\%$ *[1]*

14.3

1 HCl is a co-product and but-1-ene is a by-product
[1]; HCl is produced in the intended reaction *[1]*;
but-1-ene is produced in a second, unintended
reaction. *[1]*

2 A has a higher atom economy *[1]*; so less waste to
dispose of *[1]*; A is carried out at a lower temperature
[1]; so less energy needed for fuel *[1]*; starting
material for A is ethene *[1]*; which can be obtained
easily from crude oil. *[1]*

3 a High temperature increases rate of reaction
[1]; but high temperature will favour
endothermic direction so reduces yield *[1]*;
450 °C (is a moderately high temperature) and
is a compromise (between these two competing
factors). *[1]*

Changing pressure has no effect on yield (because
there are equal numbers of moles on each side of
the equation) *[1]*; high pressure would increase
rate *[1]*; but would be expensive, because of need
for expensive equipment / energy. *[1]*

Catalyst increases rate without need for
increased pressure / temperature. *[1]*

  b Reaction removes toxic CO *[1]*; produces
valuable H_2 *[1]*; although it releases CO_2 which
is a greenhouse gas *[1]*; and requires fuel to

be burnt which also releases CO_2 *[1]*; overall benefits outweigh the risks. *[1]*

15.1

1 $N_2 + O_2 \rightarrow 2NO$ *[1]*. NO is oxidised to NO_2 which dissolves in rain water to make acid rain. *[1]*

2 Carbon monoxide is toxic to humans because it binds irreversibly to haemoglobin, reducing the uptake of oxygen by the blood. *[1]* A possible equation is $2CH_4 + 3O_2 \rightarrow 2CO + 4H_2O$ *[1]*, but other balanced equations forming CO and H_2O (possibly together with CO_2 and/or C) could be credited.

3 There is more complete combustion. *[1]*

15.2

1 Hydrogen has a low density and may easily escape. It is highly flammable. *[1]*

2 Advantage: renewable source, carbon neutral. Disadvantage: may use up farmland which could be used for food; may promote deforestation. *[2]*

3 It is made by fermenting sugar. Sugar cane only grows in certain areas. *[1]*

15.3

1 Sun emits visible and ultraviolet, Earth emits infrared *[1]*; wavelength of radiation emitted by Sun is smaller than that emitted by Earth *[1]*; frequency of radiation emitted by Sun is greater than that emitted by Earth. *[1]*

2 Infrared window is the range of wavelengths / frequencies not absorbed by water vapour *[1]*; sulfur hexafluoride absorbs infrared radiation with a wavelength / frequency from within this range. *[1]*

3 Infrared radiation from the Earth is absorbed by bonds in the water molecule *[1]*; some of this energy is re-emitted in all directions, including back to Earth *[1]*; greater evaporation means that the concentration of water vapour will increase *[1]*; a higher concentration of water vapour means that more energy is absorbed and re-emitted in this way *[1]*; temperature of the Earth increases even more. *[1]*

16.1

Go further

There are 16 duplet combinations of the 4 RNA bases.

Summary questions

1 a Consists of 2 chains of nucleotides *[1]*; twisted into a double helix structure *[1]*; sugar and phosphate molecules bond together to form a 'backbone' *[1]*; complementary pairs of bases on the inside of the molecule bond together by hydrogen bonds. *[1]*

   b RNA (usually) is single stranded not double stranded *[1]*; RNA has a uracil base instead of a thymine *[1]*; RNA has a ribose sugar instead of a deoxyribose. *[1]*

2 Alanine–Leucine–Glutamic acid–Valine *[1]*

3 mRNA: carries information from the nucleus to the ribosome *[1]*; sequence of codons on mRNA determines sequence of amino acids in protein. *[1]*

tRNA: delivers amino acid to the ribosome *[1]*; anticodon on tRNA recognises / bonds specifically to a codon on mRNA. *[1]*

17.1

1 A molecule with more than one functional group. *[1]*

2 a addition reactions: alkenes, carbonyls. *[1]*

   b substitution reactions: haloalkanes, alcohols. *[1]*

   c oxidation reactions: primary and secondary alcohols, aldehydes. *[1]*

3 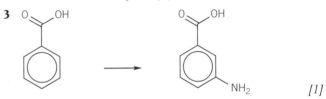 *[1]*

amine and carboxylic acid *[1]*; the amine would act as a proton acceptor so will react with acids, and it could act as a nucleophile to form an amide with acyl chlorides *[1]*; the carboxylic acid would act as a proton donor so will react with bases, and it could be converted to an acyl chloride with $SOCl_2$, and it could undergo esterification with alcohols. *[1]*

17.2

1 a $C_3H_7OH + [O] \rightarrow C_2H_5CHO + H_2O$ *[1]*

   b $C_3H_7OH + 2[O] \rightarrow C_2H_5COOH + H_2O$ *[1]*

   c $C_3H_7OH \rightarrow C_3H_6 + H_2O$ *[1]*

   d $C_3H_7OH + HCl \rightarrow C_3H_7Cl + H_2O$ *[1]*

   e $C_3H_7OH + HCOOH \rightarrow C_3H_7OOCH + H_2O$ *[1]*

2 but-1-ene $\rightarrow$ butan-1-ol *[1]* $\rightarrow$ butanoic acid *[1]*; conditions for the first step are reaction with H_2O at 300 °C with a catalyst of H_3PO_4 *[1]*; conditions for the second step are refluxing with acidified potassium dichromate. *[1]*

3 haloalkane $\rightarrow$ alcohol *[1]* $\rightarrow$ ketone *[1]* $\rightarrow$ cyanohydrin *[1]*; the ketone is butanone *[1]*; the alcohol is butan-2-ol *[1]*; the haloalkane is 2-bromobutane or 2-chlorobutane *[1]*; the conditions are OH^- ions for making the alcohol from the haloalkane *[1]*, oxidation with acidified potassium dichromate for making the ketone *[1]*, and reaction with HCN for making the cyanohydrin. *[1]*

Answers to practice questions

Chapter 1

1 B *[1]*, **2** D *[1]*, **3** C *[1]*, **4** C *[1]*, **5** D *[1]*

6 a Mol NaOH = (15.75/1000) × 0.500
= 7.875 × 10⁻³ mol *[1]*

   b Mol HCl = mol NaOH (because 1:1 ratio in the equation) = 7.875 × 10⁻³ mol. *[1]*

   c Concentration of HCl = moles/(vol/1000)
= 7.875 × 10⁻³/(10.0/1000) = 0.7875 *[1]*
= 0.788 mol dm⁻³ to 3 s.f. *[1]*

   d Concentration HCl in undiluted solution
= 0.788 × dilution factor = 7.88. *[1]*

     Dilution factor is 10, because 25 cm³ was used to make up 250 cm³ of diluted solution *[1]*
= 0.788 × 10 = 7.88 mol dm⁻³.

     Mass of 7.88 mol = 7.88 × M_r(HCl)
= 7.88 × 36.5 = 287.62.

     So concentration in g dm⁻³ = 288 (to 3 or more s.f.). *[1]*

   e Take 50 cm³ of 2.00 mol dm⁻³ NaOH. *[1]*

     Measure out using a 50 cm³ volumetric pipette / 2 × 25 cm³ volumetric pipette. *[1]*

     Add water up to the mark. *[1]*

     Fine detail: ONE of: add water dropwise when close to mark, invert solution several times to ensure mixing, observe level of liquid in pipette or flask at eye level. *[1]*

7 a Test tube connected a tube to a gas syringe OR inverted burette / measuring cylinder in trough of water *[1]*. Magnesium carbonate labelled AND heat shown being applied to magnesium carbonate *[1]*, apparatus shown as airtight (for gas syringe), OR delivery tube shown in appropriate place under burette. *[1]*

   b 2.00 g MgCO₃ = 2.00/84.3 = 0.082 30 mol *[1]*.
Volume of CO₂ formed = 0.082 30 × 24.0
= 1.98 dm³ *[1]*. Answer given to 2 or more s.f.

   c v = nRT/p *[1]* = 1 × 8.314 × 210/600 = 2.91 (m³) *[1]* = 2910 dm³. *[1]*

Chapter 2

1 A *[1]*, **2** C *[1]*, **3** B *[1]*, **4** D *[1]*

5 a i **Two or more** (lighter) nuclei **join** together, to form a **single heavier** nucleus. *[2]*

    ii To overcome the repulsion *[1]* between the **positive** nuclei. *[1]*

   b i $^1_1H + ^2_1H \rightarrow ^3_2He$ *[1 for each H isotope]*

    ii $^{12}_6C$ *[1: numbers, 1: symbol]*

iii The Sun is not hot enough / does not have high enough pressure (at centre) / is not a heavyweight star.

     OR

     Other stars are hotter / have a higher pressure (at centre) / are heavyweight stars. ALLOW that there is no hydrogen left in these stars. *[1]*

6 a i 2 electrons in sub-shell A, 2 electrons in sub-shell B.

     [1 mark: 4 electrons in total, 1 mark: electrons in correct sub-shells]

     Electrons shown with opposite spins in sub-shell A, AND electrons in separate orbitals in sub-shell B (ignore direction of spins). *[1]*

    ii 10   *[1]*

   iii A = 3s, B = 3p

     [2 marks, ALLOW 1 mark if s and p both correct but numbers are wrong]

   iv s-orbitals are spherical, p-orbitals are dumbbell shaped. ALLOW suitable sketch. *[1]*

     OR p-orbitals have two regions of electron density, s-orbitals only have one.

     OR p-orbitals have no electron density close to centre of atom.

   b

Species	Electron configuration
Na⁺	1s²2s²2p⁶ IGNORE 3s⁰ *[1]*
F, ALLOW O⁻ Ne⁺ etc. *[1]*	1s²2s²2p⁵
Cr	[Ar] 4s² *[1]* 3d⁴ *[1]* in either order

     [4]

   c i p-orbitals / sub-shells have the highest energy OR are the outermost electrons. *[1]*

    ii 5p³. *[1]*

Chapter 3

1 B *[1]*, **2** D *[1]*, **3** B *[1]*,

4 C *[1]*, **5** A *[1]*

6 a i NH₄⁺ and Cl⁻ ions shown alternating in rows *[1]* at least 2 rows shown, with ions of opposite charges shown as nearest neighbours throughout. *[1]*

    ii There is electrostatic attraction *[1]* between + and – charged ions. *[1]*

   b i One of the bonds has both electrons from the same atom *[1]* both come from the N atom. *[1]*

ii Electron pairs / groups repel each other. *[1]*

To get as far apart as possible / to minimise repulsion. *[1]*

4 groups / pairs of electrons around N atom. *[1]*

Tetrahedral arrangement (of electrons and hence tetrahedral shape of molecule). *[1]*

Bond angle is 109½°. *[1]*

c Triple bond between N atoms *[1]*. Correct number of electrons on each N atom AND lone pair shown on each N atom. *[1]*

Chapter 4

1 A *[1]*, **2** C *[1]*, **3** B *[1]*, **4** C *[1]*, **5** B *[1]*

6 a $E = m\,c\,\Delta T = 100 \times 4.18 \times 24 = 10\,032$ (J) *[1]*, correct values for m and ΔT *[1]*. Correct evaluation of E (with e.c.f.).

b i Moles of ethanol = 1.10/46 = 0.023 91 *[1]* enthalpy change per mole =10 032/0.023 91 = 418 600 J OR 418.6 kJ *[1]*. $\Delta_c H = -419\,kJ\,mol^{-1}$ *[2]*

ii Heat loss (to the surroundings) / incomplete combustion of ethanol / evaporation of ethanol during the experiment. *[ANY 2]*

7 a i Bonds broken = $3C \times C + 8C \times H + 1C=O + 6\,O=O$ *[1]*. Bonds formed = $8\,C=O + 8\,O-H$ *[1]*. Energy required to break bonds = 8 138, energy released by forming bonds = (–)10 152 *[1]*. Enthalpy of combustion = 8 138 – 10 152 = $-2104\,kJ\,mol^{-1}$. *[1]*

ii Bond enthalpies are averages / are not the same as the actual values in the molecule. *[1]* Bond enthalpies can only be used for molecules in gaseous state ALLOW butanone and water are in the liquid state under standard conditions. *[1]*

b (Enthalpy change will be similar) *[1]* because the number and type of bonds in butanone are the same as butanal *[1]* so the same (number and type of) bonds are broken and formed when both molecules react. *[1]*

8 C *[1]* **9** D *[1]* **10** D *[1]* **11** C *[1]*

12 a Heat loss to the surroundings ACCEPT not done under standard conditions *[1]*

b i *Top line*: $2Na^+(g) + SO_4^{2-}$ (g) *[1]*

Bottom line $2Na^+(aq) + SO_4^{2-}$ (aq)/Na_2SO_4(aq) *[1]*

Top LH arrow: Lattice enthalpy / $\Delta_{LE}H$

Top RH arrow: sum of enthalpy change of hydration ($\Delta_{hyd}H$) of (positive and negative) ions

Bottom arrow: enthalpy change of solution

All 3 arrows correct = *[1]*

ii sum of enthalpy change of hydration ($\Delta_{hyd}H$) of (positive and negative) ions = $\Delta_{LE}H$ + enthalpy change of solution = –1944.0 – 2.5 = $-1946.5\,kJ\,mol^{-1}$ *[1]*

Sum of enthalpy changes of hydration = 2 × –406 + $\Delta_{hyd}H$ [SO_4^{2-}] *[1]*

$\Delta_{hyd}H$ [SO_4^{2-}] = –1946.5 + 812 *[1]* = $-1134.5\,kJ\,mol^{-1}$ *[1]*

13 a Water is polar / has a dipole / partial charges *[1]*

Water molecules form ion-dipole bonds to ions *[1]*

Energy released from forming bonds is enough to compensate for energy required to break bonds in lattice OR strength of bonds formed is similar to strength of bonds in lattice *[1]*

b i $\Delta_{surr}S = -\left(-\dfrac{13100}{298}\right) = +44.0\,J\,K^{-1}\,mol^{-1}$ *[1]*

$\Delta_{tot}S = -204.8 + 44.0 = -160.8$ *[1]*

This is negative so process is not feasible *[1]*

ii (No) $\Delta_{surr}S$ can be made more positive by lowering the temperature *[1]*

If T is low enough, $\Delta_{surr}S$ will be large enough to compensate for –ve $\Delta_{sys}S$ *[1]*

T required for this would be (well) below 273 K (so water would not be liquid at this T) *[1]*

Chapter 5

1 C *[1]*, **2** B *[1]*, **3** B *[1]*, **4** C *[1]*,

5 D *[1]*, **6** A *[1]*, **7** B *[1]*

8 a i Positive and negative charges at opposite ends of the bond / uneven distribution of charge in the bond. *[1]*

ii F and Cl are both more electronegative *[1]* than C. *[1]*

iii F has a greater negative charge than the Cl atoms *[1]*; centre of negative charge is not in the centre of the molecule / is not in the same place as the centre of positive charge / charges are arranged unsymmetrically. *[1]*

iv Permanent dipole–permanent dipole bonds. *[1]*

b i Random movement of electrons *[1]*; results in an unequal distribution of charge *[1]*. This induces a dipole on a neighbouring molecule *[1]*; the two dipoles /molecules attract each other. *[1]*

ii CCl_4 has more electrons than CCl_3F *[1]* so more change of an instantaneous dipole arising *[1]* instantaneous dipole–induced dipole bonds are stronger *[1]* (more significant than) additional permanent dipole–permanent dipole bonds between CCl_3F molecules *[1]* overall, more energy needed to separate CCl_4 molecules. *[1]*

9 C *[1]* **10** B *[1]* **11** D *[1]*

12 a i sulfonic acid / sulfonate *[1]*

ii $-NH_3^+$ (accept any valid protonated amine group e.g. $-NH_2R^+$) *[1]*

b i Random movement of electrons creates an uneven distribution of charge / instantaneous dipole *[1]*

This induces a dipole on a neighbouring molecule *[1]* There is an (electrostatic) attraction between the two dipoles / two molecules *[1]*

ii Long chains / regular arrangement of chains / close packing of chains (any two) *[2]*

c bond between H atom in O-H group and O or N in fibre *[1]*

lone pair on O pointing to H *[1]*

180° bond angle around H atom *[1]*

δ– and δ+ charges on O-H and N-H bonds *[1]*

13 a i amine *[1]*

ii amine group has a lone pair of electrons on a small (electronegative) atom OR a δ+ H atom *[1]* can bond to a δ+ H (in the receptor site) OR a lone pair of electrons on a small (electronegative) atom (in the receptor site) *[1]*

b i contains a chiral centre / C atom bonded to 4 different groups *[1]*

ii At least one molecule drawn with CH_3, H, and NH_2 groups shown in a 3-dimensional / tetrahedral arrangement around the chiral carbon (at the end of the chain) *[1]*

two molecules are clearly mirror images *[1]*

Chapter 6

1 D *[1]*, **2** A *[1]*, **3** A *[1]*, **4** D *[1]*, **5** D *[1]*

6 a i Ozone absorbs high frequency / high energy ultraviolet radiation *[1]* this causes skin cancer/ eye cataracts (if it reaches the Earth's surface). *[1]*

ii CFC molecules diffuse from troposphere to stratosphere / are not broken down in the troposphere *[1]* in the stratosphere, high frequency / energy uv light breaks the C−Cl bond *[1]* Cl radicals are formed *[1]* these act as a catalyst for the break down of ozone. *[1]*

b i Homolytic *[1]*

ii A Species with an unpaired electron. *[1]*

iii Energy required to break one bond
= 305 × 1000 *[1]* / 6.02 × 1023
= 5.07 × 10⁻¹⁹ J *[1]*

iv $f = \dfrac{E}{h}$ *[1]* $= \dfrac{5.07 \times 10^{-19}}{6.63 \times 10^{-34}} = 7.65 \times 1014\,Hz$ *[1]*

v O atom combines with O_2 to form O_3. *[1]*

7 a i Bonds vibrate *[1]* with more energy *[1]*

ii Peak at 2 510–3 490 indicates O−H in carboxylic acid *[1]* no O−H in propanal *[1]* peak at 1 715 indicates C=O in carboxylic acid *[1]* C=O in aldehyde would be seen at 1720–1740. *[1]*

b i Peak at 74 is molecular ion / unfragmented propanoic acid *[1]*. Peak at 75 is due to presence of a ^{13}C atom. *[1]*

ii Peak at 57 = $C_3H_5O^+$ (loss of OH) *[1]*. Peak at 45 = $COOH^+$. *[1]* (1 max if no + charges on either species).

8 B *[1]*   **9** C *[1]*   **10** D *[1]*   **11** D *[1]*

12 $CH_3COOCH_2CH_3$ *[1]*

δ = 1.2 is $CH_2\underline{CH_3}$ AND 2H on C atom adjacent to CH_3 environment *[1]*

δ = 2.0 is $\underline{CH_3}CO$, no H atoms on C adjacent to CH_3 environment *[1]*

δ = 4.2 is $O\underline{CH_2}$, 3 H on C atom adjacent to CH_2 environment *[1]*

ACCEPT any unambiguous way of identifying environments e.g. labels on structure

13 Electrons are excited to higher energy level *[1]* absorb light energy / electromagnetic radiation *[1]* energy gap between energy levels greater in benzene *[1]* because less extensive delocalised system in benzene *[1]* benzene absorbs in uv region, disperse red absorbs in visible region / complementary wavelengths to red NOT red *[1]*

Chapter 7

1 C *[1]*,   **2** D *[1]*

3 a pH 8–10 due to the presence of OH⁻ ions. *[1]*

b H⁺ ions from the acid will react with OH⁻ ions, removing them from the system *[1]*; the position of equilibrium will move to the right to replace the OH⁻ ions. *[1]*

4 a i **Step 1:** increasing temperature moves position of equilibrium to the right as the forward reaction is endothermic *[1]*; increasing pressure moves position of equilibrium to the left as there are fewer molecules on the left *[1]*; catalyst does not affect the position of equilibrium. *[1]*

Step 2: increasing temperature moves position of equilibrium to the left as the forward reaction is exothermic *[1]*; increasing pressure moves position of equilibrium to the right as there are fewer molecules on the right *[1]*; catalyst does not affect the position of equilibrium. *[1]*

ii **Step 2:** is exothermic *[1]* so a high temperature moves the position of equilibrium to the left, making less product *[1]*; in step 1 a higher temperature moves the position of equilibrium to the right, making more product. *[1]*

iii **Step 2:** has fewer molecules on the right than on the left *[1]* so a higher pressure moves the position of equilibrium to the right, making more product *[1]*; in step 1 a higher temperature moves the position of equilibrium to the left, making less product. *[1]*

b i $K_c = \dfrac{[CH_3OCH_3][H_2O]}{[CH_3OH]^2}$ *[2]*

ii $K_c = \dfrac{(0.20) \times (0.20)}{(0.050)^2} = 16$

[1 mark for correct numbers in expression; 1 mark for identifying there are no units for K_c in this expression.]

5 D *[1]*   **6** C *[1]*   **7** C *[1]*   **8** D *[1]*

9 a $K_c = \dfrac{[C_2H_4][H_2O]}{[C_2H_5OH]}$ *[1]*

b $mol\,dm^{-3}$ *[1]*

10 a $K_c = \dfrac{[SO_3]^2}{[SO_2]^2[O_2]}$ *[1]*

b i Increasing pressure moves the position of equilibrium to the right but does not affect the magnitude of K_c. *[1]*

ii Adding a catalyst does not affect the position of equilibrium or the magnitude of K_c. *[1]*

iii Increasing the temperature moves the position of equilibrium to the left and decreases the magnitude of K_c. *[1]*

c $K_c = \dfrac{[SO_3]^2}{[SO_2]^2[O_2]}$

$= \dfrac{(1.80 \times 10^{-2}\,mol\,dm^{-3})^2}{(7.45 \times 10^{-3}\,mol\,dm^{-3})^2 \times (3.62 \times 10^{-3}\,mol\,dm^{-3})}$

$K_c = \dfrac{3.24 \times 10^{-4}\,mol^2\,dm^{-6}}{(5.55 \times 10^{-5}\,mol^2\,dm^{-6}) \times (3.62 \times 10^{-3}\,mol\,dm^{-3})}$

$= 1610$ *[1]* $mol^{-1}\,dm^3$ *[1]* (3sf)

11 a $K_{sp} = [Pb^{2+}][S^{2-}]$ *[1]*

b The maximum concentration of S^{2-} ions is $K_{sp} / [Pb^{2+}]$ *[1]*

$= 1.3 \times 10^{-28}\,mol^2\,dm^{-6} / 1.14 \times 10^{-14}\,mol\,dm^{-3}$
$= 1.14 \times 10^{-14}\,mol\,dm^{-3}$. *[1]*

12 a Retention time is the time taken for a component to emerge from the column in gas-liquid chromatography. *[1]*

b The area under the octane peak will be three times the area under the decane peak. *[1]*

c Mass spectrometry will confirm the relative molecular mass of each component. *[1]*

Chapter 8

1 B *[1]*,   **2** A *[1]*,   **3** C *[1]*,   **4** C *[1]*

5 a $2.5 \times 10^{-4}\,mol$ *[1]*

b $2.5 \times 10^{-4}\,mol$ *[1]*

c $0.0127\,mol\,dm^{-3}$ *[1]*

6 a $0.005\,mol$ *[1]*

b $0.01\,mol$ *[1]*

c $0.355\,mol\,dm^{-3}$ *[1]*

7 a i $\dfrac{25.4 \times 0.1}{1000}$ *[1]* $= 0.002\,54$ *[1]*

ii $0.00254 \times 0.5 = 0.001\,27$ *[1]*

iii $0.001\,27$ *[1]*

iv $\dfrac{0.00127 \times 1000}{5}$ *[1]* $= 0.254\,mol\,dm^{-3}$ *[2]*

b 0.254×51.5 *[1]* $= 13.1\,g\,dm^{-3}$. *[2]*

8 D *[1]*   **9** B *[1]*   **10** B *[1]*   **11** C *[1]*   **12** D *[1]*

13 A weak acid on its own can dissociate on addition of alkali, but does not have a sufficiently high concentration of [salt] to accept enough protons on addition of acid. *[1]*

14 pH = 4.2 *[1]*. The buffer minimises pH changes *[1]* on addition of small quantities of acid or alkali. Addition of H^+ moves the position of equilibrium to the left *[1]*, and addition of OH^- moves the position of equilibrium to the right *[1]*. This minimises the change in $[H^+]$. *[1]*

Chapter 9

1 A *[1]*,   **2** B *[1]*,   **3** C *[1]*,

4 C *[1]*,   **5** D *[1]*,   **6** B *[1]*

7 $2\,I^- + 4\,H^+ + MnO_2 \rightarrow I_2 + 2\,H_2O + Mn^{2+}$ *[2]*

8 Nitrogen is oxidised; iodine is reduced. *[2]*

9 a From colourless *[1]* to red/brown. *[1]*

b Loss of electrons / increase in oxidation state. *[1]*

c $2Cl^- \rightarrow Cl_2 + 2e^-$ *[1]*

d $Cl_2 + 2Br^- \rightarrow Br_2 + 2Cl^-$ *[1]*

e Chlorine oxidises the bromide ions *[1]*; chlorine is reduced. *[1]*

10 D *[1]*   **11** C *[1]*   **12** B *[1]*   **13** D *[1]*   **14** D *[1]*

15 a VO_2^+ is +5 and VO^{2+} is +4. *[2]*

b Reduced because its oxidation state goes from +5 to +4. *[1]*

c $VO_2^+ + 2H^+ + 1e^- \rightleftharpoons VO^{2+} + 1H_2O$ *[1 mark for H_2/ H_2O balancing, 1 mark for electron balancing]*

d $E_{cell} = 1.76\,V$ *[1]*

e $2VO_2^+ + 4H^+ + Zn \rightleftharpoons 2VO^{2+} + 2H_2O + Zn^{2+}$ *[1]*

f Zinc has a more negative $E^\ominus$ than all of the vanadium half-cells so it can reduce all the species to V^{2+}. *[1]*

g V^{3+} *[1]*; as $E^\ominus$ for Pb/Pb^{2+} is less negative than $E^\ominus$ for V^{3+}/V^{2+}, so Pb cannot reduce V^{3+}. *[1]*

16 a $Fe \rightarrow Fe^{2+} + 2e^-$ *[1]*

b $O_2 + 2H_2O + 2e^- \rightarrow 4OH^-$ *[1]*

c $Fe^{2+} + 2OH^- \rightarrow Fe(OH)_2$ *[1]*

d $Zn \rightarrow Zn^{2+} + 2e^-$ *[1]*

Chapter 10

1 C *[1]*, **2** A *[1]*, **3** A *[1]*, **4** D *[1]*, **5** D *[1]*

6 a Heterogeneous *[1]*

b More particles will have energy greater than the activation enthalpy *[1]*; particles collide more frequently and the reaction is faster. *[1]*

c It has a greater surface area *[1]* so collisions are more frequent. *[1]*

d It provides an alternative pathway *[1]* of lower activation enthalpy. *[1]*

e Reactants are adsorbed onto the catalyst *[1]*; bonds in the reactant break *[1]*; new bonds form *[1]*; products detach from the catalyst surface. *[1]*

7 a Homogeneous as it is in the same state as the reactant. *[1]*

b NO forms an intermediate with ozone *[1]* and provides an alternative pathway *[1]* of lower activation enthalpy. *[1]*

c Higher temperature means particles move faster/collide with greater energy. *[1]* Higher proportion *[1]* of molecules collide with energy greater than the activation enthalpy. *[1]*

8 D *[1]* **9** D *[1]* **10** B *[1]* **11** B *[1]* **12** B *[1]*

13 a The rate would double. *[1]*

b The rate would go up by a factor of eight – doubled due to doubling [M] and quadrupled due to doubling [N] *[1]*.

c $k = \dfrac{rate}{[M][N]^2}$ *[1]* $= \dfrac{0.025}{(0.1) \times (0.1)^2} = 25$ *[1]*; units are $mol^{-2}dm^6s^{-1}$ *[1]*

d k increases as temperature increases. *[1]*

e Successive half-lives of [M] are equal *[1]* and successive half-lives of [N] are not equal. *[1]*

14 a First order with respect to P and first order with respect to Q. *[2]*

b Taking tangents at time $t = 0$ / measuring successive half-lives. *[1]*

c Not supported as the slow step contains P only. *[1]*

Chapter 11

1 D *[1]*, **2** D *[1]*, **3** D *[1]*, **4** A *[1]*,

5 A *[1]*, **6** D *[1]*, **7** C *[1]*, **8** D *[1]*,

9 A *[1]*, **10** C *[1]*

11 a Bromine vapour is toxic; it spreads far because it is volatile; bromine is corrosive. *[Max 2 marks]*

b Instantaneous dipole–induced dipole bonds *[1]* are stronger between bromine molecules *[1]* as the M_r is greater/larger number of electrons *[1]* more energy is needed to break them (so a higher b.p.) *[1]*

12 Mg^{2+} and O^{2-} ions have high charge densities *[1]*. They attract each other strongly and it takes a lot of energy to separate them *[1]*.

13 a $CaCO_3 \rightarrow CaO + CO_2$ *[1]*

b $MgCO_3$ *[1]* because Mg^{2+} is a smaller ion which polarises the carbonate ion more. *[1]*

c Calcium oxide *[1]*

d $CaO + 2HA$ *[1]* $\rightarrow CaA_2 + H_2O$ *[1]*

Chapter 12

1 C *[1]*, **2** A *[1]*, **3** C *[1]*, **4** B *[1]*, **5** C *[1]*

6 It must be an alkene as it reacts with bromine and produces an addition polymer *[1]*. It must have four carbon atoms. *[1]* It must have different groups on each carbon of the C=C bond as it has E/Z isomers. *[1]* It is but-2-ene. *[1]*

7 a $C_{10}H_{22}$ *[1]*

b There are four groups of electrons around each carbon atom *[1]* which repel *[1]* as far as possible *[1]*.

c C_3H_6 is propene and C_7H_{14} is heptane. *[2]*

d C_3H_6 has no isomers. C_7H_{14} can show chain isomerism. *[2]*

e Electrophilic addition. *[1]*

f 1,2-dibromopropane. *[1]*

g The lone pair on the oxygen of the water molecule can attack the carbocation to produce $CH_2BrCH(OH)CH_3$. *[1]*

8 D *[1]* **9** A *[1]* **10** A *[1]* **11** B *[1]* **12** B *[1]*

13 a model A has 6 delocalised electrons *[1]* model B has 3 C-C pi bonds / alternating C-C and C=C bonds *[1]*

b *Level of response mark scheme:*

Evidence from bond lengths (from X-ray diffraction):

bond lengths equal
bond lengths intermediate between C-C and C=C
fits A
does not fit B (because C-C longer than C=C)

Evidence from pattern of chemical reactions

takes part in (mostly) substitution reactions
fits A as substitution maintains delocalised structure does not fit B as C=C would take part in addition reactions

Evidence from thermochemical data

enthalpy of hydrogenation is less negative than predicted / not 3 x value for cyclohexene

fits A as bonding in benzene is stronger than 3 x C=C / delocalisation increases stability

does not fit B as 3 separate C=C bonds / no extra stability

[5–6 marks]

Discusses 3 types of evidence
All evidence described in detail
Clearly describes how evidence links to bonding

[3–4 marks]

Discusses 2 types of evidence
Most evidence described in detail OR
discusses 3 types of evidence but in less detail
Some success in linking evidence to bonding

[1–2 marks]

Discusses 1 type of evidence
Describes this in some detail OR
Discusses 2 types of evidence but in little detail
Limited success in linking evidence to bonding

14 a i conc nitric and conc sulfuric acid *[1]*

   ii Sn and HCl *[1]*

   b i $C_6H_5–N^+\equiv N$ *[1]*   ii C_6H_5OH *[1]*

   c Reaction 1: below 55°C *[1]* to avoid multiple nitrations. *[1]*

   Reaction 2: below 5°C *[1]* to slow down decomposition of diazonium compound. *[1]*

Chapter 13

1 C *[1]*,   2 D *[1]*,   3 A *[1]*,   4 D *[1]*,

5 B *[1]*,   6 D *[1]*,   7 A *[1]*,   8 D *[1]*

9 The observations are:

   a Violet colour. *[1]*

   b No reaction. *[1]*

   c No reaction. Phenols need acid anhydrides for esterification. *[1]*

10 a Orange. *[1]*

   b It is a tertiary alcohol *[1]* as it has not been oxidised. *[1]*

   c 2-methylpropan-2-ol [1]; $CH_3C(CH_3)(OH)CH_3$. *[1]*

   d $CH_3CH_2CH_2CH_2OH$ (butan-1-ol) *[1]* and $CH_3CH(CH_3)CH_2OH$ (2-methyl-propan-1-ol). *[1]*

   e Oxidation of butan-1-ol would give butanal, $CH_3CH_2CH_2CHO$ *[1]* under mild conditions, and butanoic acid, $CH_3CH_2CH_2COOH$ *[1]* under reflux. Oxidation of 2-methyl-propan-1-ol would give 2-methylpropanal, $CH_3CH(CH_3)CHO$ *[1]* under mild conditions, and 2-methylpropanoic acid, $CH_3CH(CH_3)COOH$ *[1]* under reflux.

11 a Carbon has a partial positive charge and the halogen has a partial negative charge *[1]* because the halogen is more electronegative. *[1]*

   b Boiling point. *[1]*

   c The nucleophile is attracted to the partial positive charge of the carbon atom; *[1]* a bond forms between the nucleophile and the carbon atom; *[1]* the C-Hal bond breaks. *[1]*

12 D *[1]*   13 B *[1]*   14 C *[1]*   15 A *[1]*

16 a $C_5H_{10}O_2$ *[1]*

   b 2 C_4H_9COOH + Na_2CO_3 → 2 C_4H_9COONa + CO_2 + H_2O *[1]*; carbon dioxide produces bubbles and turns limewater cloudy. *[1]*

   c C_4H_9COOH + NH_3 → $C_4H_9CONH_2$ + H_2O *[1]*; primary amide. *[1]*

17 a $H_2NCH_2CONHCH(CH_3)CONHCH(CH_2OH)COOH$ *[1]*

   b If acid hydrolysis is used, amine salts of the amino acids are produced, but if alkaline hydrolysis is used, carboxylate salts are produced. *[1]*

   c Only two of the amino acids have a chiral centre *[1]*; as they have four different groups around a carbon atom *[1]*; gly does not display optical isomerism as it does not have a carbon with four different groups. *[1]*

   d A spot of the product mixture is placed on the paper *[1]*; the paper is placed in a solvent, and the solvent rises to the top of the paper, separating the components *[1]*; R_f values are calculated and compared to reference data , or use spots of the three amino acids and compare heights. *[1]*

18 a 2-heptanone is a ketone and benzaldehyde is an aldehyde. *[1]*

   b $CH_3COCH_2CH_2CH_2CH_2CH_3$ *[1]*

   c Only benzaldehyde will react with Fehling's solution *[1]*; a red precipitate will be formed. *[1]*

   d Only benzaldehyde will react with Tollens' reagent *[1]*; a silver mirror will be formed. *[1]*

   e Both will react with HCN *[1]*. $CH_3C(CN)(OH)CH_2CH_2CH_2CH_2CH_3$ *[1]* and $C_6H_5CH(OH)(CN)$ *[1]*

Chapter 14

1 C *[1]*,   2 D *[1]*,   3 A *[1]*,   4 D *[1]*

5 a Acidified *[1]* dichromate *[1]* reflux *[1]*

   b Propanone: ketone *[1]*; propan-2-ol: (secondary) alcohol *[1]*

   c 96.7% *[1]*

   d 0.1 mol *[1]*

   e 0.1 mol *[1]*

   f 5.8 g *[1]*

   g 45% *[1]*

6 a 42.2% *[1]*

b Amount of C_2H_5Br = 0.05 mol *[1]*;
expected amount of C_2H_5OH = 0.05 mol *[1]*;
expected mass of C_2H_5OH = 2.30 g *[1]*;
percentage yield = 62%. *[1]*

7 A *[1]*   **8** D *[1]*   **9** B *[1]*   **10** C *[1]*   **11** C *[1]*

12 a $(4 \times 30) \times 100 / ((4 \times 30) + (6 \times 18)) = 52.6\%$ *[1]*
low atom economy means that large mass of waste is formed *[1]*

b i co-product = H_2O *[1]* additional product formed as a result of the intended reaction *[1]*

ii by-product = NO_2 *[1]* product formed as a result of an unintended reaction *[1]*

c i higher temperature is not used as reaction is exothermic *[1]* higher temperature favours endothermic direction AND reduces yield *[1]*

ii higher pressure would decrease yield *[1]* equilibrium would favour LHS AND fewer moles of gas *[1]* higher pressure increases costs *[1]* more energy needed to compress gas OR expense of thick-walled reaction vessels *[1]* lower pressure decreases rate *[1]* 5 atm is a compromise / provides acceptable rate without decreasing yield significantly (or reverse argument) *[1]*

Chapter 15

1 C *[1]*,   **2** D *[1]*,   **3** A *[1]*,
4 B *[1]*,   **5** D *[1]*,   **6** B *[1]*

7 a $C_{10}H_{22} + 15\frac{1}{2} O_2 \rightarrow 10CO_2 + 11H_2O$ *[1]*

b $NO + \frac{1}{2}O_2 \rightarrow NO_2$ *[1]*

c $SO_2 + \frac{1}{2}O_2 \rightarrow SO_3$ *[1]*

d $SO_3 + H_2O \rightarrow H_2SO_4$ *[1]*

8 It is made from plants *[1]* which can be regrown *[1]*; it is carbon neutral *[1]*; it produces less CO, C_xH_y, SO_2, and particulates than from a diesel engine *[1]*; it is biodegradable. *[1]* [max 4 marks]

9 D *[1]*   **10** C *[1]*   **11** C *[1]*   **12** D *[1]*   **13** A *[1]*

14 Molecules of carbon dioxide absorb infrared radiation. *[1]* Infrared radiation is emitted by the Earth's surface. *[1]* Bonds vibrate with greater energy. *[1]* Some of this energy is radiated back towards Earth. *[1]* Some of the energy is passed on to other molecules by collisions. *[1]* Higher concentration of CO_2 means that a greater fraction / proportion of radiation is absorbed. *[1]*

Chapter 16

1 D *[1]*   **2** A *[1]*   **3** B *[1]*   **4** D *[1]*   **5** B *[1]*   **6** A *[1]*

7 a The hydrogen bonds between the (complementary) base pairs break AND the two strands separate. *[1]*
New RNA nucleotides bond to the exposed bases. *[1]*
The newly attached nucleotides are joined together, creating a new strand of RNA. *[1]*

b i mRNA / messenger RNA *[1]*

ii codon (IGNORE any reference to amino acid encoded) *[1]*

iii anti-codon *[1]*

iv peptide (ACCEPT amide) *[1]*

c Codons consist of 3 bases. *[1]*
There are more than 20 combinations / 64 combinations of bases in codons *[1]* ACCEPT any number >20. *[1]*

Chapter 17

1 D *[1]*   **2** A *[1]*   **3** A *[1]*   **4** A *[1]*   **5** A *[1]*

6 Ketone, amine, alcohol, ether *[4]*; ketones can be reduced to alcohols or react with HCN *[1]*; amines will react with acids or can form amides *[1]*; alcohols can react with carboxylic acids to form esters. *[1]*

7 **Step 1:** CH_3CH_2CHO to $CH_3CH_2CH_2OH$; reduction with $NaBH_4$. *[1]* **Step 2:** $CH_3CH_2CH_2OH$ to $CH_3CH_2CH_2Br$; nucleophilic substitution with HBr. *[1]* **Step 3:** $CH_3CH_2CH_2Br$ to $CH_3CH_2CH_2CN$; nucleophilic substitution with HCN. *[1]* An extra step after Step 1 could involve dehydrating the alcohol to form an alkene, then reacting the alkene with HBr to form $CH_3CH_2CH_2Br$.

8 **Step 1:** C_6H_6 to $C_6H_5NO_2$; nitration with concentrated nitric acid and concentrated sulfuric acid at <55°C. *[1]*

Step 2: $C_6H_5NO_2$ to $C_6H_5NH_2$; reduction with tin and concentrated hydrochloric acid. *[1]*

Synoptic questions

1 Aluminium chloride is manufactured from aluminium metal. It has several uses, for example as a catalyst in several important reactions of benzene. Aluminium can be formed from the reaction of aluminium with hydrogen chloride at about 700°C.

Reaction A: $2\,Al + 6\,HCl \rightarrow 2\,AlCl_3 + 3\,H_2$

a A small-scale pilot study of this process was carried out.

   i The yield in this pilot study was found to be 95%. Calculate the mass of aluminium needed to produce 10 kg of aluminium chloride. *(2 marks)*

   ii Calculate the atom economy of reaction A to 3 significant figures. *(1 mark)*

b A second reaction that could be used to form aluminium chloride is shown below:

   Reaction B: $2Al + 3Cl_2 \xrightarrow{700°C} 2AlCl_3$

   Use the principles of green chemistry to compare the environmental impact of reactions A and B. *(2 marks)*

c Aluminium chloride is a solid ionic compound at temperatures below 180°C. However, above this temperature the compound is converted into a liquid **covalent** compound with the formula Al_2Cl_6.

   i What can you predict about the electrical conductivity of aluminium chloride at room temperature, and at temperatures above 180°C? Explain your answer. *(2 marks)*

   ii The bonding of the covalent molecule Al_2Cl_6 can be represented by the diagram shown below:

   Draw a dot-and-cross diagram to show the bonding in this compound. *(2 marks)*

d Aluminium chloride is used as a catalyst in a range of reactions, known as Friedel–Crafts reactions, which are important in synthesis.

   i Which reaction is a Friedel–Crafts reaction? Explain the importance of these reactions in synthesis. *(1 mark)*

   ii Give the structure of the organic reactant used in reaction 2. *(1 mark)*

   iii Identify the **types** and **mechanisms** of the reactions that occur in reaction 1 AND reaction 2. *(2 marks)*

   iv Name the **type** of product formed in reaction 2. *(1 mark)*

   v State the reagent and conditions required for reaction 3. *(1 mark)*

2 Sodium carbonate exists as a hydrated salt, $Na_2CO_3.xH_2O$.

a Some students carry out an experiment to find calculate the value of x. They heat some hydrated sodium carbonate in a crucible and record the following results:

	Mass / g
1. Mass of crucible	10.36 g
2. Mass of crucible + hydrated salt	14.77 g
3. Mass of crucible + salt after 5 minutes heating	11.99 g
4. Mass of crucible + salt after a further 2 minutes heating	11.99 g

   i What was the purpose of taking reading 4 (mass of crucible + salt after a further 2 minutes heating)? *(1 mark)*

   ii Calculate the value of x in the formula $Na_2CO_3.xH_2O$. *(4 marks)*

b Magnesium carbonate also exists in hydrated forms. However it is not possible to use the same method to find the formula of magnesium carbonate as anhydrous magnesium carbonate undergoes thermal decomposition at a low temperature.

   i Write an equation for the thermal decomposition of anhydrous magnesium carbonate. *(1 mark)*

   ii Explain why anhydrous magnesium carbonate decomposes at a lower temperature than anhydrous sodium carbonate. *(2 marks)*

   iii A titration method is used to find the formula of **hydrated** magnesium carbonate.

   1.69 g of a sample of hydrated magnesium carbonate was dissolved in water and the volume made up to 250 cm³.

   25.0 cm³ of this solution was titrated against 0.100 mol dm⁻³ HCl. The average titre was 24.50 cm³.

   Calculate the value of x. *(4 marks)*

c Carbonate rocks can be heated to high temperatures in the interior of the Earth, releasing carbon dioxide. Some of this carbon dioxide reaches the atmosphere and acts as a greenhouse gas.

   Describe how carbon dioxide acts as a greenhouse gas. *(6 marks)*

d Carbon dioxide from the atmosphere and carbonate ions from rocks are involved in a series of processes involving water that form a weakly acidic solution:

Equation 1: $CO_2(g) + aq \rightleftharpoons CO_2(aq)$

Equation 2: $CO_2(aq) + H_2O(l) \rightleftharpoons H^+(aq) + HCO_3^-(aq)$

Equation 3: $HCO_3^-(aq) \rightleftharpoons H^+(aq) + CO_3^{2-}(aq)$

   i Explain how a decrease in the pH of the ocean can result in the release of carbon dioxide into the atmosphere. *(4 marks)*

   ii A mixture of sodium carbonate and sodium hydrogencarbonate can also be used as a buffer solution.

      Calculate the pH of a buffer solution formed by mixing equal volumes of $0.20\,mol\,dm^{-3}$ $NaHCO_3$ and $0.40\,mol\,dm^{-3}$ Na_2CO_3.

      $Ka\,[HCO_3^-] = 4.8 \times 10^{-11}\,mol\,dm^{-3}$. *(2 marks)*

e Carbon dioxide is also converted very rapidly into hydrogen carbonate ions by the enzyme carbonic anhydrase.

This catalyses the reaction shown in Equation 2, above.

The optimum temperature of the enzyme is 37°C and the optimum pH is 7.5.

   i Sketch the likely pattern in activity when the temperature is altered from 20°C to 65°C. *(2 marks)*

   ii Explain why changing the pH from 7.5 to 6.5 produces a significant drop in enzyme activity. *(2 marks)*

3 Propanone reacts with iodine in the presence of an acid catalyst

$CH_3COCH_3 + I_2 \rightarrow CH_3COCH_2I + HI$

a An experiment was carried out in which the concentration of iodine was measured at regular time intervals. The results were plotted on a graph

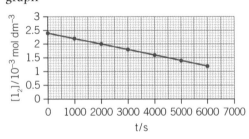

   i Describe how these data could have been obtained experimentally. *(4 marks)*

   ii State how the rate of reaction changes during the experiment, and explain how this pattern enables you to deduce that the order with respect to iodine is zero. *(2 marks)*

b Further experiments were carried out in which the rate was measured with different concentrations of reactant.

Experiment	$[H+]$ / $mol\,dm^{-3}$	$[CH_3COCH_3]$ / $mol\,dm^{-3}$	Rate / $mol\,dm^{-3}\,s^{-1}$
1	1.00	0.80	5.6×10^{-4}
2	2.00	0.80	1.2×10^{-3}
3	2.00	0.40	5.6×10^{-4}
4	1.50	0.40	4.2×10^{-4}

   i The overall rate equation is:

      rate = $k[H^+][CH_3COCH_3]$.

      Explain how the orders with respect to H^+ and propanone can be deduced from these data. *(2 marks)*

   ii Calculate a value for k for this reaction, including its units. *(2 marks)*

c The following mechanism has been proposed:

   i Complete the mechanism by adding the missing curly arrows and the formula of product X. *(4 marks)*

   ii Explain how this mechanism is consistent with the rate equation shown in **b**. *(3 marks)*

d During the reaction, a solution of hydrogen iodide is formed.

   i A solution of HI is known as hydroiodic acid, which is a strong acid.

      Calculate the pH of $0.200\,mol\,dm^{-3}$ hydroiodic acid. *(1 mark)*

ii A student tries to form a solution of hydroiodic acid by passing hydrogen iodide gas into water.

To form hydrogen iodide, he adds concentrated sulfuric acid to solid sodium iodide. Steamy fumes of hydrogen iodide are formed but also a purple vapour.

Identify the purple vapour and state **one** other gaseous product formed in the reaction. *(2 marks)*

iii Describe a better way of preparing hydrogen iodide from sodium iodide. *(1 mark)*

e If hydrogen iodide gas is heated in a sealed vessel an equilibrium is set up $2HI \rightleftharpoons H_2 + I_2$

0.0200 mol of hydrogen iodide are heated to 600 K in a vessel of volume 1 dm³. At equilibrium, 0.0090 mol of I_2 is formed.

i Use this information to calculate a value of K_c. Give your answer to an appropriate number of significant figures. *(3 marks)*

ii At 700 K, K_c has a value of 50. Explain what you can deduce about the sign of ΔH for this process. *(2 marks)*

f Hydrogen fluoride has properties that differ significantly from those of hydrogen iodide. For example it has a much higher boiling point, it is a weak acid when dissolved in water and HF does not decompose even when heated to very high temperature.

A student tries to estimate the enthalpy change for the formation of hydrogen fluoride. He uses the following bond enthalpies data:

$E(H–H) = +436 \, kJ \, mol^{-1}$

$E(H–F) = +568 \, kJ \, mol^{-1}$

$E(F–F) = +158 \, kJ \, mol^{-1}$

i Calculate a value for the $\Delta_f H$ [HF]. *(2 marks)*

ii The student knows estimates of the enthalpy change using bond enthalpies are often different to the true values. Discuss whether the value you estimated in **i** is likely to be significantly different to the true value. *(2 marks)*

iii A solution of hydrogen fluoride in water is a weak acid.

K_a [HF] = $5.6 \times 10^{-4} \, mol \, dm^{-3}$

Calculate a value for the pH of 0.100 mol dm⁻³ hydrofluoric acid. Comment on the validity of any assumptions that you make. *(4 marks)*

iv Use ideas about the structure and bonding in hydrogen fluoride to discuss reasons for the differences in properties between hydrogen fluoride and other hydrogen halides *(6 marks)*

4 1,2-dichloroethene, CHCl=CHCl, is the starting material for the production of a range of organic molecules, including polymers.

a i Explain why this molecule can exist as a pair of stereoisomers. *(2 marks)*

ii Bonds in organic molecules are described as π (pi) or σ (sigma) bonds. State the number of each type of bond present in this molecule:

Number of π bonds: _____ Number of σ bonds: _____ *(1 mark)*

b The table below gives details of three reactions, A–D of 1,2-dichloroethene.

Complete the table with the missing reagents and structures.

Reaction	Reagents	Structure of product
A	**i**	CH_2ClCH_2Cl
B	Bromine liquid	**ii** ....
C	**iii**	$CHClOHCH_2Cl$
D	Mixture of aqueous bromine and sodium chloride solution	**iv** "............ or ...." two possible products

(4 marks)

c i Give the full structural formula of the repeat unit of the polymer formed from 1,2-dichloroethene. *(1 mark)*

ii Explain what type of polymerisation has occurred in the formation of this polymer. *(1 mark)*

d A reaction can also occur with HF to form a fluorinated compound, $CHClFCH_2Cl$. This compound has been used as a solvent. It can also exist as a pair of stereoisomers.

i Give the systematic name of this molecule. *(1 mark)*

ii Explain why this molecule can exist as a pair of stereoisomers. You should include 3-dimensional structures to illustrate your answer. *(3 marks)*

e The fluorinated compound has been linked to the destruction of ozone in the stratosphere.

i Explain why the presence of ozone in the stratosphere is significant for human health. *(2 marks)*

The destruction of ozone is thought to occur because of the formation of Cl radicals. These react with ozone in a catalytic cycle.

ii Complete these equations to show the processes that destroy ozone in the stratosphere:

$Cl + O_3 \rightarrow$ _____ + _____

_____ + _____ $\rightarrow Cl + O_2$ *(2 marks)*

iii Name the type of radical reaction that is occurring in this catalytic cycle. Give a reason for your answer. *(2 marks)*

f The highest frequency radiation that reaches the troposphere is described as UVA. This has a wavelength range of 315 nm to 400 nm ($1 nm = 10^{-9} m$).

The bond energy of a C–Cl bond in $CHClFCH_2Cl$ is estimated to be $340 kJ mol^{-1}$.

Discuss whether UVA radiation could cause the production of Cl radicals from $CHClFCH_2Cl$ molecules in the troposphere. *(5 marks)*

5 Carboxylic acids and esters are found in many natural products. For example, lactic acid, a white water-soluble solid, and the ester ethyl lactate, a colourless liquid with a boiling point of 155°C, are both found in wine.

lactic acid   ethyl lactate

a i Give the systematic name of lactic acid. *(1 mark)*

ii State the reagents and conditions used to form ethyl lactate from lactic acid. *(2 marks)*

b A student attempts to separate and purify a sample of ethyl lactate from the reaction mixture.

i Suggest how the liquid ester would be separated from the reaction mixture. *(1 mark)*

ii Give the name of a substance used to remove water from the sample of the ethyl lactate. *(1 mark)*

c The purified product was analysed using infrared spectroscopy, along with a sample of lactic acid.

Explain how infrared spectroscopy could be used to show that lactic acid has been successfully converted into ethyl lactate. *(3 marks)*

d When lactic acid is heated with a phosphoric acid catalyst it is converted into a new liquid compound, X.

The empirical formula is $C_6H_{10}O_5$ and the molecular ion of the mass spectrum occurs at an M/z value of 162.

The infrared spectrum has a broad peak between 3200 and 3600 cm^{-1} and a peak at 1790 cm^{-1}.

The ^{13}C nmr spectrum has 3 peaks at $\delta = 220$, $\delta = 85$, and $\delta = 35$.

The proton NMR spectrum has 3 peaks. The details of these are shown below:

Chemical shift	Splitting	Relative number of H in the environment
1.8	doublet	3
4.2	quartet	1
11.5	singlet	1

Use the information provided to deduce the structure of the molecule X, explaining how the structure links to the information. *(6 marks)*

e Lactic acid can also form a polymer, poly(lactic acid).

i Draw out the repeat unit of this polymer and explain the type of polymerisation that has occurred. *(2 marks)*

ii Suggest why poly(lactic acid) is regarded as a 'greener' product than polymers such as nylon. *(1 mark)*

f Another carboxylic acid found in living organisms is ethanedioic acid, $H_2C_2O_4$, found in the leaves of plants such as rhubarb. Salts of ethanedioic acid, such as calcium ethanedioate, are also found naturally, and the ethanedioate ion is able to act as a bidentate ligand in several metal complexes.

i Write a balanced equation for the formation of calcium ethanedioate from the reaction between calcium hydroxide and ethanedioic acid. *(1 mark)*

ii Draw out the full structural formula of the ethanedioate ion and explain why it can act as a bidentate ligand. *(3 marks)*

iii Ethanedioate forms an 6-coordination complex ion with Fe^{3+} ions. Deduce the formula of this complex ion. *(2 marks)*

iv Draw out a 3-dimensional structure of this complex ion. *(2 marks)*

g Ethanedioc acid can act as a reducing agent. For example, it is able to reduce a solution of acidified manganate ions.

Electrode potential data for the two half-equations involved in this process are given below:

$$MnO_4^-(aq) + 8H^+(aq) + 5e^- \rightarrow Mn^{2+}(aq) + 4H_2O(l)$$
$$E^{\ominus} = +1.52V$$

$$2CO_2(g) + 2H+ (aq) + 2e^- \rightarrow H_2C_2O_4(aq)$$
$$E^{\ominus} = -0.49V$$

i Use the electrode potential data to describe why ethandioic acid is able to reduce a solution of acidified manganate ions. *(3 marks)*

ii Describe what you would see during this reaction. *(2 marks)*

iii An impure sample of ethanedioic acid with a mass of 0.98 g was dissolved to form 250 cm^3 of solution.

25.0 cm^3 of the solution was titrated with 0.0200 $mol dm^{-3}$ acidified MnO_4^- solution.

The average titre was 19.10 cm^3.

Calculate the % purity of the ethanedioic acid. Give your answer to an appropriate number of significant figures. *(5 marks)*

Answer to synoptic questions

1 a i mol $AlCl_3 = \dfrac{10{,}000}{130.5} = 76.6$ [1] mol Al required

$= 76.6 \times \dfrac{100}{95}$ AND

mass $= 80.66 \times 24 = 1.935\,kg$ [1]

ii $\dfrac{(2 \times 130.5) \times 100}{(2 \times 130.5 + 2)} = 99.2\%$ [1]

b ANY two from: similar energy requirements / more waste produced in B / similar raw materials used (aluminium ore and sea water / rocksalt) / chlorine is more hazardous than hydrochloric acid [2]

c i does not conduct below 180°C AND ions are not free to move [1] does not conduct above 180°C AND no ions / charged particles present [1]

ii

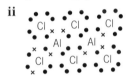

2 × dative covalent bond [1] all other details [1]

d i reaction 1 AND forms C–C bonds [1]

ii CH_3CH_2COCl [1] (ACCEPT full or skeletal formula)

iii electrophilic substitution [1] nucleophilic addition [1]

iv cyanohydrin [1]

v any named strong acid AND reflux [1]

2 a i to ensure that all water had been removed [1]

ii moles of anhydrous salt $= \dfrac{1.63}{106} =$

$0.01538\,mol$ [1] moles of water $= \dfrac{2.78}{18} =$

0.1544 [1] $x = \dfrac{0.1544}{0.01538} = 10$ [1] ecf from

incorrect calculations of moles above [1]

b i $MgCO_3(s) \rightarrow MgO(s) + CO_2(g)$ [1]

ii magnesium has a higher charge density / smaller radius [1] has a greater ability to polarise the carbonate ion [1]

iii moles HCl $= 2.45 \times 10^{-3}$ [1] mol $MgCO_3$

in $25.0\,cm^3 = \dfrac{2.45 \times 10^{-3}}{2}$ AND

moles $MgCO_3$ in $250\,cm^3 = \dfrac{2.45 \times 10^{-2}}{2} =$

$0.01225\,mol$ [1] M_r of magnesium carbonate

$= \dfrac{1.69}{0.01225} = 137.8$ [1] total M_r of water

molecules $= 137.8 - 84.3 = 53.5$ AND

$x = \dfrac{53.5}{18} = 3.0$ (to 2 s.f.) [1]

c absorbs infrared radiation [1] emitted by Earth [1] bonds vibrate with greater energy [1] reradiates this energy AND some is radiated back to Earth [1] passes on energy to other molecules by collision [1] temperature of atmosphere increases [1]

d i $[H^+]$ increases [1] $[CO_2]$ in equation 2 increases [1] to keep K_c constant [1] $[CO_2(aq)]$ in equation 1 also increases, resulting in increase in $[CO_2(g)]$ [1]

ii $[H^+] = K_a \times [HA] / [A^-] = 4.8 \times 10^{-11} \times \dfrac{0.20}{0.40} =$

$2.4 \times 10^{-11}\,mol\,dm^{-3}$ [1] pH $= -\log 2.4 \times 10^{-11} =$ 10.6 [1]

e i graph showing maximum activity at 37°C AND rapid fall above 37°C [1] exponential increase between 20 and 37 [1]

ii pattern of charges in enzyme change / NH_2 becomes NH_3^+ / COO^- becomes COOH [1] affects ability of active site to bind to substrate OR alters shape of active so substrate cannot fit in [1]

3 a i Extract small samples at regular time intervals [1] quench by cooling / adding a base [1] titrate against sodium thiosulfate [1] concentration can be calculated from the value of the titre [1]

OR use a colorimeter with appropriate filter AND zero with distilled water [1] measure absorbance at regular time intervals [1] measure absorbance of known concentrations of iodine [1] plot graph of absorbance vs conc AND read off concentration [1]

ii Rate of reaction remains constant [1] so rate is not affected by changes in $[I_2]$ (hence order is zero w.r.t. I_2) [1]

b i Comparing expt 1 + 2 AND rate doubles as $[H^+]$ doubles [1] Comparing expts 2 + 3 AND rate halves as [propanone] halves [1]

ii $k = $ rate / $[H^+]$[propanone] $= \dfrac{5.6 \times 10^{-4}}{1 \times 0.8} =$

7.0×10^{-4} [1] units $= dm^3\,mol^{-1}\,s^{-1}$ [1]

c i 6 correct curly arrows $= $ [4], 4 or 5 $= $ [3], 3 $= $ [2], 2 $= $ [1] X $= H^+$ [1]

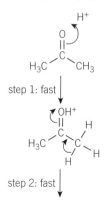

ii step 2 is the rate-determining step [1] propanone and H^+ appear before the rate-determining step so will be 1st order [1] I_2 appears after the rate-determining step so will be zero order [1]

d i pH = $-\log 0.2 = 0.7$ [1]

ii iodine [1] hydrogen sulfide [1]

iii add phosphoric acid (to solid sodium iodide) [1]

e i equilibrium concentration of HI =

$0.020 - 0.090 = 0.002$ [1] $K_c = \dfrac{0.009^2}{0.002^2}$ ecf from

incorrect calculation of [HI] or if [HI] taken as 0.020 [1] $K_c = 20$ ecf [1]

ii ΔH is positive AND K_c has increased [1] K_c increases with T for endothermic reactions OR increasing temperature favours the exothermic direction [1]

f i (reaction is $\frac{1}{2}H_2 + \frac{1}{2}F_2 \rightarrow HF$) Bonds formed = $568\,kJ\,mol^{-1}$, bonds broken = $\frac{1}{2} \times 436 + \frac{1}{2} \times$ $158 = 297\,kJ\,mol^{-1}$ [1] ΔH = broken − formed = $-271\,kJ\,mol^{-1}$ [1]

ii (unlikely to be different) average bond enthalpies are not used (these are the only compounds that contain these bonds) [1] all molecules are in the gaseous state [1]

iii $[H^+] = \sqrt{(5.6 \times 10^{-4} \times 0.050)} = 5.29 \times 10^{-3}$ mol dm^{-3} [1] pH = 2.28 [1] (2 s.f. or more)

Assumptions are that 1. $[HA]_{eqm} = [HA]_{initial}$ AND 2. $[H^+] = [A^-]$ [1]

Assumption 1 is not justified as $[H^+]$ is relatively large in comparison to [HA] 10% of [HA] value [1] IGNORE any comments about assumption 2

iv Points to be made:
• H–F bonds are relatively strong,
• because F atoms are small / bond lengths are short,
• explains stability of H–F at high temperatures,
• and may also be a factor in being a weaker acid,
• H–F bonds are very polar,
• because F is very electronegative,

• hydrogen bonding is possible between H–F molecules,
• explains high boiling point,
• also may be a factor in being a weaker acid

5–6 marks:
Discusses all three properties
Describes three features of structure and bonding
Relates all properties clearly to features of structure and bonding 3–4 marks:
Discusses two properties
Describes two features of structure and bonding
Relates some properties to features of structure and bonding 1–2 marks:
Discusses one property
Describes one features of structure and bonding
Has limited success in relating some properties to features of structure and bonding

4 a i Contains a C=C bond that cannot rotate [1] there are two different groups attached to each C atom (of the double bond) [1]

ii 1 pi bond AND 5 sigma bonds [1]

b i H_2 AND Pt catalyst at RTP OR Ni catalyst at moderately high T and P / 150°C and 5 atm [1]

ii CHClBrCHClBr [1]

iii phosphoric acid at high T and P-300°C and 60 atm OR conc. Sulfuric acid followed by water [1]

iv CHClBrCHCl$_2$ OR CHClBrCHClBr [1]

c i [1]

ii Addition AND no small molecule is lost during polymerisation [1]

d i 1,2-dichlorofluoroethane [1] IGNORE number before fluorine

ii (One of the carbons) has four different groups around it [1] so can exist as non-superimposable mirror images [1]

[1] must show correct 3-dimensional geometry

e i absorbs <u>high frequency / high energy</u> UV radiation [1] which would cause skin cancer / eye damage / sunburn [1]

ii ClO + O$_2$ in 1st equation [1] ClO + O in 2nd equation [1]

iii propagation [1] radicals on both sides of the equation [1]

f Energy needed to break one C–Cl bond =
$\frac{340000}{6.02 \times 10^{23}}$ = 5.64 × 10⁻¹⁹ J [1]

Minimum frequency required = $\frac{E}{h}$ =

$\frac{5.64 \times 10^{-19}}{6.63 \times 10^{-34}}$ = 8.52 × 10¹⁴ Hz [1]

Frequency of shortest wavelength of UVA = $\frac{c}{\lambda}$ =

$\frac{3.0 \times 10^8}{315 \times 10^{-9}}$ = 9.52 × 10¹⁴ Hz [1]

Frequency of longest wavelength of UVA =

$\frac{3.0 \times 10^8}{400 \times 10^{-9}}$ = 7.50 × 10¹⁴ Hz [1]

So only part of the UVA range is able to cause the C–Cl bond to break [1]

5 a i 2-hydroxypropanoic acid [1]

ii ethanol [1] conc sulfuric acid AND reflux [1]

b i distill off at 155°C [1]

ii any <u>anhydrous</u> salt e.g. anhydrous sodium sulfate [1]

c New product will not have a (broad) peak at 2500–3300 cm⁻¹ [1] because O–H has reacted (to form ester group) [1] Peak at 1700–1725 cm⁻¹ (due to C=O in carboxylic acid) will now be at 1735–1750 (due to C=O in ester) [1]

d *The following analysis of data could be expected:*

Mass spectrum:
M_r of empirical formula unit = 162
So molecular formula = empirical formula

Infrared: Broad peak between 3200 and 3600 indicates OH in alcohol 1790 indicates CO in acid anhydride? No very broad peak 2500–3200 / no peak between 1720-1725 indicates no COOH

NMR
3 carbon environments
δ = 220 due to C=O, δ = 85 due to C–OH, δ = 35 due to CH₃–C
3 proton environments
δ = 1.8 = CH₃–R
δ = 4.2 = CH–O
δ = 11.5 therefore is OH

5–6 marks:
Correctly deduces structure
Analyses all types of data (mass spectrum, infrared spectrum, nmr)
Explains in detail how each feature of the molecule relates to the data 3–4 marks:

Correctly deduces most feature of the structure
Analyses most types of data
Explains how most features of the molecule relate to the data 1–2 marks:

Deduces some features of the molecule
Analyses some of the data *Explains how some features of the molecule relate to the data*

e i

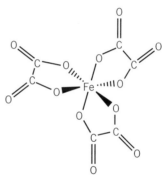

[1] condensation polymerisation AND a small molecule / water is lost [1]

ii lactic acid can be obtained from renewable natural sources AND raw materials for nylon come from crude oil [1]

f i $Ca(OH)_2 + H_2C_2O_4 \rightarrow Ca(C_2O_4) + 2H_2O$ [1]

ii

Correct structure (see above) [1] Possesses two lone pairs of electrons [1] Which it can donate / use to form bonds to one metal ion [1]

iii $[Fe(C_2O_4)_3]^{3-}$ formula [1] charge [1]

iv Structure is:

Octahedral arrangement of bonds around central Fe (with correct 3-dimensional representation) [1] Three C_2O_4 groups attached in a geometrically feasible arrangement [1]

g i ethanedioic acid system has a more negative electrode potential than manganate system [1]

ethanedioic acid donates electrons / is oxidised [1]

manganate accepts electrons / is reduced [1]

ii purple colour disappears [1] bubbling [1]

iii moles of manganate = 19.1 × $\frac{0.02}{1000}$ = 3.82 × 10⁻⁴ [1] moles of $H_2C_2O_4$ = 9.55 × 10 [1] mass of $H_2C_2O_4$ in 250 cm³ = 9.55 10⁻³ × 90 = 0.8595 g [1] % ethanedioic acid in sample = 0.8595 × $\frac{100}{0.98}$ = 87.7 [1] answer to 2 s.f. = 88 [1]

For Reference

List of constants

Molar gas volume $= 24.0\,\text{dm}^3\,\text{mol}^{-1}$ at RTP

Avogadro constant, $N_A = 6.02 \times 10^{23}\,\text{mol}^{-1}$

Specific heat capacity of water, $c = 4.18\,\text{J}\,\text{g}^{-1}\,\text{K}^{-1}$

Planck constant, $h = 6.63 \times 10^{-34}\,\text{J}\,\text{Hz}^{-1}$

Speed of light in a vacuum, $c = 3.00 \times 10^8\,\text{m}\,\text{s}^{-1}$

Ionic product of water, $K_w = 1.00 \times 10^{-14}\,\text{mol}^2\,\text{dm}^{-6}$ at 298 K

1 tonne $= 10^6\,\text{g}$

Arrhenius equation: $k = Ae^{-E_a/RT}$ or $\ln k = -E_a/RT + \ln A$

Gas constant, $R = 8.314\,\text{J}\,\text{mol}^{-1}\,\text{K}^{-1}$

Characteristic infrared absorptions in organic molecules

Bond	Location	Wavenumber / cm^{-1}
C—H	Alkanes	2850–2950
	Alkenes, arenes	3000–3100
C—C	Alkanes	750–1100
C=C	Alkenes	1620–1680
aromatic C=C	Arenes	Several peaks in range 1450–1650 (variable)
C=O	Aldehydes	1720–1740
	Ketones	1705–1725
	Carboxylic acids	1700–1725
	Esters	1735–1750
	Amides	1630–1700
	Acyl chlorides and acid anhydrides	1750–1820
C—O	Alcohols, ethers, esters and carboxylic acids	1000–1300
C≡N	Nitriles	2220–2260
C—X	Fluoroalkanes	1000–1350
	Chloroalkanes	600–800
	Bromoalkanes	500–600
O—H	Alcohols. phenols	3200–3600 (broad)
	Carboxylic acids	2500–3300 (broad)
N—H	Primary amines	3300–3500
	Amides	*ca.* 3500

Periodic table

Key
atomic number
Symbol
name
relative atomic mass

(1)	(2)	(3)	(4)	(5)	(6)	(7)	(0)

Group 1	2	3	4	5	6	7	8	9	10	11	12	13	14	15	16	17	18
1 **H** hydrogen 1.0																	2 **He** helium 4.0
3 **Li** lithium 6.9	4 **Be** beryllium 9.0											5 **B** boron 10.8	6 **C** carbon 12.0	7 **N** nitrogen 14.0	8 **O** oxygen 16.0	9 **F** fluorine 19.0	10 **Ne** neon 20.2
11 **Na** sodium 23.0	12 **Mg** magnesium 24.3											13 **Al** aluminium 27.0	14 **Si** silicon 28.1	15 **P** phosphorus 31.0	16 **S** sulfur 32.1	17 **Cl** chlorine 35.5	18 **Ar** argon 39.9
19 **K** potassium 39.1	20 **Ca** calcium 40.1	21 **Sc** scandium 45.0	22 **Ti** titanium 47.9	23 **V** vanadium 50.9	24 **Cr** chromium 52.0	25 **Mn** manganese 54.9	26 **Fe** iron 55.8	27 **Co** cobalt 58.9	28 **Ni** nickel 58.7	29 **Cu** copper 63.5	30 **Zn** zinc 65.4	31 **Ga** gallium 69.7	32 **Ge** germanium 72.6	33 **As** arsenic 74.9	34 **Se** selenium 79.0	35 **Br** bromine 79.9	36 **Kr** krypton 83.8
37 **Rb** rubidium 85.5	38 **Sr** strontium 87.6	39 **Y** yttrium 88.9	40 **Zr** zirconium 91.2	41 **Nb** niobium 92.9	42 **Mo** molybdenum 95.9	43 **Tc** technetium	44 **Ru** ruthenium 101.1	45 **Rh** rhodium 102.9	46 **Pd** palladium 106.4	47 **Ag** silver 107.9	48 **Cd** cadmium 112.4	49 **In** indium 114.8	50 **Sn** tin 118.7	51 **Sb** antimony 121.8	52 **Te** tellurium 127.6	53 **I** iodine 126.9	54 **Xe** xenon 131.3
55 **Cs** caesium 132.9	56 **Ba** barium 137.3	57–71 lanthanoids	72 **Hf** hafnium 178.5	73 **Ta** tantalum 180.9	74 **W** tungsten 183.8	75 **Re** rhenium 186.2	76 **Os** osmium 190.2	77 **Ir** iridium 192.2	78 **Pt** platinum 195.1	79 **Au** gold 197.0	80 **Hg** mercury 200.6	81 **Tl** thallium 204.4	82 **Pb** lead 207.2	83 **Bi** bismuth 209.0	84 **Po** polonium	85 **At** astatine	86 **Rn** radon
87 **Fr** francium	88 **Ra** radium	89–103 actinoids	104 **Rf** rutherfordium	105 **Db** dubnium	106 **Sg** seaborgium	107 **Bh** bohrium	108 **Hs** hassium	109 **Mt** meitnerium	110 **Ds** darmstadtium	111 **Rg** roentgenium	112 **Cn** copernicium		114 **Fl** flerovium		116 **Lv** livermorium		

57 **La** lanthanum 138.9	58 **Ce** cerium 140.1	59 **Pr** praseodymium 140.9	60 **Nd** neodymium 144.2	61 **Pm** promethium 144.9	62 **Sm** samarium 150.4	63 **Eu** europium 152.0	64 **Gd** gadolinium 157.2	65 **Tb** terbium 158.9	66 **Dy** dysprosium 162.5	67 **Ho** holmium 164.9	68 **Er** erbium 167.3	69 **Tm** thulium 168.9	70 **Yb** ytterbium 173.0	71 **Lu** lutetium 175.0
89 **Ac** actinium	90 **Th** thorium 232.0	91 **Pa** protactinium	92 **U** uranium 238.1	93 **Np** neptunium	94 **Pu** plutonium	95 **Am** americium	96 **Cm** curium	97 **Bk** berkelium	98 **Cf** californium	99 **Es** einsteinium	100 **Fm** fermium	101 **Md** mendelevium	102 **No** nobelium	103 **Lr** lawrencium